Advances in
Carbohydrate Chemistry and Biochemistry

Volume 31

Advances in Carbohydrate Chemistry and Biochemistry

Editors

R. STUART TIPSON

DEREK HORTON

Volume 31

1975

ACADEMIC PRESS   New York   San Francisco   London

A Subsidiary of Harcourt Brace Jovanovich, Publishers

ACADEMIC PRESS, INC.
111 Fifth Avenue, New York, New York 10003

United Kingdom Edition published by
ACADEMIC PRESS, INC. (LONDON) LTD.
24/28 Oval Road, London NW1

LIBRARY OF CONGRESS CATALOG CARD NUMBER: 45-11351

ISBN 0–12–007231–9

PRINTED IN THE UNITED STATES OF AMERICA

CONTENTS

Hewitt Grenville Fletcher, Jr. (1917–1973)

CORNELIS P. J. GLAUDEMANS

Deamination of Carbohydrate Amines and Related Compounds

J. MICHAEL WILLIAMS

The Reaction of Ammonia with Acyl Esters of Carbohydrates

MARIA E. GELPI AND RAÚL A. CADENAS

Chemistry and Biochemistry of Apiose

RONALD R. WATSON AND NEIL S. ORENSTEIN

Specific Degradation of Polysaccharides

BENGT LINDBERG, JÖRGEN LÖNNGREN, AND SIGFRID SVENSSON

Chemistry and Interactions of Seed Galactomannans

IAIN C. M. DEA AND ANTHONY MORRISON

The Interaction of Homogeneous, Murine Myeloma Immunoglobulins with Polysaccharide Antigens

CORNELIS P. J. GLAUDEMANS

Bibliography of Crystal Structures of Carbohydrates, Nucleosides, and Nucleotides, 1973

GEORGE A. JEFFREY AND MUTTAIYA SUNDARALINGAM

LIST OF CONTRIBUTORS

Numbers in parentheses indicate the pages on which the authors' contributions begin.

RAÚL A. CADENAS, *Departamento de Química, Facultad de Agronomía, Universidad de Buenos Aires, Av. San Martín 4453, Buenos Aires, Argentina* (81)

IAIN C. M. DEA, *Unilever Research Laboratories, Colworth House, Sharnbrook, Bedfordshire MK44 1LQ, England* (241)

MARIA E. GELPI, *Departamento de Química, Facultad de Agronomía, Universidad de Buenos Aires, Av. San Martín 4453, Buenos Aires, Argentina* (81)

CORNELIS P. J. GLAUDEMANS, *Laboratory of Chemistry, National Institute for Arthritis, Metabolism, and Digestive Diseases, National Institutes of Health, Bethesda, Maryland 20014* (1, 313)

GEORGE A. JEFFREY,[*] *Department of Crystallography, University of Pittsburgh, Pittsburgh, Pennsylvania 15260* (347)

BENGT LINDBERG, *Department of Organic Chemistry, Arrhenius Laboratory, University of Stockholm, S-104 05 Stockholm, Sweden* (185)

JÖRGEN LÖNNGREN, *Department of Organic Chemistry, Arrhenius Laboratory, University of Stockholm, S-104 05 Stockholm, Sweden* (185)

ANTHONY MORRISON, *Unilever Research Laboratories, Colworth House, Sharnbrook, Bedfordshire MK44 1LQ, England* (241)

NEIL S. ORENSTEIN,[**] *Department of Microbiology, Harvard School of Public Health, Boston, Massachusetts 02115* (135)

MUTTAIYA SUNDARALINGAM, *Department of Biochemistry, College of Agricultural and Life Sciences, University of Wisconsin, Madison, Wisconsin 53706* (347)

SIGFRID SVENSSON,[***] *Department of Organic Chemistry, Arrhenius Laboratory, University of Stockholm, S-104 05 Stockholm, Sweden* (185)

RONALD R. WATSON,[†] *Department of Microbiology, University of Mississippi Medical Center, Jackson, Mississippi 39216* (135)

J. MICHAEL WILLIAMS, *Department of Chemistry, University College, Singleton Park, Swansea SA2 8PP, Wales, United Kingdom* (9)

[*] Present address: Department of Chemistry, Brookhaven National Laboratory, Upton, Long Island, New York 11973.

[**] Present address: Department of Pathology, Massachusetts General Hospital, 2 Hawthorne Place, Boston, Massachusetts 02114.

[***] Present address: Department of Clinical Chemistry, University Hospital, S-221 85 Lund, Sweden.

[†] Present address: Department of Microbiology, Indiana University School of Medicine, 1100 West Michigan Street, Indianapolis, Indiana 46202.

PREFACE

In this thirty-first volume of *Advances*, Williams (Swansea) surveys the deamination of carbohydrate amines and related compounds, updating earlier discussions by Peat (Vol. 2), Shafizadeh (Vol. 3), and Defaye (Vol. 25). Gelpi and Cadenas (Buenos Aires) provide a comprehensive treatment of the reaction of ammonia with acyl esters of carbohydrates; their article greatly extends that by Deulofeu (Vol. 4). A chapter by Watson (Jackson, Miss.) and Orenstein (Boston, Mass.) brings the article by Hudson (Vol. 4) on the chemistry and biochemistry of apiose up to date. Lindberg, Lönngren, and Svensson (Stockholm) discuss the specific, chemical degradation of polysaccharides in an article that updates that by Bouveng and Lindberg (Vol. 15) and complements that by Marshall on their enzymic degradation (Vol. 30). The extensive literature on the chemistry and interactions of seed galactomannans is surveyed by Dea and Morrison (Sharnbrook, England), thus adding to our previous articles on the chemistry of a variety of polysaccharides. Glaudemans (Bethesda, Md.) provides an interesting discussion on the interaction of homogeneous, murine myeloma immunoglobulins with polysaccharide antigens, and also describes the career of the late H. G. Fletcher, Jr. In a continuation of our series of bibliographic articles on carbohydrate structures that have been ascertained by crystallographic methods, Jeffrey (Pittsburgh) and Sundaralingam (Madison, Wis.) treat those structures definitively established in 1973, and list all of those determined satisfactorily before 1970.

The Subject Index was compiled by L. T. Capell.

Kensington, Maryland   R. STUART TIPSON
Columbus, Ohio   DEREK HORTON
June, 1975

ix

Advances in
Carbohydrate Chemistry and Biochemistry

Volume 31

1917–1973

HEWITT GRENVILLE FLETCHER, Jr.

1917–1973

Hewitt Fletcher had a blackboard in the basement of his home. Written upon it were the formulas of compounds related to his current interests in chemical research. There, during a break from labors at his work bench, or from sighting a handgun ("please aim carefully, the line of fire runs only one foot past the hot-water heater"), he and his friends would discuss chemistry. He did this, not out of any forced sense of duty, but simply because chemistry was fun, and, like everything else in his life, was done energetically and with much enjoyment. Dr. Fletcher radiated thoroughness and dependability, and the qualities we traditionally expect from the people from "Down-East." His grandfather was Headmaster of the Normal School in Castine, Maine. Dr. Fletcher's father, Hewitt Grenville Fletcher, Senior, went to Amherst College and Harvard University, and later settled in Watertown, Massachusetts, with his wife, Frances Mitchell, and practised law in nearby Boston.

Hewitt, Jr., was born on May 28th, 1917, and grew up in Watertown. The only child in the family, he spent much time with his mother's parents. Despite the harmful possibilities of receiving too much attention in such a situation, he grew up tough and self-reliant, with a distrust of frills and slickness. Helen Thayer, his high-school chemistry teacher at Watertown, was impressed with Hewitt from the start. The summer before he was to take chemistry, Hewitt asked to have the textbook to be used. When the term started, there was nothing to teach him; he had devoured the contents of his book.

Before taking chemistry in high school he had, while in junior high school, set up a laboratory for himself in the basement of his parents' home. Here he experimented, keeping a notebook in which he meticulously recorded his observations and deductions. One day, he prepared sulfur dioxide, and the gas drove his mother's bridge-party out of the house. After a stern lecture by his parents concerning his chemical experimentation, he built a sort of fume-hood, to lead the noxious gases out of the basement into the open.

1

Unfortunately, the fumes killed his mother's best roses, and again he was made aware of his mother's displeasure. One day, he went to Helen Thayer and told her that "chemistry is at a low ebb at home today." He had asked his father to buy him a small lecture-bottle of a corrosive gas from a chemical-supply house. His father complied and took the material to his Boston law-office, where, somehow, the cylinder began to leak; one whole wing of the building had to be evacuated in the dead of winter. It is typical of the family that none of these mishaps dampened his parents' encouragement of Hewitt's scientific curiosity.

When he was a senior in high school he wanted to study organic chemistry. This was not taught, so Helen Thayer and he studied it together. The reader may wonder if he were not more bookish and industrious than is healthy for a young boy: far from it! When he became a senior in high school, his father bought him a sailboat, an original Morse Friendship Sloop called the "Lulu Belle." This was no dinghy, but a 27-foot boat of beautiful traditional design, with bunks for two. His father had done this so that his son might "learn to cope." Hewitt's first cruise was from Boston, Massachusetts, to Bangor, Maine, an ocean cruise of nearly 200 miles. As the Lulu Belle had no auxiliary engine, he was always subject to the vagaries of the wind, and spent many a night at sea, either because there was too little wind or too much for him to sail safely to shore. He also became quite well acquainted with thick New England fogs. Needless to say that Hewitt, Senior, often paced the seawall, worried about his son's being out and overdue.

Hewitt disliked dances or social affairs, but, apparently, graciously attended those he could not avoid. He did not particularly like athletics while in high school, although he was physically strong, and extremely fond of hiking and mountaineering. At 13, Hewitt met his wife-to-be, Ann Winter. Helen Thayer recalls that, on one particularly calm day, Hewitt took Ann and her for a canoe trip, from Hull, around Boston Light and outer Brewster Island, a round trip of some 20 miles over open water. Another time, during the Christmas vacation, he launched his canoe one evening in Boston Harbor, and paddled among the festively lit boats.

Following high school, he proceeded to the Massachusetts Institute of Technology in Boston for undergraduate study. During this time, Hewitt explored England by bicycle. His uncanny memory and gift for observation would allow him, some thirty years later, to recall details and lead his family to out-of-the-way places of interest when they all visited England. In 1939, he obtained his Bachelor of Science degree, followed, in 1942, by his Doctor of Philosophy degree under Dr. R. C. Hockett. While in graduate school, he had married Ann in 1940, and the Fletchers' first child, Bradford, was born in 1942. At that time, Dr. Fletcher was an instructor at M.I.T., teaching organic chemistry and doing research as well. By this time, he had already published a few papers. It is interesting that one of his first papers reviewed the life of Augustin-Pierre Dubrunfaut, an early sugar-chemist. This typifies another of his interests: history. Later in his life, one of the most prized attractions of his home would become his magnificent collection of rare and old books; these ranged from books printed in the liberal Low Countries for religiously persecuted groups of the 17th and 18th centuries to books dealing with the history of science. During much of 1943 and 1944, Dr. Fletcher worked with Robert C. Hockett on the oxidation of polyalcohols with lead tetraacetate, and, in 1945, he collaborated with Dr. R. Max Goepp, Jr., on a number of structural problems in the field of hexitol anhydrides.

In 1945, their second son, R. Theodore, was born. Later that year, the family moved to Bethesda, Maryland, where Dr. Fletcher began work under the direction of Dr. C. S. Hudson at the then National Institute of Health. Hudson had already been at the N.I.H. (formerly called the Hygienic Laboratory) for sixteen years, and having been a student of Van't Hoff's, he had brought his knowledge of stereochemistry to bear on the problems of optical rotation and the chemistry of anomers. Dr. Fletcher entered into this collaboration with characteristic energy. His interest centered originally around synthetic methods involving the cyclitols and anhydrohexitols. During this period, Drs. Robert K. Ness, Fletcher, and Hudson were the first to use lithium aluminum hydride to prepare 1,5-anhydrohexitols from the corresponding fully acetylated hexosyl halides. Then, together with Drs. Roger W. Jeanloz and Hudson, Dr. Fletcher began his well known work on the chemistry of ribose, and developed, for synthetic use, intermediates, such as tri-O-benzoyl-D-ribopyranosyl bromide, more stable than those hitherto available. Hudson's work on the relationship between structure and optical rotation also continued in collaboration with Dr. Fletcher. In all, they published some twenty papers together. During this time, Hewitt and Ann's third son, Peter Grenville, was born in 1949.

It was obvious that Dr. Fletcher had become Hudson's right-hand man, and, when Hudson retired in 1951, Dr. Fletcher became Chief of the Section on Carbohydrates. With his excellent technician Mr. Harry W. Diehl, and in collaboration with Dr. Ness, he embarked on a program of synthetic carbohydrate chemistry involving the development of protecting groups, as well as on the elucidation of structure of natural products, such as stevioside, the very sweet glycoside from the "Herb of Paraguay." It was at this time that Dr. Fletcher and his collaborators developed the preparation of tri-O-benzoyl-D-ribofuranosyl bromide, a stable intermediate of great utility in the synthesis of important glycosides and nucleosides containing the D-ribofuranosyl group. While investigating the chemistry of these derivatives, they found that, when hydrolyzed, this bromide yields 1,3,5-tri-O-benzoyl-α-D-ribofuranose which, on treatment with hydrogen halide, affords 3,5-di-O-benzoyl-D-ribofuranosyl halide. This compound lacks a participating group at O-2, thus opening a pathway to the synthesis of the difficultly accessible glycosides and N-glycosyl derivatives of α-D-ribofuranose.

In the late fifties, the N.I.H. started the Visiting Program; under this, young scientists were invited to work for one or two years at the Institutes under the direction of staff members. Dr. Fletcher actively participated in this program. It was as a Visitor in Dr. Fletcher's laboratory that Dr. Donald L. MacDonald became interested in the preparation of phosphoric esters of carbohydrates, and his work eventually led to the MacDonald method of synthesizing glycosyl phosphates.

Also, in the late fifties, Dr. Fletcher turned to the synthesis of nucleosides, and he developed the first direct synthesis of the two anomers of 9-(2-deoxy-D-erythro-pentofuranosyl)adenine. In 1961, he (with Dr. Robert Barker) started his long and thorough research into the use of benzyl ethers as nonparticipating protecting-groups in carbohydrate synthesis. This work, which culminated in the subsequent publication of some twenty papers on this use of the benzyl group, was nearest to Dr. Fletcher's heart. It led to a facile synthesis for spongoadenosine, a nucleoside of great interest in cancer therapy, to a study of the mechanism of solvolysis of aldosyl halides, and to the preparation of a large number of extremely useful intermediates, as well as to a new method for O-debenzylation employing boron trifluoride.

Other avenues of research were not ignored, however, as is indicated by the preparation of the first furanose-related glycal and the study of inversions in the cyclitol series by treatment of their esters with liquid hydrogen fluoride. Then, in the mid-sixties, Dr. Fletcher began to turn to the chemistry of acylamidodeoxyhexoses. Together with Drs. Nevenka Pravdic and T. D. Inch, he prepared 1,2-oxazoline derivatives of such carbohydrates. This led to a method for the selective transglycosylation of 2-acetamido-2-deoxy-β-D-gluco-pyranosides in which the acetylated glycoside is treated with a solution of zinc chloride in butyl acetate–butyl alcohol. The butyl glycoside is formed by way of the 1,2-oxazoline intermediate. This method has the exciting potential for selective cleavage at positions in polysaccharides where 2-acetamido-2-deoxy sugars having a 1,2-*trans* configuration are located. Dr. Fletcher's death prevented him from further developing this possibility, in which he had a real interest.

Well over thirty postdoctoral Fellows and Visiting Scientists passed through Dr. Fletcher's Section over the years. The association he had with them and the sheer enjoyment he derived from working with them were perhaps his greatest pleasures. He considered this aspect of the proper continuation of science to be one of his greatest achievements. He was an excellent teacher, with a superior research philosophy, and was of unshakable integrity. These associates, who are now scattered all over the world, have expressed the fact that, to them, their years at the National Institutes of Health were their most formative ones.

To go back just a little, to 1958: when Dr. Fletcher went alone to Missoula, in Montana, to work for six weeks at the regional laboratories of the N.I.H., he caught Rocky Mountain spotted fever, but, fortunately, his friends on the scene, the Perrines, nursed him back to health. His family joined him, and then continued with him on a lecture tour out west, during which he availed himself of the opportunity of visiting Professor Gobind Khorana's laboratory. Typically, the family camped out most of the time. There was, presumably, a fair amount of trout fishing, for outdoor activities were another of Hewitt's passions. When his children were young, he was a Scout master in the local Boy Scout troop. He loved canoeing (the family had several canoes in the barn near their house) or an enjoyable weekend of hunting in Virginia. "Due to urgent business of the highest calibre, I shall be out of the laboratory," he would inform his associates. The deer he shot would be dressed by the local butcher, and, for the fortunate few, a dish of venison at the Fletchers's was a real treat.

The association between Drs. Fletcher and Pravdić, which started when she was a Visiting Scientist at the Institute in Bethesda, continued after she returned to Yugoslavia. Together they investigated the chemistry of unsaturated 2-acetamido-2-deoxy-D-glucose derivatives, and, with Emmanuel Zissis, the preparation of lactones from these sugars, work stemming from Dr. Fletcher's interest in Tay–

Sachs disease. This is a disorder in fat metabolism wherein a "hexosaminidase" (better called an acetamidodeoxyhexosidase) is missing. The lack of this enzyme causes failure of the metabolism of ganglioside G_{M2}, leading to an uncontrolled increase of this material in the body. It was known that hexosaminidases are inhibited by 2-acetamido-2-deoxy-D-glucono-1,4-lactone, but there was a virtual hiatus in the literature on the synthesis and chemistry of 2-acylaminoaldonic acids, so Dr. Fletcher began by thoroughly investigating the chemistry of the oxidation of aldoses at C-1. He unraveled the many products obtained by these oxidations, and the interconversions of these products. Glycals also exert an inhibitory effect on glycosidases, and so Dr. Fletcher began a study of the 2-acetamidoglycals as well, and he was still actively engaged in these researches to the end.

In the early sixties, the Fletchers turned again to one of Hewitt's oldest loves: sailing. Dr. Fletcher had known Dr. Louis Long, Jr., since his graduate-school days. They shared an enthusiasm for the sea, and consequently, the two families spent many a summer sailing together off the coast of Maine. Hewitt had many interests, but sailing was probably his most fervent one. Thus, it is not surprising that, in the late sixties, he decided to have a boat built for his family. It was a happy day in 1969 when they commissioned their 35-ft auxiliary ketch "Aspara." With it, the family, but more often Hewitt and Ann alone, explored the myriad tributaries and inlets of the Chesapeake Bay, or made long cruises, such as the one to Maine and back. It was from his contact with Nature, and from the great pleasure Hewitt derived from sailing, that he drew a large part of the strength needed to bear his illness so gallantly in the last year and a half of his life.

By the early seventies, his children had grown up. Bradford married, obtained his Ph.D. in literature, and then chose a life of teaching; Hewitt delighted in the resulting grandchildren. Ted, too, married; he is a chemist, and now works at the N.I.H. Their youngest son, Peter, chose a career in teaching in secondary school after he finished college, and is a member of the staff of the Friends School in Sandy Spring, Maryland.

This obituary would be incomplete without some indication of the many contributions made by Dr. Fletcher to the administration of Science, but it would be mechanical and difficult to mention *all* the honors he received. He was a member of the American Chemical Society and The Chemical Society (London). He was on many National committees, including the Science Advisory, Post Office, National Research Council's Pioneering Research, the Chemical Literature, and the NIH Library Committee. He received the C. S. Hudson Award of the Division of Carbohydrate Chemistry of the American Chemical Society in 1968. He was on the Board of Editors of the *Journal of Organic Chemistry*, as well as the Editorial Advisory Board of *Carbohydrate Research*. He was Chairman of the first Gordon Research Conference on the Chemistry of the Carbohydrates in 1964, and was an invited lecturer at many international meetings. Dr. Fletcher was author or coauthor of more than 170 scientific papers.

Hewitt died, on October 19, 1973, of pulmonary metastases from a sarcoma in the leg. He was a man of character, one of the most distinguished carbohydrate chemists in the United States, and a wonderful and generous companion. He was truly an inspiration to all who knew him, and we shall miss him sorely.

CORNELIS P. J. GLAUDEMANS

APPENDIX

The following are the names of scientists who published with Dr. H. G. Fletcher, Jr.: R. Allerton, L. Anderson, J. B. Ames, R. Barker, A. K. Bhattacharya, B. Coxon, H. W. Diehl, M. T. Dienes, G. R. Findlay, I. Franjic, K. W. Freer, C. P. J. Glaudemans, R. M. Goepp, Jr., M. Haga, A. Hasegawa, R. Harrison, E. J. Hedgley, R. C. Hockett, A. G. Holstein, C. S. Hudson, T. D. Inch, T Ishikawa, R. W. Jeanloz, R. L. Kaufman, D. Kiely, H. R. Kirshen, L. H. Koehler, H. Kuzuhara, D. L. MacDonald, R. Montgomery, R. K. Ness, C. Pedersen, T. D. Perrine, J. R. Plimmer, N. Pravdić, Y. Rabinsohn, H. E. Ramsden, N. K. Richtmyer, W. L. Salo, E. L. Sheffield, Jr., C. F. Snyder, S. Soltzberg, C. M. Sponable, J. D. Stevens, S. Tejima, E. Vis, H. B. Wood, Jr., R. C. Young, B. Zidovec, M. Zief, and E. Zissis.

DEAMINATION OF CARBOHYDRATE AMINES
AND RELATED COMPOUNDS

By J. Michael Williams

*Chemistry Department, University College, Swansea SA2 8PP,
Wales, United Kingdom*

I. Introduction

Early deamination studies, notably those of P. A. Levene and his collaborators, were primarily carried out in attempts to elucidate the structures of 2-amino-2-deoxy-D-glucose, -D-galactose, and -D-mannose, and their aldonic acids. However, the reactions were not entirely helpful, because only in one amine was the amino group replaced by a hydroxyl group (with inversion); other amines gave rearrangement products. There were also conflicting reports concerning the structure of 2,5-anhydro-D-mannose (chitose), the product obtained from 2-amino-2-deoxy-D-glucose. By 1919, structures had been deduced for the deamination products, but the reactions could not be fully described until the structures of the amines had been confirmed by unambiguous synthesis. The reactions were better understood when they were classified as nucleophilic dis-

9

placements by Peat[1] (although he did not use this terminology), and when conformational principles were applied.[2]

The discovery, during the past two decades, of numerous naturally occurring amino sugars and aminodeoxycyclitols, and interest[3-5] in the neighboring-group and rerrangement reactions that can accompany displacement reactions of carbohydrate derivatives, have led to increasing interest in deamination of amino sugars.

The early work was conducted without the aid of chromatography and other modern analytical methods, which later made detailed analysis of product mixtures possible. Nevertheless, even in some later work, product analysis was inadequate, and this fact has to be taken into account when assessing the significance of the results. Some aspects of carbohydrate amine deamination have been discussed in previous Volumes of this Series. Peat[1], in 1946, summarized much of the early work; Shafizadeh,[2] in 1958, rationalized the known results by using conformational principles; and Defaye,[6] in 1970, discussed deaminations leading to 2,5-anhydrides of sugars and related compounds. Part of a comprehensive survey[7] of the chemistry of amino sugars is devoted to a discussion of their deamination reactions.

In the present article, an attempt is made to rationalize the different kinds of behavior of carbohydrate amines, including the cyclitol amines, upon deamination. In addition, the potential use and limitations of these reactions in synthesis and in elucidation of structure are discussed. Methods of deamination that do not involve nitrosation and related processes are not considered.

II. THE NATURE OF THE DEAMINATION REACTION

It was suggested[8] in 1932 that carbonium ions are intermediates in the conversion of a primary aliphatic amine into nitrogen-free products by the action of nitrous acid. In the weakly acidic solutions normally used for deamination of aliphatic amines, the rate-determining

(1) S. Peat, Advan. Carbohyd. Chem., 2, 37–77 (1946), especially pp. 60–64.

(2) F. Shafizadeh, Advan. Carbohyd. Chem., 13, 43–61 (1958).

(3) L. Goodman, Advan. Carbohyd. Chem., 22, 109–175 (1967).

(4) D. H. Ball and F. Parrish, Advan. Carbohyd. Chem. Biochem., 24, 167–197 (1969).

(5) B. Capon, Chem. Rev., 69, 469–493 (1969).

(6) J. Defaye, Advan. Carbohyd. Chem. Biochem., 25, 181–194 (1970).

(7) D. Horton, in "The Amino Sugars," R. W. Jeanloz, ed., Academic Press, New York, N. Y., 1969, Vol. IA, pp. 1–211.

(8) F. C. Whitmore and D. P. Langlois, J. Amer. Chem. Soc., 54, 3441–3447 (1932).

step is N-nitrosation of the amine by nitrous anhydride or a related species, such as nitrosyl chloride.[9] By analogy with diazotization of aromatic amines, the unstable N-nitrosoamine[10] is presumed to rearrange to an unstable, diazonium ion,[11] which is converted into products directly, or by way of a carbonium ion. A key feature of the reac-

$$R-NH_2 \xrightarrow[-H^{\oplus}]{NOX} R-NH-NO \longrightarrow R-N=N-OH \longrightarrow R-N_2^{\oplus}$$

$$\downarrow -N_2$$

$$\text{products} \longleftarrow R^{\oplus}$$

tion is that the product-forming steps occur at rates that are normally greater than, or comparable to, the rates of interconversion of conformers. Thus, the reaction is normally controlled by the ground-state, conformational equilibrium (corollary of the Curtin–Hammett principle[13]), in contrast to solvolysis of alkyl halides or alkyl sulfonates, which may occur by way of conformations that are unimportant in the ground state.

A characteristic feature of deamination reactions is that rearrangement occurs to a greater extent, and more products are often formed, than in related solvolyses. This behavior has been related to the low activation-energies involved in the conversion of the diazonium ion into products, there being smaller differences in the activation energies for the competing pathways than for solvolysis.[14] That the carbonium ion intermediates in deamination and solvolysis are different was suggested[18] in 1952; heterolysis of a high-energy diazonium ion was assumed to give a high-energy carbonium ion, because a stable,

(9) J. H. Ridd, *Quart. Rev.* (London), **15**, 418–441 (1961).

(10) N-Nitrosomethylamine has been observed in solution at $-80°$; see E. Müller, H. Hais, and W. Rundel, *Chem. Ber.*, **93**, 1541–1552 (1960).

(11) For a discussion of evidence for the existence of alkanediazonium ions as species possessing a finite lifetime, see Ref. 12. It has been suggested that diazonium ions are not mandatory intermediates; see later discussion (p. 14).

(12) J. T. Keating and P. S. Skell, in "Carbonium Ions," G. A. Olah and P. von R. Schleyer, eds., Wiley–Interscience, New York, N. Y., 1970, Vol. II, pp. 608–616.

(13) E. L. Eliel, N. L. Allinger, S. J. Angyal, and G. A. Morrison, "Conformational Analysis," Wiley, New York, N. Y., 1965, pp. 28–31.

(14) This point was made in connection with the decomposition of diazoesters by way of diazonium ions.[15] See also, Refs. 16 and 17.

(15) R. Huisgen and H. Reimlinger, *Ann.*, **599**, 161–182 (1956).

(16) A. Streitwieser, Jr., and W. D. Schaeffer, *J. Amer. Chem. Soc.*, **79**, 2888–2893 (1957).

(17) A. Streitwieser, Jr., *J. Org. Chem.*, **22**, 861–869 (1957).

(18) L. S. Ciereszko and J. G. Burr, Jr., *J. Amer. Chem. Soc.*, **74**, 5431–5436 (1952).

neutral molecule, namely, nitrogen, is released. Such ions, described as relatively unsolvated,[19] "hot," or vibrationally excited[20] carbonium ions, were expected to react less discriminatingly than the ("normal") solvated, carbonium ions formed in solvolysis. The rearrangements that commonly occur in deamination are highly stereospecific; for example 1,2-shifts commonly involve the migration of an atom that is trans and coplanar to the nitrogen atom of the diazonium ion in its favored conformation. Neighboring-group participation may also result in the formation of an unrearranged product with complete retention of configuration. Such results have often been discussed in terms of *concerted* reactions of diazonium ions, which, Streitwieser and Schaeffer[16] proposed, are the branching point in deamination reactions. The activation energy for heterolysis of the alkanediazonium ion is presumed to be very low,[20] and, according to the Hammond postulate,[21] there should be relatively little bond-*formation* in the transition state. The relative unimportance of the "migratory aptitude" of a neighboring group in deamination is consistent with a nonconcerted mechanism.[19]

Attention has been drawn to the importance of the counter-ion.[22-26] When first formed, the diazonium ion is, presumably, part of an ion-pair such as 1, which is then converted into an ion-pair (2) in which the ions are initially separated by the nitrogen molecule (see Scheme 1). This nitrogen-separated ion-pair has been described as a vibra-

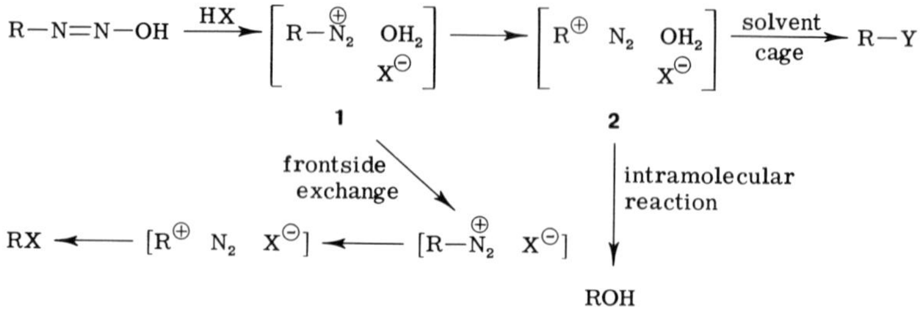

Scheme 1

(19) D. J. Cram and J. E. McCarty, *J. Amer. Chem. Soc.*, **79**, 2866–2875 (1957).
(20) E. J. Corey, J. Casanova, Jr., P. A. Vatakencherry, and R. Winter, *J. Amer. Chem. Soc.*, **85**, 169–173 (1963).
(21) G. S. Hammond, *J. Amer. Chem. Soc.*, **77**, 334–338 (1955).
(22) R. Huisgen and H. Reimlinger, *Ann.* **599**, 183–202 (1956).
(23) R. Huisgen and C. Rüchardt, *Ann.*, **601**, 1–21 (1956).
(24) R. Huisgen and C. Rüchardt, *Ann.*, **601**, 21–39 (1956).
(25) E. H. White and J. E. Stuber, *J. Amer. Chem. Soc.*, **85**, 2168–2170 (1963).
(26) For a summary, see Ref. 27.

tionally excited ion-pair[27] in which the cation is less well solvated than that formed in solvolysis. Such poorly solvated carbonium ions correspond to the "hot" ions already mentioned.

The stereospecificity of such rearrangements as 1,2-shifts has been attributed to carbonium ions whose dissymmetry is maintained by the counter-ion.[27,28] However, generalizations that disregard the structure of the amine are probably invalid, and it seems reasonable to differentiate between, on the one hand, primary, and on the other hand, secondary and tertiary alkanediazonium ions. Concerted reactions are most likely for primary alkanediazonium ions;[29,30] two results cited[30] in support of this suggestion are (a) the 85% net inversion of configuration (corrected[30] for racemization by way of diazonium ion–diazoalkane equilibration) that occurs in the deamination of optically active (+)-butylamine-1-d in glacial acetic acid,[16] and (b) the preponderant (>85%)[31] inversion of configuration at the migration terminus in the deamination of (+)(R)-neopentylamine-1-d to give products of methyl migration.[32]

Alkanediazonium ions may lose a proton to give a diazoalkane when loss of nitrogen to form a carbonium ion is unfavorable. Product formation by way of diazoalkanes can be detected by deuterium labelling, and has been shown to be most likely for primary alkanedia-

$$R-CH_2-N_2^{\oplus} \rightleftharpoons R-CH=\overset{\oplus}{N}=\overset{\ominus}{N} \xrightarrow{\text{DX}} R-CHD-N_2^{\oplus} \rightleftharpoons R-CD=\overset{\oplus}{N}=\overset{\ominus}{N}$$

$$\Big\updownarrow \text{DX}$$

$$R-CD_2-N_2^{\oplus}$$

zonium ions in nonpolar solvents.[33] Diazoalkanes are stable products of deamination when such electron-withdrawing groups as carbonyl are present adjacent to the nitrogen-bearing carbon atom.[34]

(27) E. H. White and D. J. Woodcock, in "The Chemistry of the Amino Group," S. Patai, ed., John Wiley and Sons, Ltd., London, 1968, p. 461–480.

(28) C. J. Collins, Accounts Chem. Res., 4, 315–322 (1971).

(29) J. H. Bayless, A. T. Jurewicz, and L. Friedman, J. Amer. Chem. Soc., 90, 4466–4468 (1968).

(30) L. Friedman, in "Carbonium Ions," G. A. Olah and P. von R. Schleyer, eds., Wiley–Interscience, New York, N. Y., 1970, Vol. II, pp. 660–668.

(31) The reliability of the optical rotation on which this figure was based has been questioned (Ref. 12, p. 591).

(32) R. D. Guthrie, J. Amer. Chem. Soc., 89, 6718–6720 (1967).

(33) L. Friedman, A. T. Jurewicz, and J. H. Bayless, J. Amer. Chem. Soc., 91, 1795–1799 (1969).

(34) G. W. Cowell and A. Ledwith, Quart. Rev. (London), 24, 119–167 (1970).

The nature of the intermediates in the deamination of secondary alkanamines seems less clear cut. Thus, Whiting and coworkers[35] believed that heterolysis of the alkanediazohydroxide (diazotic acid) yields the carbonium ion in one step. It has also been suggested, by analogy with the behavior of aromatic diazohydroxides, that the intermediate formed depends on the configuration of the diazohydroxide, the *syn* isomer (having the nitrogen lone-pair *trans* to the hydroxyl group) giving the diazonium ion, and the *anti* isomer (having the alkyl group *trans* to the hydroxyl group) being converted directly into the carbonium ion. The *anti* configuration was considered to be the more likely for alkanediazohydroxides.[36]

The discovery by C. J. Collins and collaborators that up to 60% retention of configuration at the migration terminus could occur in 1,2-shifts clearly ruled out *concerted* reactions of secondary alkanediazonium ions. These results were explained by postulating that the lifetime of the intermediate carbonium ion is such that partial rotation about a carbon–carbon bond can occur prior to migration of an alkyl group.[28] An alternative suggestion has been advanced by Friedman,[37] who proposed that *cis* migration might occur by way of an eclipsed conformer.

The products formed in deamination depend on the lifetimes of the different ion-pair species, which, in turn, depend on the structure of the amine and the nature of the solvent. The counter-ion in such reactions can conveniently be varied by using N-nitrosoamides as starting materials. These compounds react to give diazonium ion-pairs in reactions that closely resemble amine deamination, and the lifetime of the ion-pairs can be studied by using ^{18}O-labelled nitrosoamides (see Scheme 2, XH = solvent or added nucleophile).

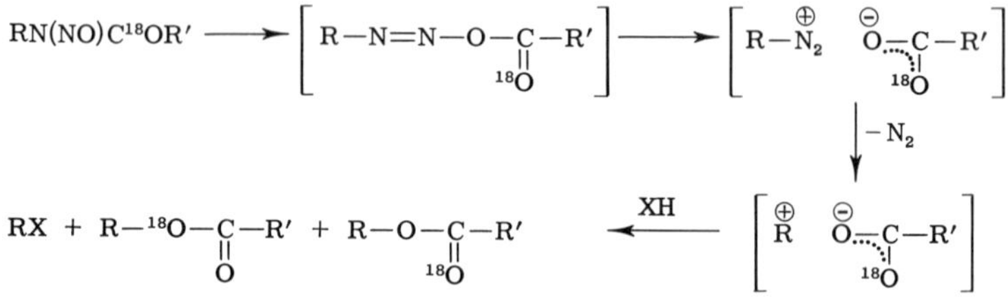

$$
RN(NO)C^{18}OR' \longrightarrow \left[R-N=N-O-\underset{\underset{^{18}O}{\parallel}}{C}-R' \right] \longrightarrow \left[R-\overset{\oplus}{N_2} \quad \overset{\ominus}{O}\underset{\underset{^{18}O}{\cdots\parallel}}{=}C-R' \right]
$$

$$
\downarrow -N_2
$$

$$
RX + R-{}^{18}O-\underset{\underset{O}{\parallel}}{C}-R' + R-O-\underset{\underset{^{18}O}{\parallel}}{C}-R' \overset{XH}{\longleftarrow} \left[\overset{\oplus}{R} \quad \overset{\ominus}{O}\underset{\underset{^{18}O}{\cdots:\parallel}}{=}C-R' \right]
$$

Scheme 2

(35) H. Maskill, R. M. Southam, and M. C. Whiting, *Chem. Commun.*, 496–498 (1965); M. C. Whiting, *Chem. Brit.*, 482–489 (1966).

(36) M. Chérest, H. Felkin, J. Sicher, F. Šipoš, and M. Tichý, *J. Chem. Soc.*, 2513–2521 (1965).

(37) See Ref. 30, p. 671.

The extent of ^{18}O-scrambling, and of inversion of configuration in the ester product resulting from intramolecular reaction, reflect the lifetime of the intermediate ion-pairs.[24,25,38]

There is little information concerning the lifetime of intermediate ion-pairs in aqueous media, but the influence of solvent composition on the stereochemistry of the substitution pathway for deaminations in aqueous acetic acid was pointed out in an important paper by Cohen and Jankowski.[39] The deamination of 9,10-*trans,trans*-2-aminodecahydronaphthalene (3) in acetic acid gave alcohol and acetate

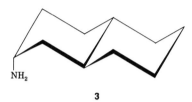

3

with preponderant (59%) retention of configuration; Table I[40] shows the yields of products as a proportion of the total substitution prod-

TABLE I[40]

Deamination of the 9,10-*trans*-2-Aminodecahydronaphthalenes

Products	Mole-% of acetic acid (+water)				
	100	75	50	25	3.4
	Yields of products				
A. From *trans,trans*[a]-2-aminodecahydronaphthalene					
trans-2-Acetoxydecahydronaphthalene	32	29	25	17	7
cis-2-Acetoxydecahydronaphthalene	39	35	29	24	26
trans-Decahydro-2-naphthol	27	31	33	32	16
cis-Decahydro-2-naphthol	2	5	13	27	51
B. From *trans,cis*[a]-2-Aminodecahydronaphthalene					
trans-2-Acetoxydecahydronaphthalene	18	15	11	6	3
cis-2-Acetoxydecahydronaphthalene	55	48	46	37	26
trans-Decahydro-2-naphthol	1	3	3	4	3
cis-Decahydro-2-naphthol	26	36	40	53	68

[a] Refers to the hydrogen atom on C-2.

(38) E. H. White and C. A. Aufdermarsh, Jr., *J. Amer. Chem. Soc.*, **83**, 1179–1190 (1961).

(39) T. Cohen and E. Jankowski, *J. Amer. Chem. Soc.*, **86**, 4217–4218 (1964).

(40) Reproduced from Ref. 27, p. 479, with the permission of John Wiley and Sons, Ltd.

ucts. In 3.4 mole-% aqueous acetic acid, inversion (77%) predominated. The stereochemistry of substitution also varied, although to
a much smaller extent, in the deamination of the corresponding
equatorial amine, 9,10-*trans,cis*-2-aminodecahydronaphthalene, there
being 81% retention in acetic acid and 94% retention in 3.4 mole-%
aqueous acetic acid. The most striking feature of these results is that,
for both amines in mixtures containing up to 50 mole-% of water, the
alcohol was formed with a much higher degree of retention than the
acetate, this presumably resulting from rapid collapse of the ion pair
corresponding to 2 in Scheme 1 (X = OAc).[41] These results probably
reflect a tendency towards the formation, in water-rich mixtures, of a
common intermediate, such as a fully-solvated carbonium ion. That
chair–chair interconversion of the intermediate carbonium ion does
not compete with reaction with nucleophilic solvent in the deamination of simple cyclohexylamines in aqueous medium is indicated by
the deamination of *cis*-2-deuteriocyclohexylamine,[42] which gave
mainly *cis*-2-deuteriocyclohexanol. That the carbonium ion had a
finite lifetime was shown by the deamination of cyclohexylamine by
^{18}O-labelled nitrous acid.[43] The reaction was conducted at or above
pH 6 (to minimize ^{18}O exchange with water), and the cyclohexanol
product contained < 10% of ^{18}O.

Solvent can have a marked effect on deamination in other respects.
For example, it is commonly found that the extent of rearrangement
is greater, the more polar the solvent;[29,44,45] this effect is attributed to
the increased lifetime of carbonium-ion species and their rearrangement to species that are more thermodynamically stable. Reactions
performed at different pH values may also give different products, as
illustrated by the deamination of 2-amino-2-deoxy-D-glucose diethyl
dithioacetal hydrochloride in solution in (*a*) aqueous acetic acid and
(*b*) hydrochloric acid (see p. 65).

Elimination to give alkenes is a major reaction-pathway in the
deamination of simple amines. The relative unimportance of elimination for amino groups equatorially attached to six-membered rings
has been attributed to the fact that, in the supposed transformation
from diazonium ion to carbonium ion, the developing carbon p-or-

(41) See also, Refs. 24 and 25.
(42) A. Streitwieser, Jr., and C. E. Coverdale, *J. Amer. Chem. Soc.*, **81**, 4275–4278
       (1959).
(43) D. L. Boutle and C. A. Bunton, *J. Chem. Soc.*, 761–763 (1961).
(44) H. Felkin, *Compt. Rend.*, **227**, 1383–1384 (1948).
(45) A similar trend has been observed for the closely related, *N*-nitrosoamide decom-
       position; see Ref. 23.

bital is never coplanar with the β-carbon–hydrogen orbital.[46] Substitution is the major reaction. However, it has now been found, by using deuterium-labelled amines, that *cis* elimination accounts for more than half of the elimination that occurs in the deamination of the epimeric 9,10-*trans*-2-aminodecahydronaphthalenes.[47] These results have been interpreted by postulating that the counter ion removes the β-proton. In a detailed study of the closely related decomposition of N-nitrosoamide derivatives of the epimeric 9,10-*trans*-2-aminodecahydronaphthalenes, the amount of *cis* elimination was found to be 84% and 94% for the equatorial and axial isomers, respectively, when α-elimination and hydride shift were taken into account.[48]

The minor products frequently formed in deaminations may result from such side reactions as reaction of the intermediate cation with nucleophiles, for example, acetate or chloride ions,[49,50] present in low concentration. Nitrite ions can give rise to nitrite esters or nitroalkanes;[50,51] the former are probably hydrolyzed prior to or during processing of the products.[52] Some alkenes have been reported to react with nitrous acid or related nitrosating agents.[19,53] Enol ethers would be expected to be more nucleophilic than simple alkenes, and it has been found that D-glucal reacts with nitrous acid under the conditions of deamination.[54]

Because of the number of oxygen functions present in carbohydrate amines, rearrangements of the semipinacol type, the formation of cyclic ethers, and neighboring-group participations are common in deamination. Most deamination studies reported have been conducted in water or water-containing solvents. There is clearly scope for studying deamination in nonaqueous solvents (solubility permitting) with alkyl nitrites[55] as nitrosating agents, and for using such al-

(46) See Ref. 12, pp. 644–646.

(47) T. Cohen and A. R. Daniewski, *J. Amer. Chem. Soc.*, **91**, 533–535 (1969).

(48) T. Cohen, A. R. Daniewski, G. M. Dub, and C. K. Shaw, *J. Amer. Chem. Soc.*, **94**, 1786–1787 (1972).

(49) S. Akiya and T. Osawa, *Yakugaku Zasshi*, **76**, 1280–1282 (1956).

(50) A. T. Austin, *Nature*, **188**, 1086–1088 (1960).

(51) A. T. Austin, *J. Chem. Soc.*, 149–157 (1950).

(52) I. T. Millar and H. D. Springhall, (Sidgwick's) "The Organic Chemistry of Nitrogen," Oxford University Press, London, 1966, p. 86.

(53) J. R. Park and D. L. H. Williams, *Chem. Commun.*, 332–333 (1969).

(54) G. Hutchinson and J. M. Williams, unpublished results.

(55) A. T. Jurewicz, J. H. Bayless, and L. Friedman, *J. Amer. Chem. Soc.*, **87**, 5788–5790 (1965).

ternative procedures as *N*-nitrosoamide thermolysis (see p. 67) and
the action of acids on triazenes.[56]

III. CYCLIC COMPOUNDS

1. Pyranose and Cyclitol Derivatives

a. Equatorial Amines. — (i) **Pyranose Derivatives.** Ledderhose[57]
was the first to treat 2-amino-2-deoxy-D-glucose hydrochloride with
nitrous acid in an attempt to convert it into a known hexose. The
syrupy, reducing, sugar product, 2,5-anhydro-D-mannose, was also
obtained by Fischer and Tiemann,[58] who named it chitose, and who
oxidized it to 2,5-anhydro-D-mannonic (chitonic) acid and 2,5-
anhydro-D-mannaric ("isosaccharic") acid. The latter acid had previously been obtained by the action of nitric acid on 2-amino-2-deoxy-
D-glucose,[59] but owing to incorrect analysis, the presence of an
anhydro ring was not recognized[60] until 1894. That this ring is a five-
membered one was indicated by the conversion of 2,5-anhydro-D-
mannaric acid into 2,5-furandicarboxylic acid and 2-furoic acid.[61] The
conversion of 2,5-anhydro-D-mannonic acid into 5-(acetoxymethyl)-
2-furoic acid by acetylation provided additional evidence for the for-
mulation of chitose as a 2,5-anhydrohexose.[62] When the deamination–
oxidation sequence was reversed, isomeric anhydrohexonic (chitaric)
and anhydrohexaric ("epi-isosaccharic") acids were obtained (by way
of 2-amino-2-deoxy-D-gluconic acid).[58] Fischer[63] suggested that the
isomeric acids are epimeric; this view was supported by Levene
and LaForge,[64] who established,[65,66] by an indirect method (summarized by Peat[1]), that chitose has the *manno* configuration. When
the configuration of 2-amino-2-deoxy-D-glucose (**4**) was confirmed by
synthesis[67] in 1939, inversion of configuration, which until then had
merely been presumed to occur in the deamination, was established.

(56) E. H. White and H. Scherrer, *Tetrahedron Lett.*, 758–762 (1961).
(57) G. Ledderhose, *Z. Physiol. Chem.*, **4**, 139–159 (1880).
(58) E. Fischer and F. Tiemann, *Ber.*, **27**, 138–147 (1894).
(59) F. Tiemann, *Ber.*, **17**, 241–251 (1884).
(60) F. Tiemann, *Ber.*, **27**, 118–138 (1894).
(61) F. Tiemann and R. Haarman, *Ber.*, **19**, 1257–1281 (1886).
(62) E. Fischer and E. Andreae, *Ber.*, **36**, 2587–2592 (1903).
(63) E. Fischer, *Ann.*, **381**, 123–141 (1911).
(64) P. A. Levene and F. B. LaForge, *J. Biol. Chem.*, **21**, 345–350 (1915).
(65) P. A. Levene and F. B. LaForge, *J. Biol. Chem.*, **21**, 351–359 (1915).
(66) P. A. Levene, *Biochem. Z.*, **124**, 37–83 (1921).
(67) W. N. Haworth, W. H. G. Lake, and S. Peat, *J. Chem. Soc.*, 271–274 (1939).

However, there was still some controversy regarding the structure of chitose. Compounds that had elemental analyses corresponding to those calculated for hexose derivatives were prepared from chitose in low or unspecified yield.[68] The D-*arabino*-hexose phenylosazone obtained from chitose by Fischer and Andreae in 1903 was attributed to the presence of unchanged amine;[62] an osazone claimed to be different from D-*arabino*-hexose phenylosazone was later shown to be, probably, a mixture of the osazones derived from D-glucose and D-arabinose, and D-glucose was considered to be the main deamination product.[69] It was not at that time realized that a substituent on C-2 could be cleaved in osazone formation. The uncertainty concerning the anhydride structure of chitose was reinforced by the report that a derivative of 2-amino-2-deoxy-D-glucose gave D-mannose after deamination and hydrolysis (see later). A crystalline diphenylhydrazone, different from the corresponding derivatives from D-glucose and D-mannose was obtained[70,71] in 1935, and a crystalline benzylphenylhydrazone[72] in 1950. Crystalline 2,5-anhydro-D-mannose diphenylhydrazone was obtained in 1954, and reconverted to the free sugar by the action of benzaldehyde. 2,5-Anhydro-D-mannose (**5**) regenerated in this way was reported to consume one equivalent of periodate and yield no formaldehyde or formic acid.[73] The preparation of crystalline di-O-trityl and di-O-*p*-tolylsulfonyl (**6**) derivatives from the product obtained by high-pressure hydrogenation of 2,5-anhydro-D-mannose, and the oxidation of the di-*p*-toluenesulfonate with lead tetraacetate to a reducing compound having an analysis corresponding to that calculated for the dialdehyde **7**, provided further evidence for the 2,5-anhydro-D-mannose structure.[73] Indeed, the optical activity of the dialdehyde di-*p*-toluenesulfonate (**7**) confirmed the *manno* configuration, because 2,5-anhydro-D-glucose would have given an optically inactive, *meso*-dialdehyde.

Grant[74] conducted a detailed study of the properties of 2,5-anhydro-D-mannose (chitose), and prepared several new derivatives that confirmed the anhydrohexose structure. The mutarotation of 2,5-anhydro-D-mannose, its conversion under mild conditions into a

(68) C. Neuberg, H. Wolff, and W. Neimann, *Ber.*, **35**, 4009–4023 (1902).
(69) L. Zechmeister and G. Tóth, *Ber.*, **66**, 522 (1933).
(70) P. Schorigin and N. N. Makarowa-Semljanskaja, *Ber.*, **68**, 965–969 (1935).
(71) See also: Y. Inoue and K. Onodera, *Nippon Nogei Kagaku Kaishi*, **25**, 45–47 (1951–52); *Chem. Abstr.* **48**, 2002f (1954).
(72) T. Nagaoka, *Tohoku J. Exp. Med.*, **53**, 21–27 (1950).
(73) S. Akiya and T. Osawa, *Yakugaku Zasshi*, **74**, 1259–1262 (1954).
(74) A. B. Grant, *New Zealand J. Sci. Technol.*, **B37**, 509–521 (1956).

dimethyl acetal, and its inability to react with the Schiff reagent led Grant to postulate the bicyclic, furanose structure **8**. However, it has

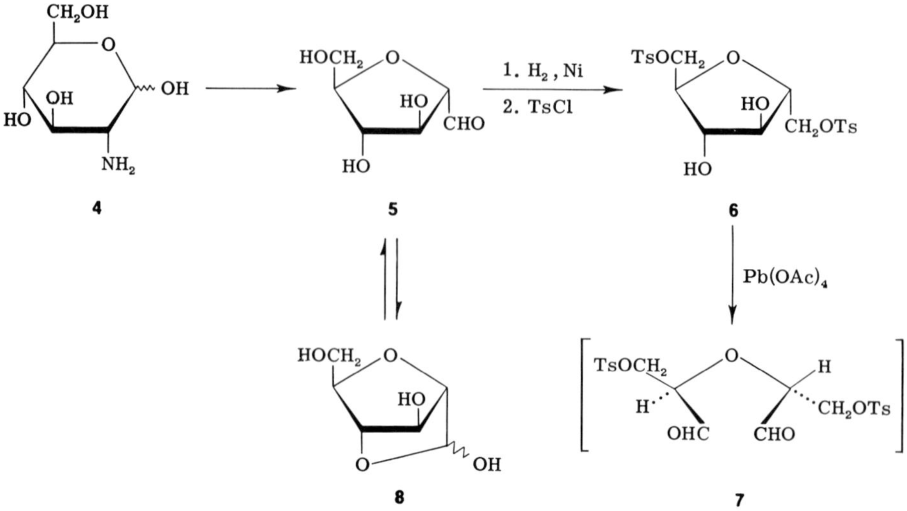

where   Ts = $SO_2C_6H_4Me$-p.

been suggested that the mutarotation of those 2,5-anhydrohexoses in which C-1 and C-6 are *trans*-related is due to (*a*) dimerization, (*b*) reversible addition of water or alcohols to the carbonyl group, (*c*) enolization, or (*d*) dehydration of the free aldehyde.[6,75] 2,5-Anhydrohexoses (such as 2,5-anhydro-D-glucose) in which C-1 and C-6 are *cis*-related can form a pyranoid ring.[76] Further confirmation of the 2,5-anhydro-D-mannose structure was obtained[79] by hydrogenation to crystalline 2,5-anhydro-D-mannitol, which was converted into optically inactive bis(2-deoxyglycerol-2-yl) ether by periodate oxidation followed by reduction with sodium borohydride (see also Ref. 80).

2,5-Anhydro-D-mannose is labile under acidic and basic conditions, and decomposition occurs when a neutral, aqueous solution is heated, or evaporated to dryness under diminished pressure at

(75) J. Defaye and S. D. Gero, *Bull. Soc. Chim. Biol.*, **47**, 1767–1775 (1965).
(76) The reported lack of mutarotation for 2,5-anhydro-D-glucose, which was tentatively identified as the product of the reaction of 2-amino-2-deoxy-D-mannose with mercuric oxide, has not been confirmed.[77,78]
(77) P. A. Levene, *J. Biol. Chem.*, **39**, 69–76 (1919).
(78) P. A. Levene, *J. Biol. Chem.*, **59**, 135–139 (1924).
(79) B. C. Bera, A. B. Foster, and M. Stacey, *J. Chem. Soc.*, 4531–4535 (1956).
(80) J. Kiss, E. Hardegger, and H. Furter, *Chimia*, **13**, 336–337 (1959).

35–40°. Such lability[74] doubtless accounts for the difficulties earlier experienced in obtaining crystalline derivatives.

Glycosides of 2-amino-2-deoxy-D-glucose also give 2,5-anhydro-D-mannose upon deamination, the aglycon being released. The first glycoside to be studied was methyl 2-amino-2-deoxy-β-D-glucopyranoside,[81] which gave a syrupy, reducing product resembling chitose. The 4,6-O-benzylidene derivative of the glycoside was first reported to give a crystalline O-benzylidene-D-mannose,[82] but this was later shown to be a 2,5-anhydro-D-mannose derivative.[83] Methyl 2-amino-2-deoxy-4,6-O-ethylidene-3-O-methyl-D-glucopyranoside gave, in 23% yield, a crystalline product tentatively identified as 2,5-anhydro-4,6-O-ethylidene-3-O-methyl-D-mannose.[84]

The deaminations of both anomers of methyl 2-amino-2-deoxy-D-glucopyranoside were compared in 1953, and the β-D anomer was found by polarimetry to react more rapidly than the α-D anomer;[85] the reasons for this difference are not clear.[86] In subsequent work, the 2,5-anhydro-D-mannose product was, in each case, reduced, and the product characterized as 2,5-anhydro-D-mannitol.[87,88] The latter was obtained in lower yield (59%) from the α-D-glycoside than (71%) from 2-amino-2-deoxy-D-glucose, and this may be due to the greater relative importance of other reaction pathways.[88] Thus, when reduction with tritiated sodium borohydride followed by radiochromatography were used to compare the deamination products of methyl 2-amino-2-deoxy-α- and -β-D-glucopyranoside, it was found that the α anomer gave a substantial proportion ($\sim 40\%$ of the total tritium) of a product having a mobility higher than that of 2,5-anhydro-D-mannitol on paper chromatograms.[89] That the second product from ethyl 2-amino-2-deoxy-α-D-glucopyranoside had an even higher mobility suggests that these products retained the aglycon, the ethyl glycoside being slightly less hydrophilic and, hence, more mobile. A likely structure for these second products is the 2-deoxy-2-C-(hydrox-

(81) J. C. Irvine and A. Hynd, *J. Chem. Soc.*, **101**, 1128–1146 (1912).

(82) J. C. Irvine and A. Hynd, *J. Chem. Soc.*, **105**, 698–710 (1914).

(83) S. Akiya and T. Osawa, *Chem. Pharm. Bull.* (Tokyo), **7**, 277–281 (1959).

(84) S. Akiya and T. Osawa, *Yakugaku Zasshi*, **74**, 1276–1279 (1954).

(85) A. B. Foster, E. F. Martlew, and M. Stacey, *Chem. Ind.* (London), 825–826 (1953).

(86) The comparison, in Ref. 85, between deamination and borate complexing is not strictly relevant, as the latter is reversible and may involve some distortion of the pyranose ring.

(87) B. C. Bera, Ph. D. Thesis, University of Birmingham, England, 1956; cited in Ref. 88.

(88) D. Horton and K. D. Philips, *Carbohyd. Res.*, **30**, 367–374 (1973).

(89) J. E. Shively and H. E. Conrad, *Biochemistry*, **9**, 33–43 (1970).

ymethyl)pentofuranosides (**9**), which would result from migration of
C-2 (followed by reduction).

Such compounds have been identified amongst the products of
deamination, followed by reduction, of methyl 2-amino-2-deoxy-α-
and -β-D-glucopyranoside.[89a] Two methyl 2-deoxy-2-(hydroxymethyl)pentofuranosides, epimeric at C-2, were detected in each
case, and the results of deuterium-incorporation experiments
suggested that the primary product was epimerized, probably during
reduction with sodium borohydride. However, in the writer's
opinion, the configurations assigned to the primary and epimerized
products are incorrect and should be reversed.

Migration of the ring-oxygen atom also occurred predominantly in
the deamination of the tetra-O-acetyl derivatives of 2-amino-2-deoxy-
α- and -β-D-glucopyranose to give 3,4,6-tri-O-acetyl-2,5-anhydro-D-
mannose (**10**), which is unstable and readily eliminates acetic acid to

$$HOCH_2 \quad O$$
$$OR$$
$$HO \quad CH_2OH$$

9 R = Me or Et

$$AcOCH_2 \quad O$$
$$AcO$$
$$CHO$$
$$AcO$$

10

form 5-(acetoxymethyl)-2-furaldehyde.[88] The triacetate (**10**) was
characterized by reduction, and deacetylation to crystalline 2,5-
anhydro-D-mannitol (44%). Small amounts of D-glucose, D-mannose,
and 2-acetamido-2-deoxy-D-glucose were detected after deacetylation. Although the use of O-acetylated derivatives did not prevent
the migration of the ring-oxygen atom in these reactions, other
rearrangements that involve migration of a carbon atom can be so
minimized (see p. 29). The closely related reactions of N-nitroso
derivatives of methyl 2-acetamido-3,4,6-tri-O-acetyl-2-deoxy-β-D-
glucopyranoside and of 2-acetamido-1,3,4,6-tetra-O-acetyl-2-deoxy-
α- and -β-D-glucopyranose are discussed in Section V,2 (see p. 67).

The unsuccessful attempt,[90] in 1911, to identify the product of
deamination of methyl 3,4,6-tri-O-acetyl-2-amino-2-deoxy-β-D-glu-
copyranoside hydrobromide was clearly due to the instability of the
triacetate **10**.

The cleavage of glycosidic linkages by means of deamination in

(89a) B. Erbing, B. Lindberg, and S. Svensson, *Acta Chem. Scand.*, **27**, 3699–3704
    (1973).
(90) J. C. Irvine, D. McNicoll, and A. Hynd, *J. Chem. Soc.*, **99**, 250 (1911).

more-complex glycosides, including polymers, is discussed in detail in Section VI (see p. 73).

The highly stereoselective rearrangements that occur in the de-amination of six-membered ring compounds containing a hydroxyl (or alkoxyl) group vicinal to an amino group are markedly dependent on conformation, as shown by the behavior of cyclohexylamine deriva-tives, such as the 2-amino-4-*tert*-butylcyclohexanols, in which one chair conformation is highly favored.[36] When the amino group is equatorially attached, the atom *trans* and coplanar to the nitrogen atom is a ring-carbon atom, and ring contraction results. The impor-tance of conformation is well illustrated by the deamination of *cis*-2-aminocyclohexanol, the two chair conformers of which are of similar free-energy; products are obtained from both conformers[91] (see Scheme 3). The ratio of the two products has been found not to cor-

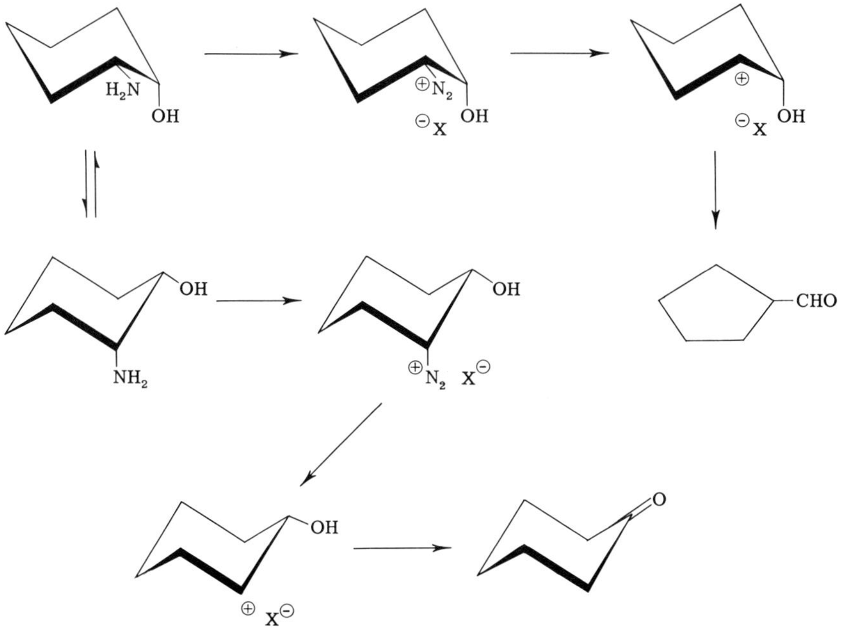

Scheme 3

respond to the ratio of the two chair conformers of the amine, and it was suggested that chair–chair interconversion occurs at an interme-diate stage, such as the diazohydroxide.[36]

A concerted reaction involving heterolytic bond-fission of the dia-

(91) G. E. McCasland, *J. Amer. Chem. Soc.*, **73**, 2293–2295 (1951).

zonium ion intermediate was proposed by Peat[1] for the deamination of 2-amino-2-deoxy-D-glucose, and a carbene intermediate was postulated by Matsushima.[92] Although a carbene can, in principle, arise from a diazoalkane, the latter, even if formed, is more likely to be reconverted into a diazonium ion by protonation. The importance of conformation in these reactions was pointed out by Shafizadeh,[2] who cited the deamination of cis-2-aminocyclohexanol. Although these reactions were represented as concerted, they may proceed through carbonium-ion pairs, as suggested by White (see p. 12). The driving force for these semipinacolic rearrangements is considered to be the formation of a resonance-stabilized cation (12). When there is

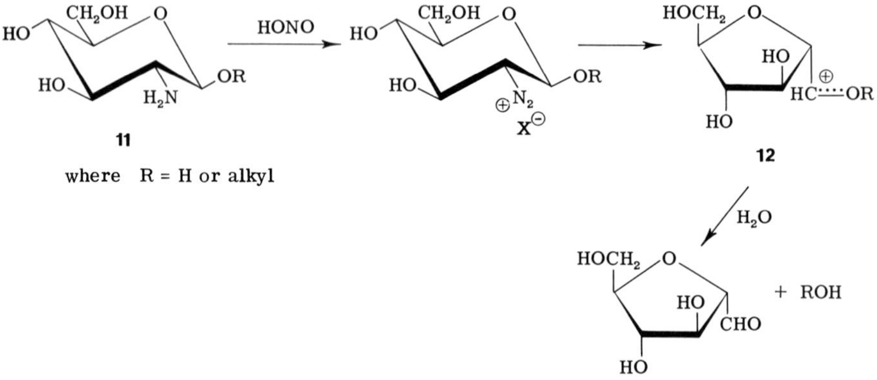

where   R = H or alkyl

no anomeric oxygen atom, as in 2-amino-1,5-anhydro-2-deoxy-D-glucitol (13), the major product of deamination does not result from migration of the ring-oxygen atom. Instead, a substitution product, namely 1,5-anhydro-D-glucitol (15), is formed, and an unidentified,

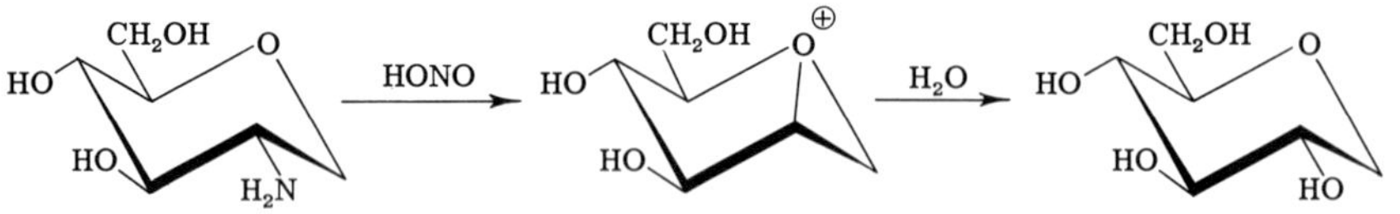

reducing product may be the compound resulting from migration of C-4. The stereospecific formation of 1,5-anhydro-D-glucitol as the major product was attributed to participation by the ring-oxygen atom, to give the bicyclic oxonium ion (14). Attack of solvent water could then occur at C-1, C-2, or C-5, and the formation of a six-mem-

(92) Y. Matsushima, Bull. Chem. Soc. Jap., 24, 144–147 (1951).

bered ring was evidently favored.[93] Similar participation of the ring-oxygen atom in the deamination of 2-amino-2-deoxy-D-glucose derivatives (11) may occur prior to cleavage of the C-1–ring-oxygen bond, as transannular overlap of the lone-pair orbital of the ring-oxygen atom can occur with the developing p-orbital on C-2. Related σ–p orbital interactions may account for the bicyclo[3.1.0]hexane derivatives formed from some cyclohexylamines having an equatorially attached amino group.[94]

2,5-Anhydro-D-mannose is also formed in the solvolysis of methyl 2-O-(p-nitrophenylsulfonyl)-α-D-glucopyranoside in water containing sodium acetate.[95]

The deamination of 2-amino-2-deoxy-D-galactose is entirely analogous to that of the D-glucose amine, and the product, 2,5-anhydro-D-talose ("chondrose"), was first obtained by Levene and LaForge,[96–98] who converted it into D-*lyxo*-hexose phenylosazone, 2,5-anhydro-D-talonic acid, and 2,5-anhydro-D-altraric(-talaric) acid derivatives. The epimeric 2,5-anhydro-D-galactonic and -D-galactaric acids were obtained by deamination of 2-amino-2-deoxy-D-galactonic acid, and the configurations of these acids, and, hence, of 2,5-anhydro-D-talose, were deduced from the optical inactivity of the anhydrogalactaric acid. 2,5-Anhydro-D-talose, which cannot be stored at room temperature,[99,100] was later characterized as its crystalline benzylphenylhydrazone[72] and as 2,5-anhydro-D-talitol[99] derivatives, and further confirmation of the structure was provided by studies of the latter compound (see Ref. 6).

Formation of neither a furanose nor a pyranose form can occur for 2,5-anhydro-D-talose, and the mutarotation reported for an aqueous solution must be due to hydration of the aldehyde group, or other processes,[6,75] or both (see p. 20).

Deamination followed by reaction with indole in the presence of hydrochloric acid was introduced in 1950 as a method for the micro-

(93) N. M. K. Ng Ying Kin, J. M. Williams, and A. Horsington, *J. Chem. Soc. (C)*, 1578–1583 (1971).

(94) A. T. Jurewicz, Ph. D. Thesis, Case Institute of Technology, Cleveland, Ohio, U. S. A., 1967; cited in Ref. 30, p. 673.

(95) P. W. Austin, J. G. Buchanan, and R. M. Saunders, *Chem. Commun.*, 146–147 (1956); *J. Chem. Soc. (C)*, 372–377 (1967).

(96) P. A. Levene and F. B. LaForge, *J. Biol. Chem.*, **18**, 123–130 (1914).

(97) P. A. Levene and F. B. LaForge, *J. Biol. Chem.*, **20**, 433–444 (1915).

(98) P. A. Levene, *J. Biol. Chem.*, **31**, 609–621 (1917).

(99) J. Defaye, *Bull. Soc. Chim. Fr.*, 999–1002 (1964).

(100) P. A. Levene and R. Ulpts, *J. Biol. Chem.*, **64**, 475–483 (1925).

determination of 2-amino-2-deoxy-D-glucose and -D-galactose.[101a,101b] The deamination products (2,5-anhydrohexoses) reacted under these conditions to give a chromophore absorbing at 492 nm. When 2-amino-2-deoxy-D-xylose and a new pentosamine (believed to be 2-amino-2-deoxy-L-arabinose) were found to give, respectively, strong and weak absorptions at 492 nm, the different behavior was correctly attributed to the presence in each amine of different proportions of the conformation possessing an equatorially attached amino group.[102] Although it was claimed that these results supported the L-*arabino* configuration postulated, the new amine was subsequently shown to be 2-amino-2-deoxy-L-ribose.[103] Despite the fact that the products of deamination of 2-amino-2-deoxypentoses have not yet been identified, the intensity of the absorption at 492 nm obtained in the indole reaction indicates that the 2,5-anhydropentose (which results from the equatorial amine) is formed from the xylose amine to a major extent, from the ribose amine to a minor extent, and from the arabinose amine[104] to some extent.

The deamination of derivatives of paromobiosamine [3-O-(2,6-diamino-2,6-dideoxy-β-L-idopyranosyl)-D-ribose (16)] was used in the elucidation of the structure of the disaccharide, although the interpretation of the results was not straightforward.[105] The deamination of the methyl glycoside (17) of this disaccharide[106] was first

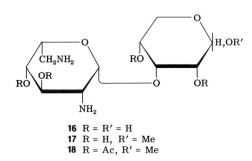

16 R = R' = H
17 R = H, R' = Me
18 R = Ac, R' = Me

(101a) Z. Dische and E. Borenfreund, *J. Biol. Chem.*, **184**, 517–522 (1950).
(101b) See also, D. Exley, *Biochem. J.*, **67**, 52–60 (1957).
(102) M. L. Wolfrom, F. Shafizadeh, and R. K. Armstrong, *J. Amer. Chem. Soc.*, **80**, 4885–4891 (1958).
(103) M. L. Wolfrom, R. K. Armstrong, and T. M. Shen Han, *J. Amer. Chem. Soc.*, **81**, 3716–3719 (1959).
(104) M. L. Wolfrom and Z. Yosizawa, *J. Amer. Chem. Soc.*, **81**, 3477–3478 (1959).
(105) T. H. Haskell and S. Hanessian, *J. Org. Chem.*, **28**, 2598–2604 (1963).
(106) Obtained from the antibiotic neomycin B.

reported[107] to give a nonreducing product that, on hydrolysis, gave D-ribose. However, in later work,[105] ribose and a trace of galactose were identified as products of the deamination of both the disaccharide **16** and its methyl glycoside (**17**). The formation of ribose was attributed to rearrangement of the diaminodideoxyhexose moiety, with concomitant cleavage of the glycosidic linkage, but cleavage of *two* glycosidic linkages was necessary to account for the formation of ribose from the methyl glycoside. In order to avoid such glycosidic cleavage, the tetra-O-acetyl derivative of the methyl glycoside was deaminated, although there was no theoretical justification for assuming that the ring contraction would be so avoided (see p. 24). The formation of ribose in the deamination of the disaccharide (**16**) indicated that at least some of the L-idopyranosyl moiety exists in the $^4C_1$ conformation (**19**) in which the bond to the 2-amino group is

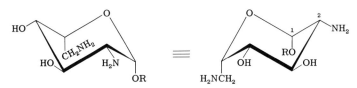

19 R = D-ribose residue

equatorial. The formation of D-ribose from the methyl glycoside suggests that the methyl riboside[108] released in the deamination was at least partially hydrolyzed during processing.

Deamination of the tetraacetate (**18**), followed by acid hydrolysis, gave at least three products, two of which were identified as ribose and L-galactose. The latter was isolated in 7% yield, and characterized as the crystalline methylphenylhydrazone. The formation of L-galactose was rationalized by postulating the participation of the 3-acetoxyl group of the $^4C_1$ conformation[109] to give a cyclic acetoxonium ion (**20**), which then suffered attack of water at C-3. However, no evidence for similar acetoxyl-group participation has been reported in the deamination of other per-O-acetylated amino sugars and inosamines (see p. 38), and such acetoxonium ions as **20** would be ex-

(107) K. L. Rinehart, Jr., A. D. Argoudelis, W. A. Goss, A. Sohler, and C. P. Schaffner, *J. Amer. Chem. Soc.*, **82**, 3938–3946 (1960).

(108) The methyl riboside may have contained some methyl ribofuranosides, which are more susceptible to acid hydrolysis.

(109) This conformation would, however, be expected to give the ring-contracted product.

pected to be hydrolyzed in aqueous medium without inversion of configuration.[110] As the tetraacetate **18** was not fully characterized, it is possible that the L-galactose could have arisen by way of an epoxide, which would be the likely deamination product for the $^1C_4$ conformation (**21**) (see Section III,1,b; p. 40) were the 3-hydroxyl

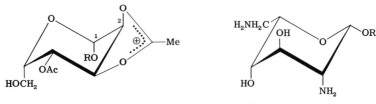

20 R = D-ribose residue   **21** R = D-ribose residue

group unesterified. Formation of an epoxide intermediate is the only way in which to account for the L-galactose isolated from the unsubstituted disaccharide.

On deamination, methyl 3-amino-3-deoxy-α-D-glucopyranoside (**22**) gave[111] a branched-chain glycofuranoside that resulted from migration of C-5. Methyl 3-deoxy-3-C-formyl-α-D-xylofuranoside[111] (**23**) was isolated in 30% yield[112] as a crystalline Schiff base, and was oxidized to a γ-lactone. The formyl-xyloside crystallized as the cyclic hemiacetal[95,113] (**24**), and a crystalline dibenzoate[93] of **24** was obtained in later work. At least one other, unidentified, minor product was

22   **23**   **24**

formed in the deamination.[112] The *manno* epimer of **22** underwent a similar rearrangement, to give, in 60% yield, methyl 3-deoxy-3-C-formyl-α-D-lyxofuranoside, whose structure was confirmed by synthesis of the derived lactone.[95] In the favored conformer of both the aminodeoxy-D-glucoside (**22**) and aminodeoxy-D-mannoside, the atoms that are trans and coplanar to the nitrogen atom are C-5 and C-1, and the tendency for C-5 migration may be attributable to the lower nucleophilicity of C-1, which is bonded to two oxygen atoms.

(110) C. B. Anderson, E. C. Friedrich and S. Winstein, *Tetrahedron Lett.*, 2037–2044 (1963), and previous papers.
(111) This was incorrectly named as a ribofuranoside in Ref. 113.
(112) S. Inoue, personal communication.
(113) S. Inoue and H. Ogawa, *Chem. Pharm. Bull.* (Tokyo), 8, 79–80 (1960).

The same products were obtained in the solvolysis of the corresponding 3-O-(p-nitrophenylsulfonyl) glycosides.[95]

There seems to be no obvious reason for the reported failure to deaminate 3-amino-3,6-dideoxy-D-mannose.[114]

The deamination of methyl 3-amino-3-deoxy-β-D-xylopyranoside (25) gave a mixture which was shown by n.m.r. spectroscopy to contain three components in the ratios of 4:1:1. The major product, methyl 3-deoxy-3-C-formyl-α-L-threofuranoside (26) was characterized as methyl 3-deoxy-3-C-(hydroxymethyl)-α-L-threofuranoside (64%) and its di-p-nitrobenzoate (55%), the structures being established by n.m.r. spectroscopy.[115] Methyl 2-deoxy-2-C-formyl-β-D-threofuranoside (27) was assumed to be a second product, because

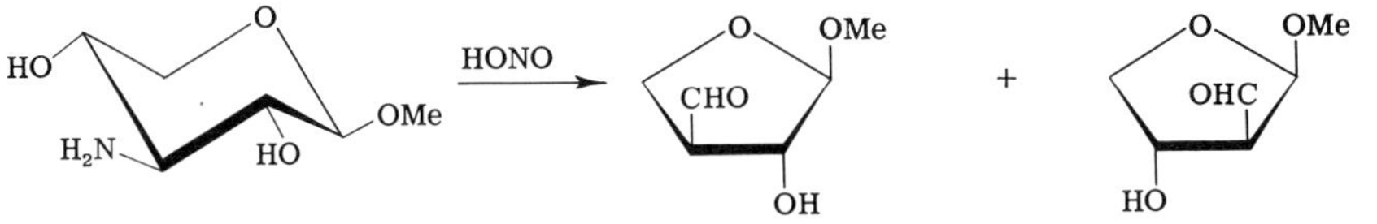

the crude deamination product was converted, by a sequence that included reduction, esterification, and glycosyl chloride formation, into four isomeric 9-[deoxy(hydroxymethyl)tetrofuranosyl]adenines, two of which were assumed to be the anomers of 9-[2-deoxy-2-C-(hydroxymethyl)tetrofuranosyl]adenine.

Such rearrangements as the foregoing can often be minimized by deaminating the per-O-acetylated amine. Such deaminations were first reported for inosamines (see p. 36), and the relative unimportance of rearrangement is presumably due to the smaller tendency of the acetoxy oxygen atom to delocalize the charge in the rearranged cation (28). In a study of the structure of 3-amino-3-deoxy-D-glucose,

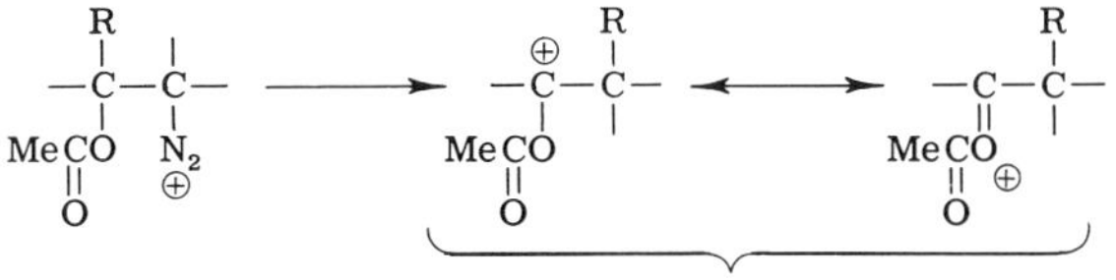

obtained from the antibiotic kanamycin, the 1,2,4,6-tetra-O-acetyl derivative was prepared, and deaminated *in situ*.[116] Acetylation, fol-

(114) J. D. Dutcher, D. R. Walters, and O. Wintersteiner, *J. Org. Chem.*, **28**, 995–999 (1963).

(115) E. J. Reist, *Chem. Ind.* (London), 1957–1958 (1967); E. J. Reist, D. F. Calkins, and L. Goodman, *J. Amer. Chem. Soc.*, **90**, 3852–3857 (1968).

(116) M. J. Cron, O. B. Fardig, D. L. Johnson, D. F. Whitehead, I. R. Hooper, and R. U. Lemieux, *J. Amer. Chem. Soc.*, **80**, 4115 (1958).

lowed by hydrolysis of the product, gave glucose (identified by paper chromatography), and α-D-glucopyranose pentaacetate was isolated in 1% yield[117] after column chromatography. Thus, the major reaction-pathway appears to be substitution with retention of configuration. However, substitution with inversion of configuration was found to occur in the deamination of the 2,4,6-triacetate hydrochloride of 1-(3-amino-3-deoxy-β-D-glucopyranosyl)uracil (**29**, R = uracil-1-yl).[118] After deacetylation, the allopyranoside (**30**, R = uracil-1-yl)

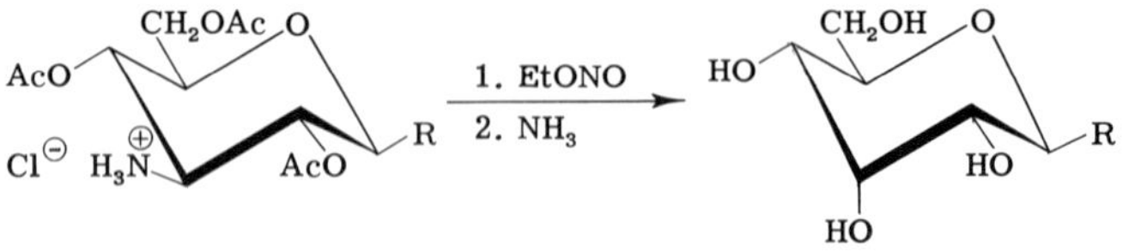

was isolated crystalline in 32% yield and its structure was established unambiguously by synthesis.[119] To account for the inversion of configuration, it was suggested that electrostatic interaction between the diazonium group and an oxygen atom in the aglycon occurs, thus resulting in the SN2 type of attack of solvent at C-3. However, the postulated interaction of diazonium group and aglycon can only occur in such highly unfavorable conformations as **31** or

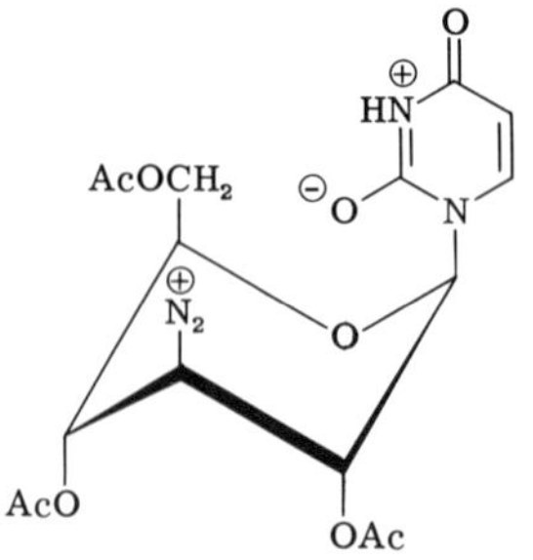

skews, and the triacetate **29** has the $^4C_1$ conformation in aqueous and dimethyl sulfoxide solutions, as shown by n.m.r. data.[120] The deaminations of the two 3-amino-3-deoxy-D-glucose derivatives are, for several reasons, difficult to compare and rationalize. The conditions

(117) M. J. Cron, personal communication.
(118) K. A. Watanabe, J. Beránek, H. A. Friedman, and J. J. Fox, *J. Org. Chem.*, **30**, 2735–2739 (1965).
(119) D. H. Warnock, K. A. Watanabe, and J. J. Fox, *Carbohyd. Res.*, **8**, 127–130 (1971).
(120) K. A. Watanabe, personal communication.

used were different; the tetraacetate was deaminated with sodium nitrite in dilute aqueous perchloric acid, and the deamination of the nucleoside **29** was effected with ethyl nitrite in a solution containing approximately equal volumes of acetic acid, water, and ethanol. Only the nucleoside starting material was well characterized, and it gave the higher yield of purified product, but in neither reaction was the crude product monitored for the epimer of the product isolated. Results obtained with O-acetylated inosamines (see p. 37) indicated that substitution with inversion of configuration is the normal result of deamination of such compounds, except when steric factors dictate otherwise. The 1,2,4,6-tetra-O-acetyl-3-amino-3-deoxy-D-glucopyranose presumably consisted mainly of the α-D anomer, and the steric effect of the axially attached, anomeric, acetoxyl group may account for the different results obtained with the tetraacetate and the nucleoside **29**.

Occurrence of ring contraction was not reported for the deamination of methyl 4-amino-4,6-dideoxy-α-D-mannopyranoside, the major product obtained after hydrolysis being tentatively identified (by paper chromatography) as a 6-deoxymannose,[121] formed by substitution with retention of configuration. In a detailed study of the deamination of methyl 4-amino-4-deoxy-α-D-glucopyranoside (**32**) four products were identified.[93,122] Methyl α-D-glucopyranoside (**33**) was shown to be the major product, and was isolated in 35% yield. Two products of rearrangement, namely methyl β-L-altrofuranoside (**34**) and methyl 3-deoxy-3-C-formyl-α-D-xylofuranoside (**24**) were isolated in 7 and 14% yields, respectively, and a third rearrangement product, namely, 4,5-anhydro-D-galactose (**35**), formed in ~10%

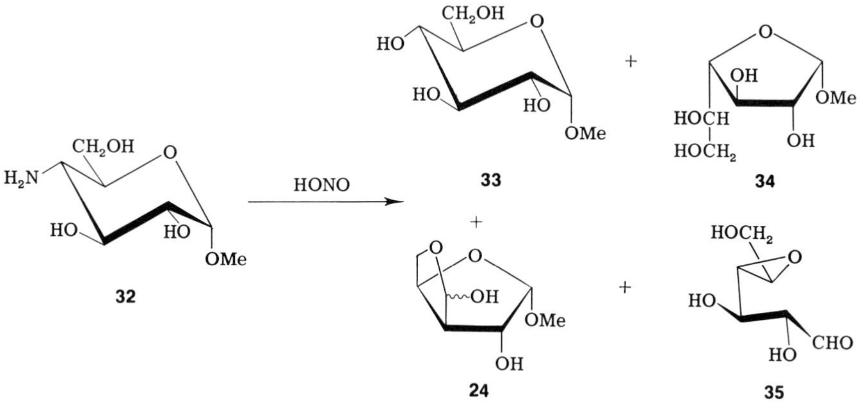

(121) C. Lee and C. P. Schaffner, *Tetrahedron Lett.*, 5837–5840 (1966).
(122) N. M. K. Ng Ying Kin, J. M. Williams, and A. Horsington, *Chem. Commun.*, 971–972 (1969).

yield, was labile and was identified as follows. When the reaction solution was de-ionized by means of ion-exchange resins, the labile component was hydrolyzed to glucose. This labile component was shown to be rapidly hydrolyzed under acidic and basic conditions, to give glucose and a trace of altrose (detected on chromatograms). 4,5-Anhydro-D-galactose was considered to be a likely deamination product and precursor of glucose, which was formed by an intramolecular cyclization (see Scheme 4, R = H) to give, initially, D-glucofuranose.

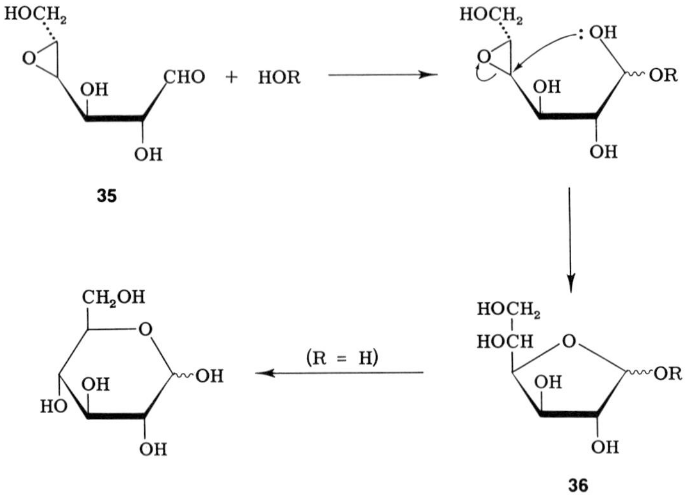

Scheme 4

Cyclization to give a six-membered ring (L-altropyranose) would be expected to be less favored in such irreversible reactions. Support for the epoxide structure was provided by "trapping" the D-glucofuranose intermediate as the methyl D-glucofuranosides (36, R = Me) by treating the neutralized, lyophilized product-mixture with cold, methanolic sodium methoxide. After chromatography, the α- and β-D-glucofuranosides were isolated in 3.4 and 6% yields, respectively.

The stereospecific formation of methyl α-D-glucopyranoside as the major product indicated that the ring-oxygen atom participates to give a bicyclic oxonium ion (37), which can then react at C-1 with solvent to give the epoxide 35, at C-4 to give the D-glucopyranoside 33, and at C-5 to give the L-altrofuranoside 34. The formation of a six-membered ring is clearly favored, as with 2-amino-1,5-anhydro-2-deoxy-D-glucitol (13). Migration of C-2 competes with participation of the ring-oxygen atom, but the product expected from such a migration is the D-ribofuranoside **38**. That the D-xylofuranoside (**24**) isolated

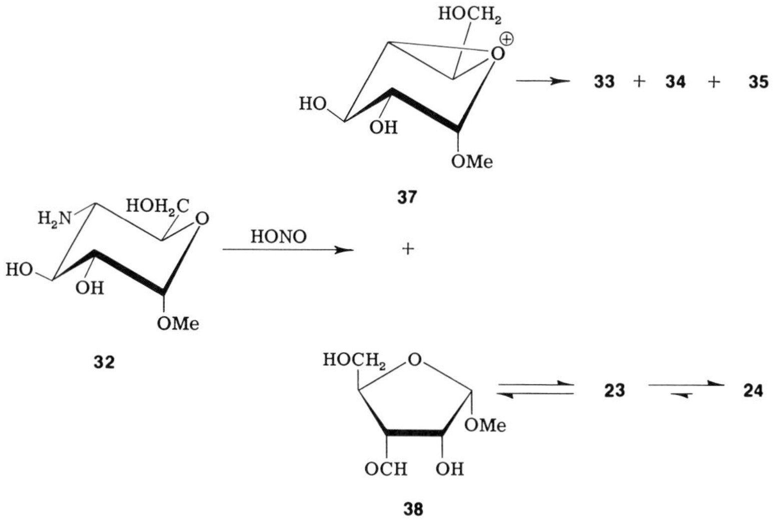

was a secondary product formed by epimerization of the D-ribofuranoside under the conditions of the reaction or the processing was indicated by the isolation, after column chromatography, of a slightly impure aldehyde which, by n.m.r. measurements, was shown to isomerize at room temperature to the D-xylofuranoside. The equilibrium between the two aldehydes is, presumably, displaced by cyclic hemiacetal formation. The D-ribofuranoside (**38**) and the D-xyloside hemiacetal (**24**) had the same R_F value on a paper chromatogram, but, with alkaline silver nitrate, reacted rapidly and slowly, respectively.

The solvolysis of methyl 4-O-(p-nitrophenylsulfonyl)-α-D-glucopyranoside in water containing sodium acetate differed significantly from the amine deamination just discussed.[123] 4,5-Anhydro-D-galactose did not survive the reaction conditions (100°) and glucose was formed, the epoxide being postulated as an intermediate. Methyl α-D-glucopyranoside (50%) and methyl β-L-altrofuranoside (8%) were also formed, but the former was not formed stereospecifically, because methyl α-D-galactopyranoside (8%) was also isolated. The absence of the product resulting from migration of C-2 can be attributed to the much stronger anchimeric effect of the ring-oxygen

(123) P. W. Austin, J. C. Buchanan, and D. G. Large, *Chem. Commun.*, 418–419 (1967).

atom, which is likely to lead to the bicyclic oxonium ion (37). The methyl α-D-galactopyranoside was considered to arise from the C-4 carbonium ion,

Participation of the ring-oxygen atom has also been postulated,[124] to account for the stereospecific formation of L-mannopyranoside (40) and D-allofuranoside (41) products in the deamination, by way of 39b, of methyl 4-amino-4,6-dideoxy-2,3-O-isopropylidene-α-L-mannopyranoside (39a). Ring contraction to give furanoside products pre-

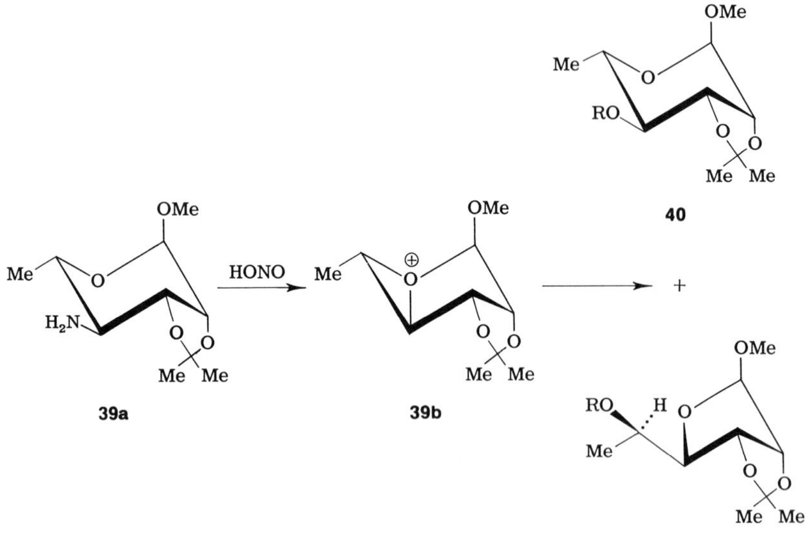

dominated, in contrast to the other deaminations, which have been postulated to occur through similar, cyclic oxonium ions. However, the fusion of a second ring to the pyranose ring may favor ring contraction. Ring contraction was not the major reaction-pathway in the deamination of methyl 4-amino-4,6-dideoxy-α-D-mannopyranoside, but a detailed product-analysis was not reported.[121]

The deamination of the pentoside methyl 4-amino-4-deoxy-2,3-O-isopropylidene-α-D-lyxofuranoside paralleled that of the mannopyranoside 39a in that substitution (with inversion) and ring contraction occurred to give α-D-lyxopyranoside (60%) and β-L-ribofuranoside (40%) derivatives as products.[124a] In contrast, there was no ring

(124) A. K. Al-Radhi, J. S. Brimacombe, and L. C. N. Tucker, *Chem. Commun.*, 1250–1251 (1970); *J. Chem. Soc. Perkin I*, 315–320 (1972).

(124a) J. S. Brimacombe, J. Minshall, and L. C. N. Tucker, *Carbohyd. Res.*, **35**, 55–63 (1974).

contraction when the 4-*p*-toluenesulfonate corresponding to the pentoside amine was submitted to nucleophilic displacements. With sodium azide in *N,N*-dimethylformamide, nucleophilic displacement (48.5%) and elimination (51.5%) resulted, presumably by way of the alternative chair or skew conformer in which the bond to the sulfonate group is axial. Ring contractions that involve migration of the ring-oxygen atoms have been observed when nucleophilic displacements have been attempted on such hexopyranosides as methyl 6-deoxy-2,3-*O*-isopropylidene-4-*O*-(methylsulfonyl)-α-D-mannopyranoside, in which a *β-trans* axial substituent at C-2 hinders the displacement reaction.[124b] The difference in behavior of the pentopyranoside and hexopyranoside 4-sulfonates is, presumably, attributable to the greater ease of interconversion of the conformers of the former esters. On the other hand, the amine deaminations are controlled by the ground-state conformations.

The deamination of the methyl glycoside (**42**) of neuraminic acid has been reported, but the products were not identified.[125]

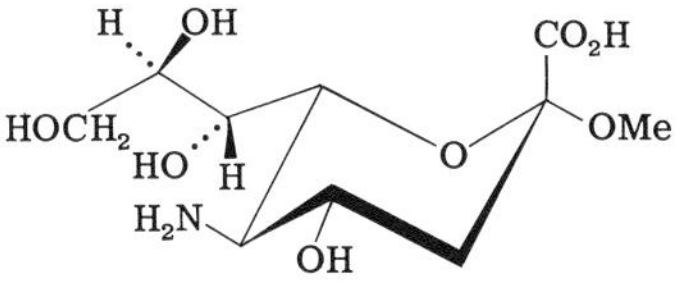

42

The foregoing results show that the deaminations of pyranose derivatives containing equatorially attached amino groups at C-2 or C-4 are dominated by participation of the ring-oxygen atom.

(ii) **Inositol Derivatives.** In the inosamine field, the first equatorial amine to be deaminated was aminodeoxy-*scyllo*-inositol, which gave the substitution product *myo*-inositol in 12% yield, together with unidentified, reducing products. The latter were presumably aldehydes (or derived hemiacetals) resulting from ring contraction.[126] At least five products were detected in a later study, but none were identified.[127] The substitution product was obtained in much higher yield when the penta-*O*-acetyl derivative of the amine hydrochloride was deaminated, *myo*-inositol hexaacetate being isolated in 71%

(124b) C. L. Stevens, R. P. Glinski, K. G. Taylor, and F. Sirokman, *J. Org. Chem.*, **35**, 592–596 (1970).
(125) E. Klenk, H. Faillard, F. Weygand, and H. H. Schöne, *Z. Physiol. Chem.*, **304**, 35–52 (1956).
(126) T. Posternak, *Helv. Chim. Acta*, **33**, 1597–1605 (1950).
(127) S. J. Angyal and J. S. Murdoch, *Aust. J. Chem.*, **22**, 2417–2428 (1969).

yield after acetylation.[128,129] The isolation of *myo*-inositol hexaacetate, in 33% yield, was also reported for the sequential deamination and acetylation of aminodeoxy-*scyllo*-inositol sulfate, obtained from glebomycin.[130]

The first per-O-acetylated derivative to be studied[131] was that (**43**) of streptamine, which gave, after deamination and acetylation, a crystalline N-acetyl-penta-O-acetyl derivative (**44**) (77% yield, before recrystallization), which was subsequently shown[132] to be the product of inversion of configuration. The use of the di-O-acetyl derivative of the bis(acetamido) aminotetradeoxyinositol **45** improved[133] the yield of substitution product isolated on deamination from less than 1% to 30%.

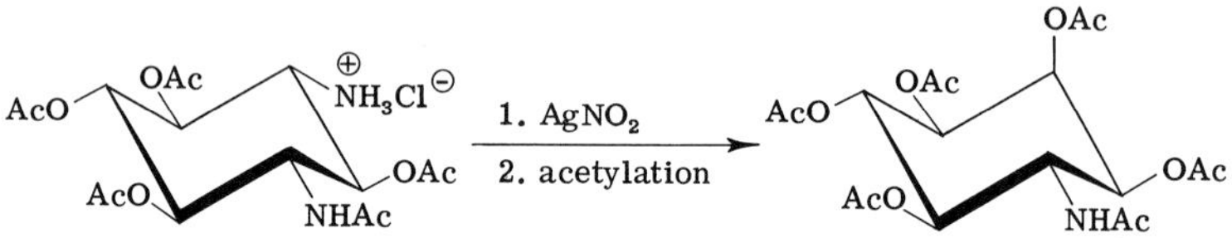

and enantiomer   and enantiomer
43   **44**

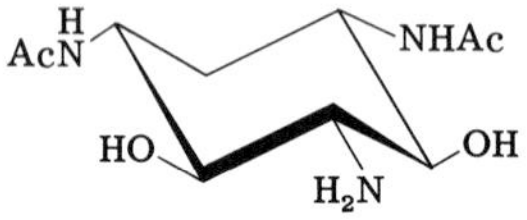

45

In all of the deaminations of inosamines just referred to, the substitution product was formed with inversion of configuration, but in no report was a product analysis given, and so, when a product was isolated in low yield, its significance is uncertain. In 1969, Angyal and Murdoch[127] made a valuable contribution to these studies when they reported the results of detailed product-analysis for several inosamine deaminations. They confirmed that rearrangement is minimized by the use of O-acetyl derivatives, and their results for equatorial amines are summarized in Table II. The crystalline O-acetyl

(128) T. Posternak, W. H. Schopfer, and R. Huguenin, *Helv. Chim. Acta,* **40,** 1875–1880 (1957).

(129) G. I. Drummond, J. N. Aronson, and L. Anderson, *J. Org. Chem.,* **26,** 1601–1607 (1961).

(130) T. Naito, *Penishirin Sono Ta Koseibusshitsu,* **15,** 373–379 (1962); *Chem. Abstr.,* **60,** 4230e (1964).

(131) O. Wintersteiner and A. Klingsberg, *J. Amer. Chem. Soc.,* **73,** 2917–2924 (1951).

(132) H. Straube-Rieke, H. A. Lardy, and L. Anderson, *J. Amer. Chem. Soc.,* **75,** 694–697 (1953).

(133) A. Hasegawa and H. Z. Sable, *Tetrahedron,* **25,** 3567–3578 (1969).

TABLE II

Deamination of O-Acetylated Inosamines having
Equatorially Attached Amino Groups[127]

Parent inosamine	Probable conformation	Inversion in substitution product (%)	Other products
Aminodeoxy-*scyllo*-	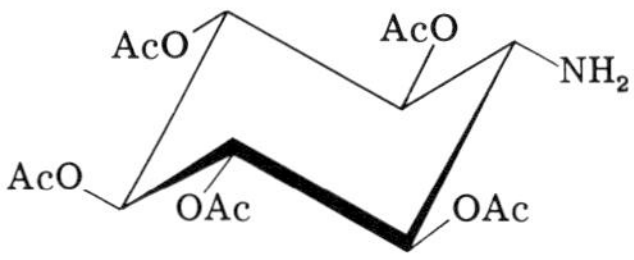	~100	
L-3-Amino-3-deoxy-*chiro*-	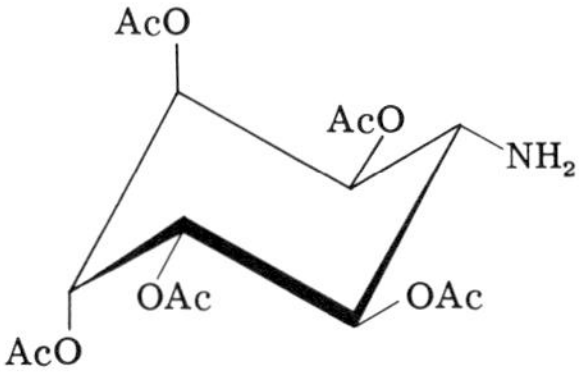	75	?
DL-4-Amino-4-deoxy-*myo*-[a]	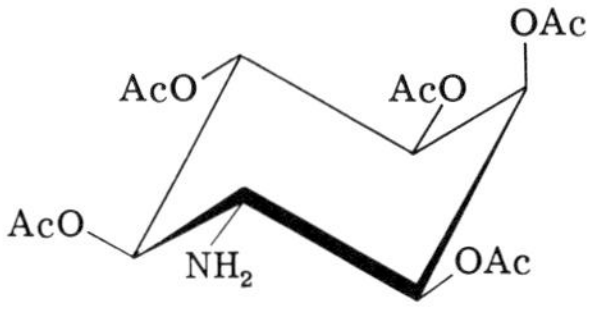	mainly inversion	?
L-1-Amino-1-deoxy-*neo*-	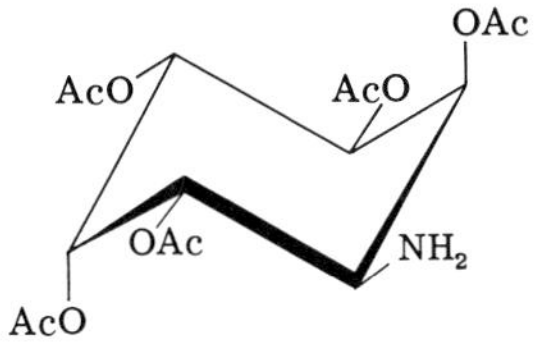	31	enol ester (?), *myo*-inositol ester
DL-1-Amino-1-deoxy-*myo*-[a]	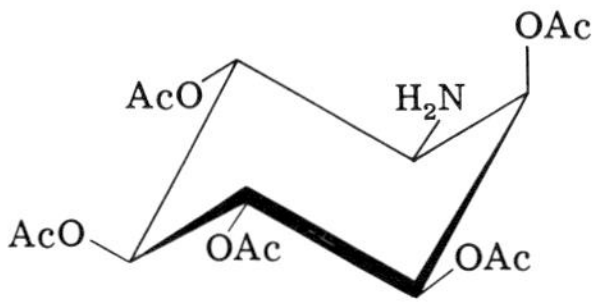	17	enol ester (?)

[a] One enantiomer is shown.

amine hydrochlorides were deaminated; alternatively, the inosamine
was acetylated with acetic anhydride–acetic acid containing perchloric acid, and then dilute hydrochloric acid and sodium nitrite were
added. Thus, the composition of the deamination medium differed in
these two experiments, there being ~30% of acetic acid present in
the latter procedure. This difference is not likely to be significant

(see p. 15), but, because the extent of inversion can vary with the solvent composition, it is desirable to use the same solvent composition in such comparative studies. The products of deamination were acetylated, and the acetates analyzed by gas–liquid chromatography (g.l.c.). Compounds having a low, relative retention-time were tentatively identified as enol esters that had resulted from elimination.[134] The percentage of inversion shown in Table II was calculated from the ratio of peak areas of the two substitution products, and so it was only a qualitative value.

There are two main points of interest. First, the extent of inversion varied over a wide range, and second, the high degree of inversion found for the aminodeoxy-*scyllo*- (46) and 3-amino-3-deoxy-*chiro*-inositol acetates was in marked contrast to the predominant retention of configuration observed in the deamination of simple, equatorial amines.

The possibility that participation of neighboring acetoxyl groups might occur in such reactions was suggested by Hasegawa and Sable,[133] but this possibility was discounted by Angyal and Murdoch,[127] who propionylated the products obtained upon deaminating penta-*O*-acetyl-aminodeoxy-*scyllo*-inositol (46) and detected, by g.l.c., only the monopropionate of 1,3,4,5,6-penta-*O*-acetyl-*myo*-inositol (47) and none of that of the 1,2,4,5,6-pentaacetate (49).

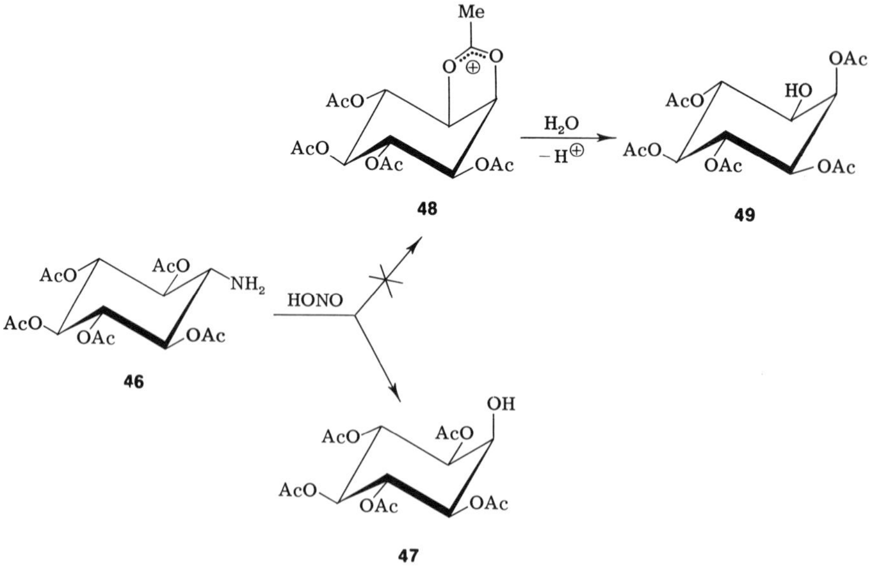

(134) Another possibility is that these compounds were unsaturated aldehydes that resulted from ring contraction during deamination followed by elimination, during acetylation at 100° in pyridine.

Angyal and Murdoch[127] argued that both pentaacetates would be expected to be formed on hydrolysis of the intermediate, acetoxonium ion 48, but King and Allbutt[135] have shown that such ions are hydrolyzed to give mainly the axial acetate. Thus, the 1,2,4,5,6-pentaacetate would be obtained[136] from 48.

These results can best be rationalized[127] by postulating that heterolysis of the diazonium ion to give a carbonium ion is less favored in amines possessing neighboring acetoxyl groups than in amines lacking such electron-withdrawing groups. Displacement of nitrogen from the diazonium ion, by concerted attack of solvent, then competes more effectively with heterolysis to a carbonium ion. Thus, inversion results, except when steric factors interfere. Displacements of the SN2 type are known to be hindered by axial substituents adjacent to the reaction center,[139] and also by substituents that are *syn*-axial to the leaving group[140] or to the attacking nucleophile.[139] The results given in Table II indicate that, in these reactions, neighboring axial substituents have a greater effect than *syn*-axial substituents.

Elimination (51) and ring-contraction (52) products were isolated in low yield after extraction of an aqueous solution of the deamination products of the amine 50 with ethyl acetate. Any triol substitution-products would, presumably, have remained in the aqueous layer (which was not examined).[141]

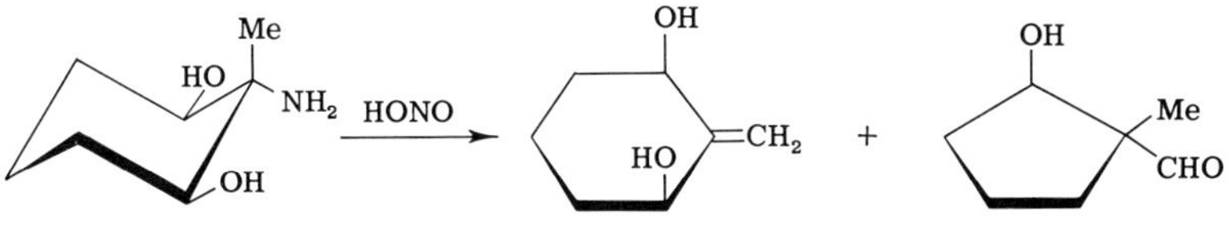

b. Axial Amines. — (i) Pyranose Derivatives. The results of deamination of the 2-amino-4-*tert*-butylcyclohexanol diastereoisomers showed that axial amines possessing vicinal hydroxyl groups give epoxides when the hydroxyl group is axial, and undergo rearrange-

(135) J. F. King and A. D. Allbutt, Can. J. Chem., 48, 1754–1769 (1970).
(136) Acetyl migration is unlikely under the reaction conditions,[137] and the equilibrium would, in any case, include much of the 1,2,4,5,6-pentaacetate.[138]
(137) S. J. Angyal, personal communication.
(138) S. J. Angyal and G. J. H. Melrose, J. Chem. Soc., 6494–6500 (1965).
(139) A. C. Richardson, Carbohyd. Res., 10, 395–402 (1969).
(140) A. K. Al-Radhi, J. S. Brimacombe, and L. C. N. Tucker, Chem. Commun., 363 (1970).
(141) K. Sakai, S. Oida, and E. Ohki, Chem. Pharm. Bull. (Tokyo), 16, 1048–1055 (1968).

ment to give ketones when the hydroxyl group is equatorial (**53**). The yields of ketone (**54**, 74%) and epoxide (76%) were significantly lower than the yields (90–98%) of ring-contracted aldehyde (**55**)

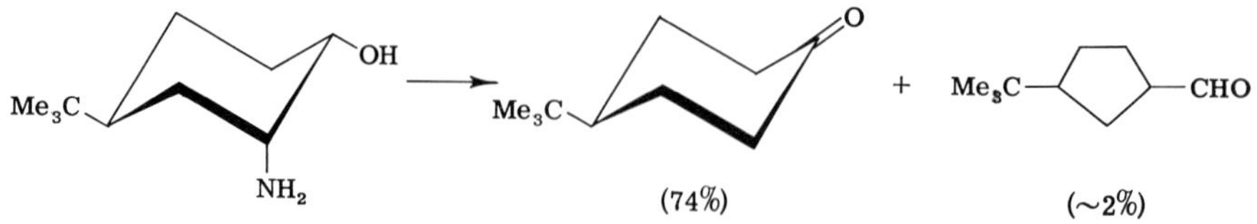

which resulted from the deamination of the two, equatorial-amine diastereoisomers. As only the pentane-soluble products were estimated, the diols were the most likely of the unidentified products.[36]

The deamination of 2-amino-2-deoxy-D-mannose was regarded by Levene[77] as "anomalous," because the product obtained after oxidation with nitric acid was D-glucaric acid. In other words, the product was not an anhydride, as in the deamination of the D-glucose and D-galactose analogs. The identification of D-glucose as the major product was later confirmed,[142,143,144a] the crystalline sugar being isolated[143] in 72% yield.

The deamination of methyl 3-amino-3-deoxy-β-D-altropyranoside was studied in 1934, but the products were not fully characterized.[144b] The syrupy product was converted into a methylated derivative that had an elemental analysis corresponding to that calculated for a methyl tetramethylhexoside. The conditions used for methylation would have opened an epoxide ring. Methyl 2,3-anhydro-4,6-O-benzylidene-α-D-mannopyranoside precipitated quantitatively from solution when the corresponding 3-amino-3-deoxyaltroside derivative was deaminated in aqueous medium.[145] Epoxide formation was likewise reported to be quantitative in the deamination of the analogous 2-amino-2-deoxyaltroside.[83,145] On deamination, 4-amino-1,6-anhydro-4-deoxy-β-D-mannopyranose also gave an epoxide, namely, 1,6:2,3-dianhydro-β-D-talopyranose, in unspecified yield.[146]

The deamination of 2-amino-2,6-dideoxy-L-talose resembles that of 2-amino-2-deoxy-D-mannose, in that substitution with inversion of configuration is the major reaction, the product being tentatively

(142) N. M. K. Ng Ying Kin and J. M. Williams, Chem. Commun., 1123–1124 (1971).
(143) D. Horton, K. D. Philips, and J. Defaye, Carbohyd. Res., 21, 417–419 (1972).
(144a) S. Hase and Y. Matsushima, J. Biochem. (Tokyo), 69, 559–565 (1971).
(144b) E. W. Bodycote, W. N. Haworth, and E. L. Hirst, J. Chem. Soc., 151–154 (1934).
(145) L. F. Wiggins, Nature, 157, 300 (1946).
(146) V. G. Bashford and L. F. Wiggins, Nature, 165, 566 (1950).

identified by paper chromatography and electrophoresis as 6-deoxy-galactose.[147] Although the absolute configuration was not established, it is extremely unlikely that inversion of configuration would occur at C-3, C-4, and C-5.

In 2-amino-1,6-anhydro-2-deoxy-β-D-glucopyranose (**56**, R = H), the oxygen atoms at C-1 and C-3 are *trans* and coplanar to nitrogen. There is similar competition between axial hydroxyl and alkoxyl (methoxyl) groups in methyl 2-amino-4,6-O-benzylidene-2-deoxy-α-D-altropyranoside, and in its deamination, epoxide formation apparently occurs to the exclusion of alkoxyl migration. However, in the 1,6-anhydride **56** (R = H), the introduction of an epoxide ring into the somewhat strained, bridged, bicyclic ring-system was relatively unfavorable, as the only products identified were those resulting from migration of oxygen from C-1 to C-2. Thus, after deamination with sodium nitrite in 90% acetic acid, and acetylation of the product, 2,6-anhydro-D-mannopyranose triacetate (**58**, R = Ac) was isolated in 10% yield.[148] The same compound was isolated in 37% yield when the di-O-acetyl derivative **56** (R = Ac) was deaminated. The 2,6-anhydro-D-mannose product was converted into various derivatives, including the known 2,6-anhydro-D-mannitol. When the hydrochloride of amine **56** (R = H) was deaminated with silver nitrite in aqueous solution, a dimeric product (**59**) was obtained after acetylation (see Scheme 5). The dimer hexaacetate **59**

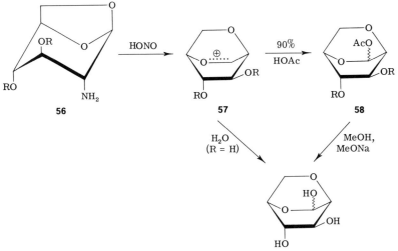

Scheme 5 (cont'd on p. 42)

(147) S. A. Barker, J. S. Brimacombe, M. J. How, M. Stacey, and J. M. Williams, *Nature*, **189**, 303 (1961).

(148) F. Micheel, W. Neier, and T. Riedel, *Chem. Ber.*, **100**, 2401–2409 (1967).

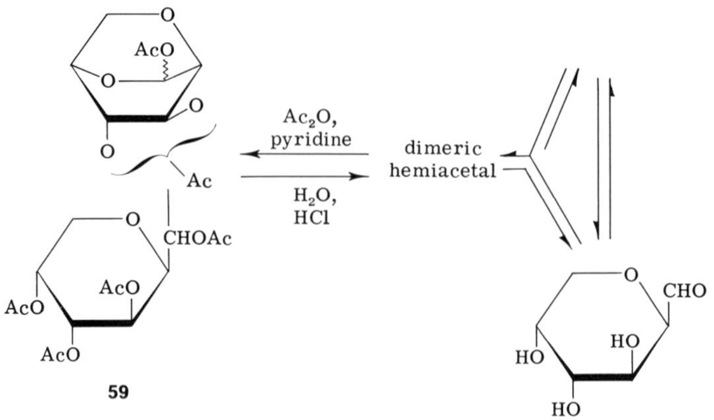

Scheme 5 (*Cont'd*)

was converted by acid hydrolysis or acid-catalyzed methanolysis into 2,6-anhydro-D-mannose and its dimethyl acetal, respectively. The dimer hexaacetate **59** was also isolated in 20% yield when 2,6-anhydro-D-mannose was acetylated with acetic anhydride in pyridine. Thus, the consequence of using 90% aqueous acetic acid (instead of water) as the solvent for the deamination was that the rearranged cation (**57**, R = H) reacted mainly with acetic acid to give 1-*O*-acetyl-2,6-anhydro-D-mannopyranose (**58**, R = H) which, on acetylation, gave the 1,3,4-triacetate (**58**, R = Ac).[148]

Substitution with inversion of configuration occurred in the deamination of methyl 4-amino-4,6-dideoxy-2,3-*O*-isopropylidene-α-L-talopyranoside in 90% acetic acid, to give the corresponding mannopyranoside derivative (32%) and its acetate (57%). A third (unidentified) product (11%) was also detected by g.l.c. analysis. The stereospecific formation of the two mannopyranosides was attributed to steric hindrance to the approach of solvent to the C-4 carbonium ion from the axial direction.[124] Likewise, the pentoside methyl 4-amino-4-deoxy-2,3-*O*-isopropylidene-β-L-ribopyranoside gave, on deamination, the D-lyxopyranoside (34%) and its acetate (66%).[124a] In contrast to these results, rearrangements were found to constitute important reaction-pathways in the deamination of two monocyclic glycopyranosides respectively possessing an axial amino group at C-4 and C-3. In deamination, methyl 4-amino-4-deoxy-α-D-galactopyranoside gave low yields of the substitution products, methyl α-D-glucopyranoside and α-D-galactopyranoside, which were isolated in 9 and 2 % yields, respectively.[142] Methyl 4-deoxy-α-D-*erythro*-hexopyranosid-3-ulose was a major product, isolated by chromatography in ~40% yield. Three other (unidentified) products were also reported.

Deamination of methyl 3-amino-3-deoxy-β-D-allopyranoside (**60**) gave at least seven products, which included the substitution products, methyl β-D-glucopyranoside (**61**, ~30%) and methyl β-D-allopyranoside (**62**, ~3%), and rearrangement products. The latter included at least two glycosiduloses (ketones), one of which was labile and isomerized during isolation by preparative paper-chromatography to another glycosidulose that was also present in the product mixture.[142] The stable glycosidulose was tentatively identified by n.m.r. spectroscopy as methyl 3-deoxy-α-L-*threo*-hexopyranosid-4-ulose (**64**), which was presumed to arise from the C-5 epimer (**63**).

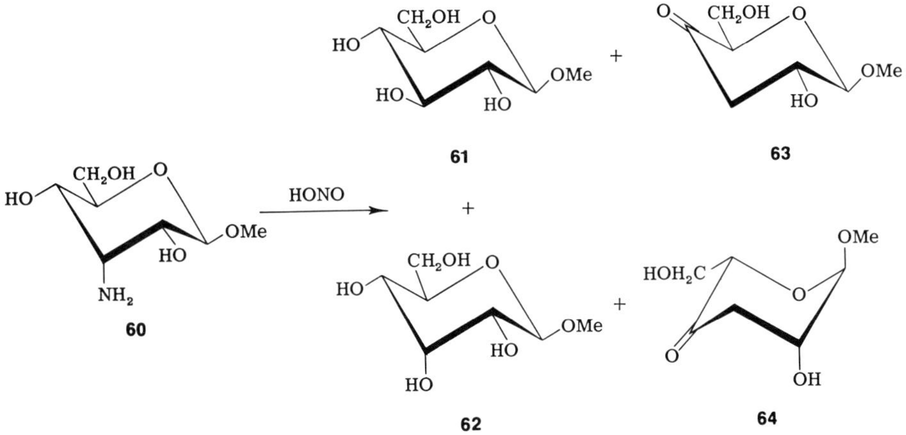

An analogous epimerization, in aqueous pyridine, of a glyco-pyranosid-4-ulose derivative has been reported.[149a] The glycosidulose **63** was, therefore, presumed to be a deamination product, and its ready epimerization to **64** suggests that it has a higher free-energy than **64**. The chair conformer shown (**63**) would be destabilized by unfavorable, dipolar interactions[149b] (anomeric effect), and the alternative chair conformer ($^1C_4$) has its hydroxymethyl group and hydroxyl group in a *syn*-diaxial relationship.

(149a) C. L. Stevens and K. K. Balasubramanian, *Carbohyd. Res.*, **21**, 166–168 (1972).
(149b) On the basis of recent measurements[149c] of conformational equilibria for cyclohexanone derivatives, the anomeric effect would be expected to be enhanced in glycopyranos-4-ulose derivatives.
(149c) R. D. Stolow and T. W. Giants, *Chem. Commun.*, 528–529 (1971).

Ring contraction also occurred in the deamination of **60** to give methyl 3-deoxy-3-C-formyl-β-D-xylofuranoside (**65**) as a minor product, isolated in 5% yield. As the 3-amino-3-deoxy-D-alloside is likely to exist mainly in the chair conformation shown (**60**), in which the bond to the amino group is axial, and as ring contractions are normally observed in the deamination of equatorial amines, the D-xyloside **65**

65

must be formed either (*a*) as a result of stereochemical leakage at the carbonium-ion stage or (*b*) from conformations in which the amino group is equatorial or, perhaps, quasi-equatorial. However, on ring contraction, the latter conformations would give an aldehyde, methyl 3-deoxy-3-C-formyl-β-D-ribofuranoside, which is the 3-epimer of the observed product (**65**). Thus, an additional step, epimerization of methyl 3-deoxy-3-C-formyl-β-D-ribofuranoside, would be required to account for the configuration of the xyloside (**65**). An analogous epimerization of the α anomer occurred in the deamination of methyl 4-amino-4-deoxy-α-D-glucopyranoside (see compounds **24** and **38**, p. 33). Ring contraction also occurred to a small extent (~2%) in the deamination of the axial amine **53**.

Because monocyclic axial amines containing a neighboring, equatorial hydroxyl group normally give ketones as major products, the reactions of 2-amino-2-deoxy-D-mannose and 2-amino-2,6-dideoxy-L-talose appear to be anomalous. This observation together with the fact that D-glucose was formed stereospecifically[142] led to the suggestion[142] that D-glucose is formed by way of 1,2-anhydro-D-glucose (**68**), which, in turn, arises by participation of the axial hydroxyl group in the α anomer (**66**) of 2-amino-2-deoxy-D-mannose. However, this possibility was rejected[143] on the grounds that the yield (≥72%) of D-glucose did not correspond to the proportion (43%) of the α anomer present in solution, and a concerted displacement of nitrogen from the diazonium ion by solvent water was suggested.[143]

Direct substitution did not occur[150] in the deamination of methyl

(150) J. W. Llewellyn and J. M. Williams, *J. Chem. Soc. Perkin I*, 1997–2000 (1973).

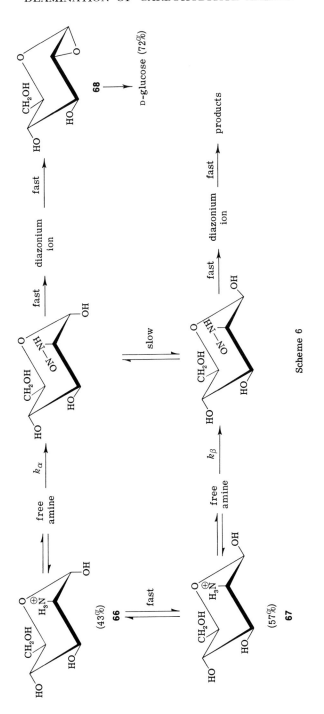

Scheme 6

2-amino-2-deoxy-α-D-mannopyranoside (**69**), which gave the ketone

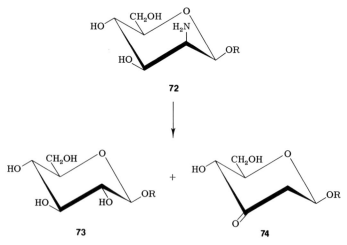

(**70**) and 2-*O*-methyl-D-glucose (**71**) in the ratio of 2:1. Both substitution and rearrangement to give a ketone occurred when 6-*O*-(2-amino-2-deoxy-β-D-mannopyranosyl)-D-glucose (**72**, R = 6-deoxy-D-glucos-6-yl) and -D-glucitol (**72**, R = 6-deoxy-D-glucitol-6-yl) were deaminated.[151,152] The relative yields of **73** and **74** were not reported, but a gas–liquid chromatogram indicated that the yields were probably comparable, provided that there were no differential losses in the

R = 6-deoxy-D-glucitol-6-yl or 6-deoxy-D-glucos-6-yl.

(151) S. Hase and Y. Matsushima, *J. Biochem.* (Tokyo), **72**, 1117–1128 (1972).
(152) S. Hase, Y. Tsuji, and Y. Matsushima, *J. Biochem.* (Tokyo), **72**, 1549–1555 (1972).

analytical procedure, which involved deionization with ion-exchange resins and trimethylsilylation.

The objection[143] to the mechanism involving participation of the axial hydroxyl group of **66** to give an epoxide intermediate was based on the premise that the product formed should be related to the anomeric composition of the amine. This would only hold were the rate of anomerization of 2-amino-2-deoxy-D-mannose much lower than the rate of N-nitrosation, or if the rates of N-nitrosation of the anomers were identical. The anomerization of 2-amino-2-deoxy-D-mannose hydrochloride is unusually rapid, and it has been suggested that this is due to the ability of the $-\overset{+}{N}H_3$ group to protonate the ring-oxygen atom.[153] The nitrosoamine is likely to anomerize at "normal" rates. The rates of N-nitrosation (the rate-determining step) of the anomers of the amine are likely to be different, the α anomer reacting the more rapidly ($k_\alpha > k_\beta$ in Scheme 6).[153a] An analogy is provided by the relative rates of deamination of methyl 2-amino-2-deoxy-α- and -β-D-glucopyranosides,[85] the slower reaction of the α anomer being attributed to steric effects resulting from the *cis* relationship of the substituents on C-1 and C-2. The difference should be even greater in the mannose anomers **66** and **67**, because the substituents on C-1 and C-2 in the α anomer are well separated. Thus, it is possible that the epoxide route to D-glucose by way of the α anomer (**66**) could be the major route, even though only 43% of the α anomer of the amine is present.

The occurrence of some substitution in the deamination of 2-amino-2-deoxy-β-D-mannopyranosides[151,152] (**72**), and its absence in the reaction of the α-D-pyranoside[150] **69**, must be due to the steric effect of the axial anomeric substituent which (in the α-D-pyranoside) hinders the approach of the nucleophile (water) to either the C-2 carbonium ion or to C-2 of the diazonium ion. The glucose and glucitol tentatively detected as minor products in the deamination of **72** (R = D-glucose residue and R = D-glucitol residue) presumably arose by way of a hydride shift of H-1 to C-2. 2-Deoxy-D-glucono-1,5-lactone (**75**) was not detected, as it would probably have

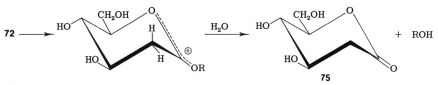

R = D-glucose residue or D-glucitol residue.

(153) D. Horton, J. S. Jewell, and K. D. Philips, *J. Org. Chem.*, **31**, 3843–3845 (1966).
(153a) Scheme 6 is reproduced from Ref. 150 by permission of The Chemical Society, London.

been hydrolyzed, and removed by the treatment with anion-exchange resin.

On deamination, 2-amino-1,5-anhydro-2-deoxy-D-mannitol gave 1,5-anhydro-2-deoxy-D-*erythro*-3-hexulose as the major product.[54] The substitution product 1,5-anhydro-D-glucitol was formed to a small extent, and inversion again predominated, as with other axial amines (**60, 72**, and methyl 4-amino-4-deoxy-α-D-galactopyranoside); this is in accord with the predominant inversion of configuration reported for deamination, in water-rich media, of simple cyclohexyla-mines,[154] and aminodecahydronaphthalenes[39] in which the amino group is axial. As discussed previously (see p. 15), the extent of in-version can vary considerably with the nature of the solvent, and steric factors appear to be important in deaminations of per-*O*-acetyl amines (see pp. 37 and 49).

(ii) **Inositol Derivatives.** On deamination, 2-amino-2-deoxy-*myo*-inositol (**76**) gave the product of substitution with inversion of config-

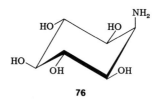

76

uration, the hexaacetate derivative being isolated in 31% yield.[126] Rearrangement also occurred, to give a ketone (probably as the major product), the phenylosazone derivative of which was isolated in 32% yield.

A low yield of substitution product, formed with inversion, was likewise isolated in the deamination of 2-amino-2-deoxy-*neo*-inositol,[155,156] L-1-Amino-1-deoxy-*chiro*-inositol was first reported to give an inositol with retention of configuration,[157] but a subsequent study[127] showed that an epoxide, L-1,2-anhydro-*myo*-inositol, is the major product (60%); L-*chiro*-inositol was also isolated, in 20% yield. Epoxide formation and rearrangement to give a ketone could be minimized by deaminating the *O*-acetylated amines, but elimination to give enol esters occurred, and so this procedure did not necessar-

(154) W. Hückel and K. Heyder, *Chem. Ber.*, **96**, 220–225 (1963).
(155) J. B. Patrick, R. P. Williams, C. W. Waller, and B. L. Hutchings, *J. Amer. Chem. Soc.*, **78**, 2652 (1956).
(156) R. L. Mann and D. O. Woolf, *J. Amer. Chem. Soc.*, **79**, 120–126 (1957).
(157) G. G. Post, Ph. D. Thesis, University of Wisconsin, Madison, Wisconsin, U. S. A., 1959; cited by S. J. Angyal and L. Anderson, *Advan. Carbohyd. Chem.*, **14**, 188 (1959).

ily improve the yield of substitution products.[127] The structure of one enol ester, formed from penta-*O*-acetyl-2-amino-2-deoxy-*neo*-inositol (79), was established by hydrogenation and deacetylation to give a known tetrol.

The deamination products obtained from several *O*-acetylated, axial amines were analyzed by g.l.c., and the extent of inversion is indicated in Table III.[127] As in Table II, the values for percent of in-

TABLE III

Deamination of *O*-Acetylated Inosamines having Axially Attached Amino Groups[127]

Parent inosamine	Probable conformation	Inversion in substitution product[a] (%)	Other products
2-Amino-2-deoxy-*myo*-	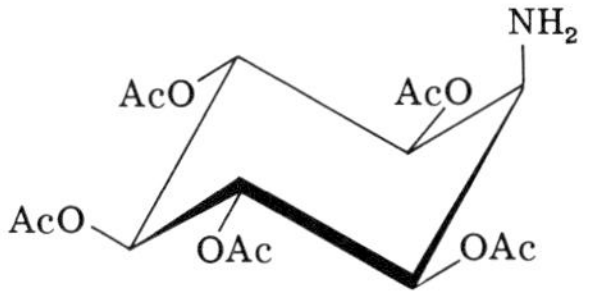 **77**	100	enol ester (?)
DL-2-Amino-2-deoxy-*epi*-	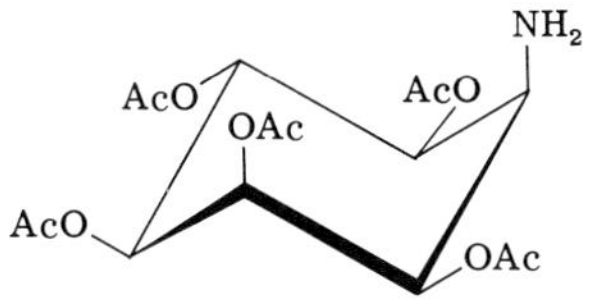 **78**[b]	100	two crystalline, enol esters
2-Amino-2-deoxy-*neo*-	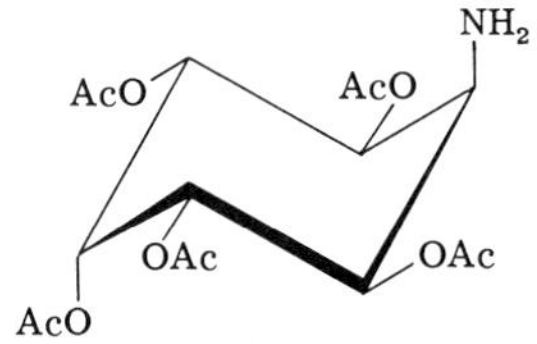 **79**	95	crystalline enol ester
L-1-Amino-1-deoxy-*chiro*-	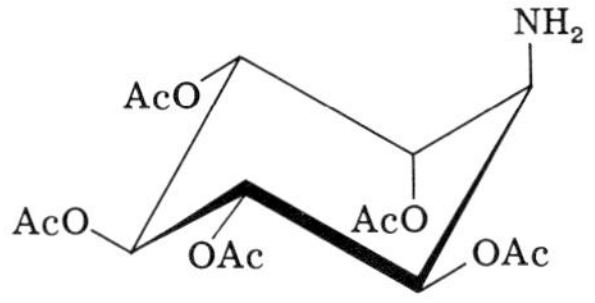 **80**	35	enol ester (?)

[a] The major substitution product was isolated crystalline in each case.
[b] One enantiomer is shown.

version are qualitative. The same trend was observed as with equatorial amines, namely, inversion predominated, except when an axial acetoxyl group was adjacent to the amino group (see p. 37). Inversion (68%) of configuration predominated in the deamination of the penta-*O*-acetyl derivative of 3-amino-3-deoxy-*muco*-inositol, which probably exists as a mixture of both chair conformers. After deacetylation, *epi*-inositol (34%) and *muco*-inositol (14%) were isolated by chromatography.[127]

The extent of inversion of simple, axial amines is much more dependent on solvent composition than for equatorial amines. It would be interesting to determine whether the deaminations of these more-complex amines are equally (or more, or less) dependent on the solvent.

Two inosamine derivatives containing axial amino groups were reported to give, on deamination, substitution products (in low or unspecified yield) with inversion[158] and retention[133] of configuration, respectively. No product analyses were reported.

c. Exocyclic Amines. — In 1887, the structure of the amine obtained from D-*arabino*-hexulose phenylosazone by reduction was inferred to be 1-amino-1-deoxy-D-fructose on the basis of its reaction with nitrous acid to give D-fructose.[159] In a related sequence, 1-amino-1,3-dideoxy-D-*erythro*-2-hexulose (**82**) (prepared, together with one molecular proportion of 1-amino-1-deoxy-D-fructose, by catalytic hydrogenation of the azine **81**, and deaminated *in situ*) gave 3-deoxy-D-*erythro*-2-hexulose (**83**, 37%).[160] Later, 1-deoxy-1-(diben-

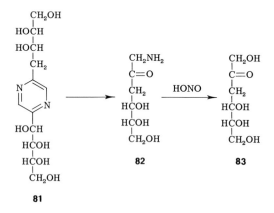

(158) M. Nakajima, A. Hasegawa, and F. Lichtenthaler, *Ann.*, **680**, 21–31 (1964).
(159) E. Fischer and J. Tafel, *Ber.*, **20**, 2566–2575 (1887).
(160) R. Kuhn, H. J. Haas, and A. Seeliger, *Chem. Ber.*, **94**, 2534–2535 (1961).

zylamino)-D-tagatose, prepared from D-galactose, was converted into D-tagatose by deamination of the intermediate, primary amine (which was not isolated).[161] These products were substitution products, but they might have resulted from several competing reactions, depending on which tautomers were present in solution. Thus, a diazoketone was a possible intermediate from the acyclic ketone, and the furanose and pyranose amines could undergo reactions involving participation of (*a*) the ring-oxygen atom to give a bicyclic, oxonium ion, or (*b*) the 2-hydroxyl group to give a 1,2-anhydride. 1-(Aminomethyl)cyclohexanol derivatives give epoxides and ring-expanded ketones, the latter normally preponderating.[162] If the pyranose forms of the 1-amino-1-deoxyhexuloses preponderated, then, clearly, ring expansion to give seven-membered ring anhydrohexuloses was not favored.

The first 6-amino-6-deoxyhexose to be studied was the D-galactose isomer, the 1,2:3,4-di-*O*-isopropylidene derivative of which was deaminated in 1:9 (v/v) acetic acid–water to give 1,2:3,4-di-*O*-isopropylidene-α-D-galactopyranose.[163] However, when the free sugar was deaminated, the syrupy product could not be characterized completely; its oxidation to galactaric (mucic) acid in 30% yield suggested the presence of some galactose. Deamination of 1,2,3,4-tetra-*O*-acetyl-6-amino-6-deoxy-D-glucopyranose gave, after hydrolysis of the product, D-glucose.[164]

On deamination, 1-amino-2,6-anhydro-1-deoxy-D-*glycero*-D-*galacto*-heptitol (**84**) gave the corresponding anhydroheptitol (**85**) in 58% yield, plus the 1-deoxy-2-heptulose **86**, isolated in syrupy (17%) and crystalline (8%) forms.[165] The rearrangement leading to the latter

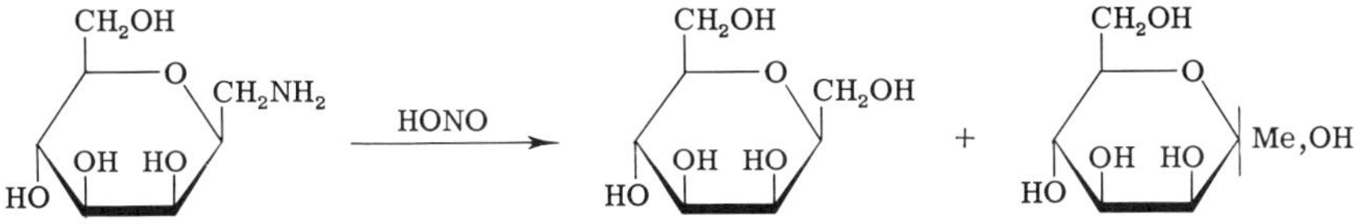

84   85   86

(161) R. Gruennagel and H. J. Haas, *Ann.*, **721**, 234–235 (1969).
(162) See, for example, R. G. Carlson and N. S. Behn, *J. Org. Chem.*, **33**, 2069–2073 (1968).
(163) K. Freudenberg and A. Doser, *Ber.*, **58**, 294–300 (1925).
(164) M. J. Cron, O. B. Fardig, D. L. Johnson, H. Schmitz, D. F. Whitehead, I. R. Hooper, and R. U. Lemieux, *J. Amer. Chem. Soc.*, **80**, 2342 (1958).
(165) J. C. Sowden, C. H. Bowers, and K. O. Lloyd, *J. Org. Chem.*, **29**, 130–132 (1964).

was incorrectly assumed to involve opening of the anhydro ring at some stage. An analogous result was reported for the D-*glycero*-L-*manno* isomer of **84,** substitution and rearrangement products being isolated in 49 and 23% (as a derivative) yields, respectively.[166] In an attempt to distinguish between the hydride-shift and elimination routes (see Scheme 7) to the rearranged 1-deoxy-2-heptulose (**87**),

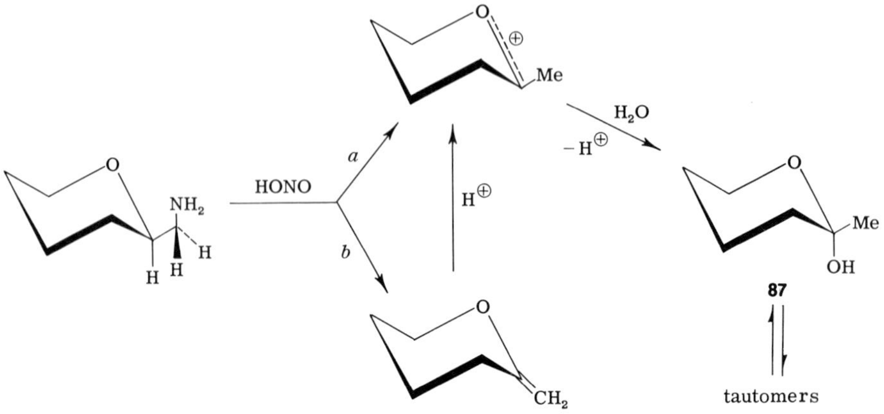

Scheme 7

deamination was conducted in deuterium oxide. The proportion of deuterium found in the total product was one sixth of that expected for route *b*, and it was concluded that both routes were, perhaps, operating, although it was admitted that hydrogen–deuterium exchange was possible for the glyculose **87** in the acyclic form.

Substitution was also the major reaction in the deamination of 1-amino-2,6-anhydro-1-deoxy-D-allitol[167] and 1-amino-2,6-anhydro-1,7-dideoxy-L-*glycero*-L-*galacto*-heptitol.[168] The rearrangement product, a 1,6-dideoxy-2-heptulose, was also isolated from the latter amine, the ratio of substitution to rearrangement being 1.6 : 1.

That rearrangement to give 1-deoxy-2-glyculoses is not the major pathway in the reactions just described can be attributed to the low proportion of the rotamer (**88**) in which the diazonium group is gauche to both C-3 and the ring-oxygen atom. The major (substitution) product may result from rearside attack of water on C-1 of the diazonium ion and/or participation of the ring-oxygen atom in ro-

(166) B. Coxon and H. G. Fletcher, Jr., *J. Amer. Chem. Soc.,* **86,** 922–926 (1964).
(167) B. Coxon, *Tetrahedron,* **22,** 2281–2302 (1966).
(168) A. Veyrières and R. W. Jeanloz, *Biochemistry,* **9,** 4153–4159 (1970).

tamer **89** to give a bicyclic oxonium ion (**91**), which is then attacked by water at C-1. Ring expansion by way of attack of water at C-2 of **91**, or by way of migration of C-3 in rotamer **90**, to give seven-membered-ring products was, evidently, relatively unfavorable. Ring ex-

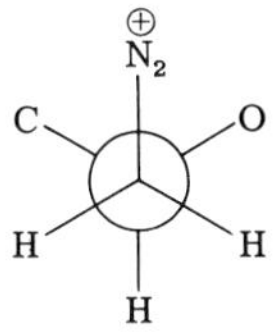

"Newman" projections viewed along the C-1 – C-2 bond.

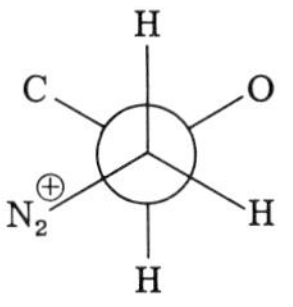

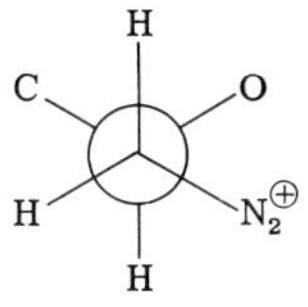

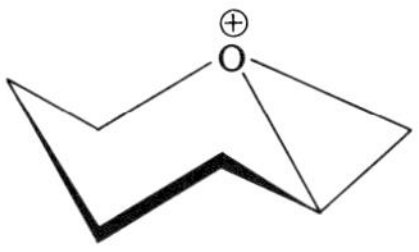

91

pansion was also not the major pathway in the deamination of two branched-chain amines, methyl 2-C-(aminomethyl)-β-L-arabinopyranoside and its *ribo* isomer, the major product from each being the corresponding *spiro*-epoxide, which was characterized by hydrolysis.[169]

In the somewhat related reaction of glycopyranosiduloses with diazomethane, *spiro*-epoxides were the major products, but reaction with diazoethane gave ring-expansion products exclusively.[170] The formation of the latter was rationalized in terms of nonbonded interactions in the transition states leading to *spiro*-epoxides. It was also considered that steric effects determine which of the two ring-expansions possible occur.

An example of selective N-nitrosation is provided by the conversion of the diamine **92** into the mono-N-nitroso compound **93** (73%)

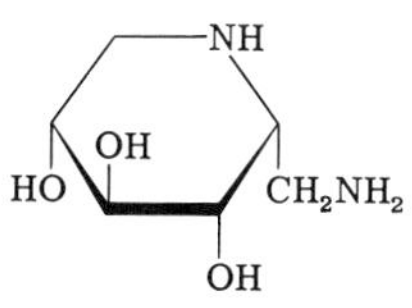

92 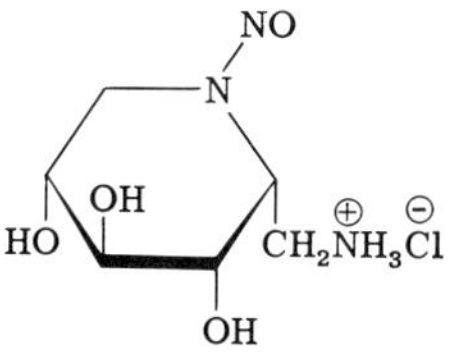   **93**

(169) S. W. Gunner, R. D. King, W. G. Overend, and N. R. Williams, *J. Chem. Soc. (C)*, 1954–1961 (1970).

(170) T. D. Inch, G. L. Lewis, and R. P. Peel, *Chem. Commun.*, 1549–1550 (1970); *Carbohyd. Res.*, **19**, 29–38 (1971).

by the action of silver nitrite in dilute hydrochloric acid solution.[171] Deamination in more weakly acid solution gave several products, the interesting, bridged compound **94** being isolated in 12% yield after

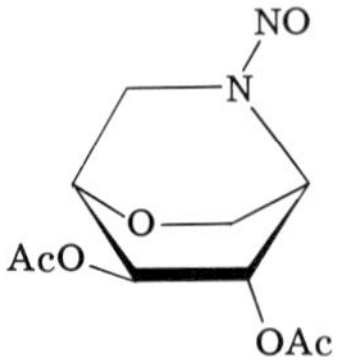

94 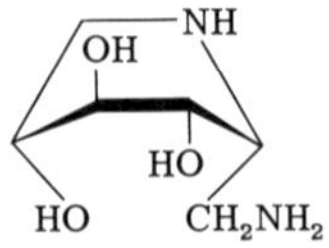   95

acetylation and chromatography. The formation of this product, which must have resulted from a boat conformation (**95**), was rationalized on the grounds that the two chair conformers of the diamine must presumably be of comparable energy, and that their interconversion must involve a boat intermediate.

2. Furanose Derivatives[172]

Almost all of the reported deaminations of furanose amines concern compounds in which the amino group is not directly attached to a ring atom. The reactions of 1,2-O-alkylidene-6-amino-6-deoxy-hexose derivatives appear to depend on the configuration of C-5. Thus, the L-idose derivative **96** gave the 3,6-anhydro-L-idose derivative (**97**), which was isolated in 80% yield,[173] whereas the D-glucose isomer (**100**) gave the 5,6-anhydride **101** in unspecified yield.[146] Because, with the L-idose amine, the reaction mixture was heated, it is tempting to speculate that 5,6-anhydro-1,2-O-isopropylidene-β-L-idofuranose was an intermediate in the formation of the 3,6-anhydride. However, the subsequent conversion of two L-idose amines[174] (**98**) and[171] (**99**) into 3,6-anhydro-L-idose derivatives on deamination at room temperature (or below) does not support this theory.

The L-idose derivative **98** gave a small proportion of the substitution product, and an attempt to improve its yield by protecting the 3-hydroxyl group by acetylation surprisingly did not prevent formation of the 3,6-anhydro ring, the substitution product being formed only in low yield. The D-glucose isomer of **98** could not be converted by

(171) H. Paulsen and E. Mäckel, *Chem. Ber.* **102**, 3844–3853 (1969).
(172) 1,4-Lactones are referred to in the discussion of acids (*see* Section IV, p. 58).
(173) H. Ohle and R. Lichtenstein, *Ber.*, **63**, 2905–2912 (1930).
(174) H. Paulsen, *Ann.*, **665**, 166–187 (1963).

deamination into the corresponding alcohol, and none of the products were identified. A related D-glucose amine (**102**) gave[175] the substitution product in 61% yield.

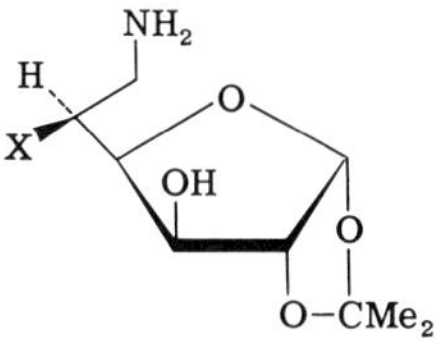

96 X = OH
99 X = NHCH₂Ph

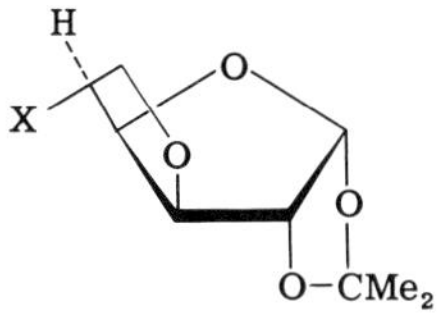

X = OH
97

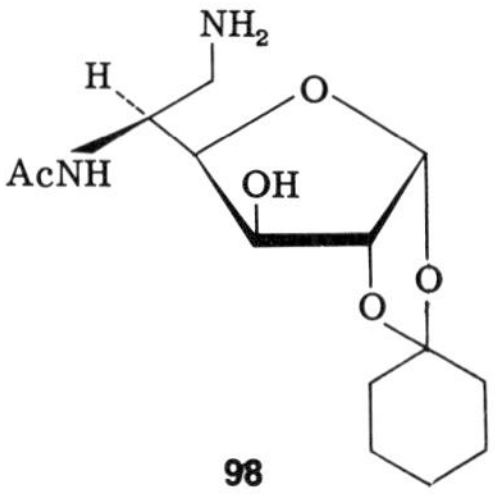

98

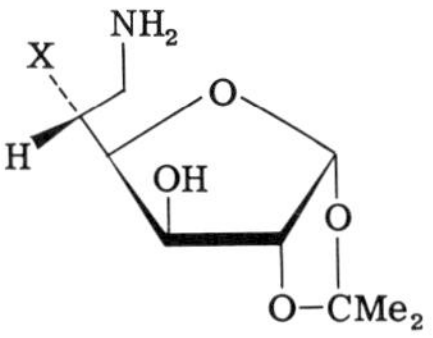

100   X = OH
102   X = SCH₂Ph

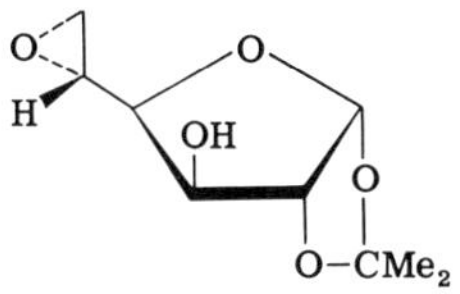

101

The difference between the behavior of the 6-amino-6-deoxy-D-
glucose and -L-idose derivatives can be rationalized as follows. In
the transition states for the formation of the 3,6-anhydro ring from the
D-*gluco* isomers, the oxygen (or hetero) atoms attached to C-4 and
C-5 are in a sterically unfavorable, *cis* relationship. The transition
states from the L-*ido* amines involve a *trans* relationship of these hetero atoms. Thus, for the D-*gluco* amine, deamination reactions other
than 3,6-anhydro ring-formation compete more effectively.

The 2,5-diamino-2,5-dideoxy derivatives of 1,4:3,6-dianhydro-D-
mannitol and -D-glucitol gave, on deamination, the same product,
namely, 1,4:3,6-dianhydro-L-iditol, in unspecified yield.[176] The product presumably arose from carbonium-ion intermediates, which
reacted predominantly with solvent on the less hindered, *exo* sides
of the V-shaped molecule.

Deamination performed in hydrochloric acid solution may result in
the formation of chlorine-containing products, as illustrated by the
deamination   of   5-amino-5-deoxy-1,2-O-isopropylidene-α-D-xylo-
furanose, which gave the 5-chloro-5-deoxy derivative (isolated in
6% yield).[49] 1,2-O-Isopropylidene-α-D-xylofuranose, the major prod-

(175) R. L. Whistler and R. E. Pyler, *Carbohyd. Res.*, **12**, 201–210 (1970).
(176) V. G. Bashford and L. F. Wiggins, *J. Chem. Soc.*, 371–374 (1950).

uct, was isolated as the di-*p*-toluenesulfonate in 28% yield. The chlorodeoxy compound was, presumably, formed by reaction of the intermediate, diazonium ion (or, possibly, a carbonium ion) with chloride ion. Another possible intermediate is the tricyclic oxonium ion **103**. Attack of nucleophiles at C-4 of the ion **103**, to give pyranose

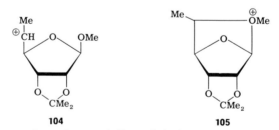

103

products, may be unfavorable due to constraints imposed by the dioxolane ring.

The deamination, in 90% aqueous acetic acid, of methyl 5-amino-5,6-dideoxy-2,3-*O*-isopropylidene-α-L-talo- and -β-D-allo-furanosides was studied, because these reactions could, in principle, proceed through the same intermediate cations as the deamination of methyl 4-amino-4-deoxy-2,3-*O*-isopropylidene-α-L-mannopyrano-side (**39a**, see p. 34).[177] Elimination occurred, in both furanoside amines, to give 5,6-dideoxy-2,3-*O*-isopropylidene-β-D-*ribo*-hex-5-enofuranoside (44% from the L-*talo* amine, 8% from the D-*allo* amine). This alkene was absent from the deamination products of the pyran-oside amine **39a**, and could possibly be regarded as diagnostic of the C-5 carbonium ion (**104**). Epimeric, furanoside derivatives re-sulting from substitution were also formed from both furanoside amines, and the products differed in relative yields, and also in that migration of the anomeric methoxyl group occurred only with the β-D-*allo* amine, to give 6-deoxy-2,3-*O*-isopropylidene-5-*O*-methyl-L-talofuranose and its 1-*O*-acetyl derivative. This behavior parallels the solvolysis of the corresponding furanoside 5-sulfonates,[178] and the oxonium ion **105** is, presumably, an intermediate. An oxo-

(177) J. S. Brimacombe and J. Minshall, *Carbohyd. Res.*, **25**, 267–269 (1972).

(178) C. L. Stevens, R. P. Glinski, K. G. Taylor, P. Blumbergs, and F. Sirokman, *J. Amer. Chem. Soc.*, **88**, 2073–2074 (1966).

nium ion such as **39b** (see p. 34) is, presumably, not an intermediate in these reactions, as pyranosides were not identified among the products.

Because of their basicity (lower than that of aliphatic amines), aromatic primary amines can be selectively nitrosated[179] in the presence of aliphatic amines at low pH. An example is provided by the deamination of 3'-amino-3'-deoxyadenosine, although the yield of the product isolated, 3'-amino-3'-deoxyinosine, was[180] only about 4%. Some 50% of the starting material remained unchanged, and hydrolysis released adenine (30%).

Deamination of methyl 6-amino-5,6-dideoxy-2,3-O-isopropylidene-β-D-*ribo*-hexofuranoside gave the corresponding alcohol as the major product. No data were given for an alkene, reported as a minor product resulting from elimination.[181] Substitution also occurred when 1-(6-amino-2,5,6-trideoxy-D-*erythro*-hexofuranosyl)-thymine was deaminated, the parent nucleoside being isolated pure in 51% yield.[182] The crystalline product isolated (in unspecified yield) from the deamination of a nucleoside, 1-(5-amino-5-deoxy-β-D-allofuranosyluronic acid)-5-(hydroxymethyl)uracil, resulted from substitution with retention of configuration.[183a] Such behavior is typical of α-amino acids (see p. 58).

Deamination of methyl 3-amino-3-deoxy-β-D-xylofuranoside gave an epoxide, methyl 2,3-anhydro-β-D-ribofuranoside, as the major product, together with some methyl β-D-xylofuranoside and methyl β-D-ribofuranoside. A similar reaction occurred when the α anomer was deaminated, the products being tentatively identified chromatographically. Migration of the methoxyl group occurred in the deamination of methyl 2-amino-2-deoxy-α-D-arabinofuranoside to give 2-O-methyl D-ribose; methyl 2,3-anhydro-α-D-ribofuranoside was also formed, as a minor product (~20%). Thus, methoxyl migration is more favorable than epoxide formation, whereas the reverse obtains for methyl 2-amino-4,6-O-benzylidene-2-deoxy-α-D-altropyranoside[183b] (see p. 40).

(179) N. Kornblum and D. S. Iffland, *J. Amer. Chem. Soc.*, **71**, 2137–2143 (1949).
(180) N. N. Gerber, *J. Med. Chem.*, **7**, 204–207 (1964).
(181) J. A. Montogomery and K. Hewson, *J. Org. Chem.*, **29**, 3436–3438 (1964).
(182) G. Etzold, G. Kowollik, and P. Langen, *Chem. Commun.*, 422 (1968).
(183a) K. Isono and S. Suzuki, *Tetrahedron Lett.*, 203–208 (1968).
(183b) D. R. Clark and J. G. Buchanan, personal communication; D. R. Clark, Ph. D. Thesis, University of Newcastle-upon-Tyne, England (1969).

IV. Acyclic Compounds

The deamination of 2-amino-2-deoxy-D-gluconic acid and -D-galactonic acid with nitrous acid was referred to in Section III,1,a (see p. 18). If those deaminations that involved treatment of the amino sugar with nitric acid are disregarded, the first deamination[58] of an aminodeoxyaldonic acid was that of 2-amino-2-deoxy-D-gluconic acid in 1894. Levene and his collaborators[65,97,184-186] subsequently studied the deamination of all eight 2-amino-2-deoxy-D-aldohexonic acids (see Ref. 1). Each 2-amino-2-deoxy-D-aldonic acid gave the corresponding 2,5-anhydro-D-hexonic acid with retention of configuration. That retention of configuration had occurred could be concluded only when the configurations of the parent hexosamines had been established, although, in 1925, Levene,[187] after a consideration of the $[M]_D$ values of the 2-amino-2-deoxyhexonic acids, stated that it is probable that deamination of these compounds occurs without Walden inversion.

The retention of configuration in these reactions was attributed by Foster[188] to participation of the carboxylic hydroxyl group, to give an unstable α-lactone, by analogy with the reaction of simple α-amino acids.[189] Attack of the 5-hydroxyl group on C-2 of the α-lactone would then result in a second inversion at this center. In such irreversible, ring closures, the formation of five-membered rings is more favorable than that of six-membered rings.[2,190]

One feature of these early studies that could not be explained was the fact that, whereas 2-amino-2-deoxy-D-mannonic acid and its 1,4-lactone give the same product (2,5-anhydro-D-mannonic acid) on deamination, 2-amino-2-deoxy-D-idonic acid and its lactone give[185] products epimeric at C-2. Defaye has pointed out[6] that the 5-hydroxyl group of 2-amino-2-deoxy-D-idono-1,4-lactone is sterically well placed to attack the rear of C-2 in the diazonium ion (or other product precursor); this would account for the observed inversion of configuration. Although the retention of configuration found on deamination of 2-amino-2-deoxy-D-mannono-1,4-lactone can be explained on the basis of a C-2 carbonium-ion intermediate, the reac-

(184) P. A. Levene and F. B. LaForge, *J. Biol. Chem.*, **22**, 331–335 (1915).
(185) P. A. Levene, *J. Biol. Chem.*, **36**, 89–94 (1918).
(186) P. A. Levene and E. P. Clark, *J. Biol. Chem.*, **46**, 19–33 (1921).
(187) P. A. Levene, *J. Biol. Chem.*, **63**, 95 (1925).
(188) A. B. Foster, *Chem. Ind.* (London), 627 (1955).
(189) P. Brewster, F. Hiron, E. D. Hughes, C. K. Ingold, and P. A. D. Rao, *Nature*, **166**, 179–180 (1950).
(190) J. A. Mills, *Advan. Carbohyd. Chem.*, **10**, 1–53 (1955), especially pp. 23–25.

tion may be more complex; for example, a 2-deoxy-2-diazolactone may be an intermediate.

During the studies by Levene and his collaborators on the deamination of aminodeoxy-D-gluconic acid derivatives, the preparation of the first diazohexonic acid derivative was accomplished. When ethyl 2-amino-4,6-O-benzylidene-2-deoxy-D-gluconate was treated with nitrous acid, the yellow 2-diazo-2-deoxy derivative separated out.[64] This product[191] was formulated as a diazirine, following Curtius's structure[193] for ethyl diazoacetate. The diazoester was found to react rapidly with hydrogen chloride or hydrogen bromide in ether, with evolution of nitrogen. Crystalline 2-chloro- and 2-bromo-2-deoxy derivatives were isolated, and tentatively assigned the D-*manno* configuration on the basis of measurements of optical rotations.[194] The diazo ester was also reduced to 2-deoxy- and 2-amino-2-deoxy-D-hexonate derivatives,[195] and hydrolyzed in 5% aqueous acetic acid to the D-gluconate derivative.[194] The preparation of the diazoester has been repeated,[196,197] and complete characterization was reported by Horton and Philips,[197] who also prepared the 3-O-acetyl derivative and the methyl ester analog. Treatment of methyl 2-amino-2-deoxy-D-gluconate hydrochloride with nitrous acid, followed by acetylation of the product, gave the diazo tetraacetate **106** in 30–50% yield, the rather low yield being due to hydrolysis, which gave methyl penta-O-acetyl-D-gluconate and -D-mannonate. Thermolysis of the deoxy-diazo-D-gluconate **106** at 100–110° gave a mixture of the *cis* and *trans* isomers of the enol acetate **107**, and photolysis in 2-propanol gave methyl 3,4,5,6-tetra-O-acetyl-2-deoxy-D-*arabino*-hexonate (**108**) in 61% yield.[197] The latter reaction provides a method for the

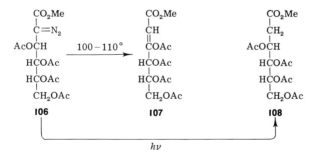

(191) The 5,6-O-benzylidene structure was corrected in Ref. 192.
(192) P. Karrer and J. Meyer, *Helv. Chim. Acta*, **20**, 407–417 (1937).
(193) T. Curtius, *J. Prakt. Chem.*, **39**, 107–139 (1889).
(194) P. A. Levene, *J. Biol. Chem.*, **53**, 449–461 (1922).
(195) P. A. Levene, *J. Biol. Chem.*, **54**, 809–813 (1922).
(196) P. W. Kent, K. R. Wood, and V. A. Welch, *J. Chem. Soc.*, 2493–2497 (1964).
(197) D. Horton and K. D. Philips, *Carbohyd. Res.*, **22**, 151–162 (1972).

conversion of an amino sugar derivative into a deoxy sugar derivative.

Other stable diazoalkanes, such as diazoketones, can also be converted into halogen derivatives, as illustrated by the conversion of 3,4-di-O-benzoyl-1-deoxy-1-diazo-D-glycero-tetrulose into the 1-bromo- and 1-chloro-deoxy derivatives by the action of hydrogen bromide and hydrogen chloride, respectively.[196]

The products obtained from the deamination of 3-amino-3-deoxyheptonic acids were not obtained pure.[198] Oxidation of the products gave acids whose calcium salts had elemental analyses agreeing with those calculated for the calcium salts of pentaric acids, and Shafizadeh[2] suggested that the latter are formed from 3-deoxy-2-heptulosonic acids, which, in turn, result from a rearrangement accompanying deamination.

The first deamination of a 1-amino-1-deoxyalditol to be reported was that of 1-amino-1-deoxy-D-glucitol, which gave 1,4-anhydro-D-glucitol in 57% yield.[199] 1-Amino-1-deoxy-D-mannitol similarly gives the 1,4-anhydride.[146] Substitution was later shown to be an important competing pathway in these reactions, the hexitols being identified, as well as the 1,4-anhydrohexitols, in the products of deamination of 1-amino-1-deoxy-D-allitol, -D-altritol, -D-iditol, and -D-gulitol.[200] The anhydrides were the only products reported for the D-talitol and D-galactitol amines.

Both substitution and cyclization occur in the deamination of 1-amino-1-deoxy-D-glycero-D-galacto-heptitol and of the D-glycero-D-gulo-heptitol isomer. Products were isolated in low yield from the former, and identification of the products from the latter amine was tentative, being based on mobilities on paper chromatograms.[201]

In a detailed study of the deamination of the four 1-amino-1-deoxy-D-pentitols (109–112), the products were analyzed by g.l.c. It was found that each aminodeoxypentitol gives three products in differing ratios.[202] In addition to the pentitol and the corresponding 1,4-anhydropentitol, there is formed a second 1,4-anhydropentitol. The latter was found to be formed by ring closure between the 5-hydroxyl group and C-2, with inversion at C-2. For example, 1-amino-1-deoxy-D-lyxitol gives 1,4-anhydro-D-lyxitol, 2,5-anhydro-D-xylitol (≡ 1,4-anhydro-L-xylitol), and D-lyxitol. Because the second 1,4-

(198) P. A. Levene and I. Matsuo, J. Biol. Chem., 39, 105–118 (1919).
(199) V. G. Bashford and L. F. Wiggins, J. Chem. Soc., 299–303 (1948).
(200) R. Barker, J. Org. Chem., 29, 869–873 (1964).
(201) H. J. F. Angus and N. K. Richtmyer, Carbohyd. Res., 4, 7–11 (1967).
(202) D. D. Heard, B. G. Hudson, and R. Barker, J. Org. Chem., 35, 464–467 (1970).

anhydropentitol is formed stereospecifically in each instance, it was concluded that 1,2-anhydropentitols are intermediates in their formation, the opening of the epoxide ring being accompanied by inversion at C-2. It was considered unlikely that the 1,4-anhydropentitols of retained configuration (**113, 115, 116,** and **114,** respectively) are formed by way of 1,2-anhydropentitols, because 1,2-epoxy-4-butanol is not cyclized to a tetrahydrohydroxyfuran under mildly alkaline conditions.

An attempt was made to rationalize the variation in the extent of cyclization in relation to the configuration of the aminodeoxypentitol.[202] The formation of the alditol probably involves the reaction of water with a diazonium ion (or, possibly, a carbonium ion), and it was presumed that this is but little affected by changes in configuration. However, cyclization involving attack of the 4-hydroxyl group on C-1 would be expected to vary with the configuration, which would affect the transition-state energy. The data in Table IV[202a]

TABLE IV[202a]

Products of Deamination of 1-Amino-1-deoxypentitols

1-Amino-1-deoxypentitol	Pentitol (%)	1,4-Anhydride (%)	2,5-Anhydride, inverted (%)
xylo	2	89	9
ribo	7	78	15
arabino	14	78	9
lyxo	20	55	24

show that ring formation becomes increasingly more difficult in the series *xylo, ribo, arabino,* and *lyxo,* although the differences are rather small. The transition states presumably resemble the cyclic products to some extent, and, on the basis of interactions between the substituents at C-2, C-3, and C-4 of the cyclic transition state, it was suggested that the ease of cyclization should decrease in the order: *arabino, xylo, ribo,* and *lyxo.* However, it was recognized that ground-state conformations might also be important in deamination reactions. The results observed can be qualitatively accounted for by considering both the ground-state and the transition-state energies. Thus, there are no unfavorable interactions in the planar, zigzag con-

(202a) Reproduced from Ref. 202, p. 464, with the permission of the American Chemical Society.

formations [203] of the *lyxo* (**109**) and *arabino* (**110**) isomers, and the cyclization of the *lyxo* isomer places all three substituents (at C-2, C-3, and C-4) in *cis* relationships (**113**). Cyclization of the *lyxo* isomer is, therefore, the least favorable. β-Eclipsed hydroxyl groups are present in the zigzag conformations of the *ribo* (**111**) and *xylo* (**112**) isomers. The latter (**112**) may be regarded as the most unfavorable, as all three

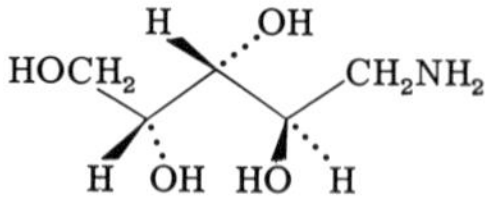

lyxo
109

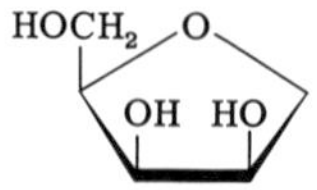

(55%)
113

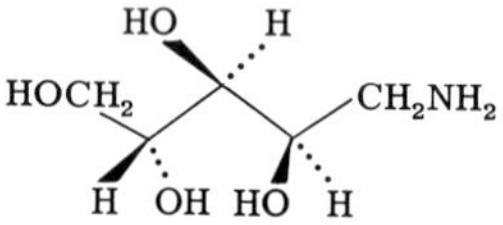

arabino
110

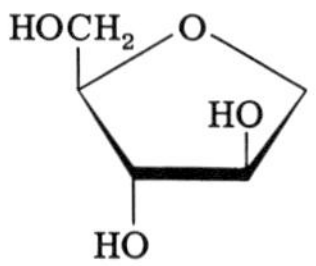

(78%)
115

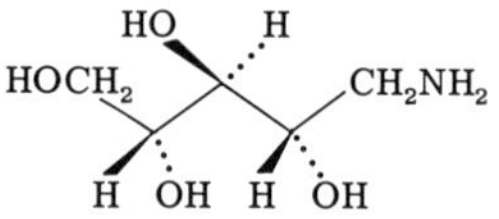

ribo
111

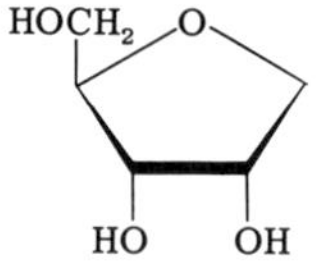

(78%)
116

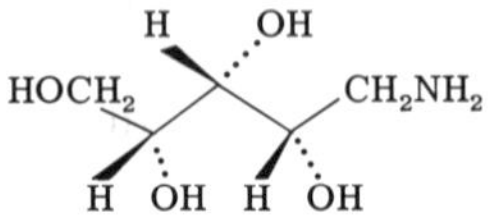

xylo
112

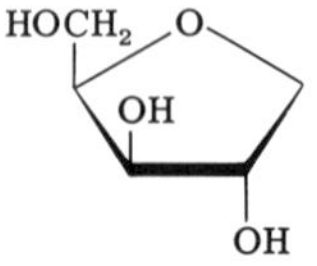

(89%)
114

of its hydroxyl groups are *gauche* disposed. The *xylo* isomer is, thus, the most likely to exist as a sickle conformer,[203] which requires less reorientation in order that cyclization may occur. The *xylo* isomer gave the highest yield of anhydride **114**. Experimental measure-

(203) P. L. Durette and D. Horton, *Advan. Carbohyd. Chem. Biochem.*, **26**, 68–73 (1971).

ments by 250-MHz n.m.r spectroscopy showed that the *ribo* and *xylo* isomers do indeed favor sickle conformations, whereas extended conformations are favored for the *arabino* and *lyxo* isomers.[204]

Ring formation in the deamination of 1-amino-1-deoxyalditols can be prevented (or minimized) by using *O*-acetyl derivatives. Thus, deamination of the *p*-toluenesulfonate (salt) of 2,3,4-tri-*O*-acetyl-1-amino-1-deoxy-D-erythritol gave erythritol,[204a] in unspecified yield, presumably after deacetylation (no deacetylation step was reported). Treatment of 2,3-di-*O*-acetyl-1-amino-4-bromo-1,4-dideoxyerythritol hydrobromide with sodium nitrate in acetic acid containing hydrogen bromide gave, in unspecified yield, 2,3-di-*O*-acetyl-1,4-dibromo-1,4-dideoxyerythritol.[204a]

A product that had an elemental analysis agreeing with that calculated for an anhydrohexitol tetraacetate was obtained by sequential deamination and acetylation of 2-amino-2-deoxy-D-glucitol.[146] In the light of subsequent reports, this product was probably a 2-deoxyhexose tetraacetate. Matsushima[92] reported that the major product of this reaction was 2-deoxy-D-*arabino*-hexose, which was isolated as its crystalline 2,2-diphenylhydrazone in 26% yield. A similar reaction occurred when a mixture of 2-amino-2-deoxy-D-ribitol and 2-amino-2-deoxy-D-arabinitol was deaminated, 2-deoxy-D-*erythro*-pentose being isolated as its benzylphenylhydrazone in 5% yield.[205] Foster[188] suggested an elimination mechanism for such rearrangements, the enol **117** being an intermediate. Another possibility (see Scheme 8) is a hydride shift to give the cation **118**.

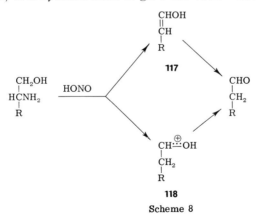

Scheme 8

(204) J. Defaye, D. Horton, and M. Muesser, *Abstr. Papers Amer. Chem. Soc. Meeting*, **168**, CARB 1 (1974).

(204a) I. Ziderman, *Carbohyd. Res.*, **18**, 323–328 (1971).

(205) Y. Matsushima and Y. Imanaga, *Bull. Chem. Soc. Jap.*, **26**, 506–507 (1953); *Nature*, **171**, 475 (1953).

Although migration of a carbon atom to give a branched-chain product is possible in the deamination of acyclic, amino sugar derivatives, no such migration was detected until a detailed study of the deamination of 2-amino-2-deoxy-D-glucitol (119) was made.[206] Five products were detected, four of which were identified as 2-deoxy-D-*arabino*-hexose (120, 46%), 2-deoxy-2-C-(hydroxymethyl)-D-*arabino*- (or *ribo*)-pentose (121, 17%, impure), 2-deoxy-D-*erythro*-3-hexulose (122, 23%), and D-mannitol (123, 5%). Except for the branched-chain sugar, these products were isolated as homogeneous fractions, in the yields indicated (after chromatography). The deamination products were also analyzed by g.l.c. after reduction with sodium borohydride, and periodate oxidation of the reduced products was used to confirm the structures.[206]

$$
\begin{array}{ccccc}
\begin{array}{l}CH_2OH\\HCNH_2\\HOCH\\HCOH\\HCOH\\CH_2OH\end{array}
& \xrightarrow{\;HONO\;}
& \begin{array}{l}CHO\\CH_2\\HOCH\\HCOH\\HCOH\\CH_2OH\end{array}
\;+\;
\begin{array}{l}CHO\\CH(CH_2OH)\\HCOH\\HCOH\\CH_2OH\end{array}
\;+\;
\begin{array}{l}CH_2OH\\CH_2\\C=O\\HCOH\\HCOH\\CH_2OH\end{array}
\;+\;
\begin{array}{l}CH_2OH\\HOCH\\HOCH\\HCOH\\HCOH\\CH_2OH\end{array}
\\[2em]
119 & & (46\%) \qquad (\sim17\%) \qquad (23\%) \qquad (5\%)\\
& & \;\;120 \qquad\;\; 121 \qquad\;\; 122 \qquad\;\; 123
\end{array}
$$

It is interesting to speculate on the reasons for the absence of carbon migration in the deamination of the 1-amino-1-deoxypentitols. Rearrangements involving migration of carbon or hydrogen account for more than 85% of the products from 2-amino-2-deoxy-D-glucitol. The marked difference in reaction pathways must be occasioned by the fact that the latter is a secondary alkanamine whereas 1-amino-1-deoxypentitols are primary alkanamines. Deamination of the latter is more likely to involve displacement of nitrogen (see Section II, p. 13). Water and a hydroxyl group are both more nucleophilic than a carbon or hydrogen atom; thus, substitution and anhydro-ring formation are more likely to occur for 1-amino-1-deoxyalditols than for alditols possessing an amino group on a nonterminal carbon atom. Carbonium ions are probable intermediates in the deamination of the latter, and it would be interesting to know whether the products from the deamination of 2-amino-2-deoxy-D-

(206) T. Bando and Y. Matsushima, *Bull. Chem. Soc. Jap.*, **46**, 593–596 (1973).

glucitol differ significantly from those given by 2-amino-2-deoxy-D-mannitol, as the ground-state conformations of the two amines are probably different.

That the delicate balance between competing pathways in deaminations can be changed by a change in pH was illustrated by the deamination of 2-amino-2-deoxy-D-glucose diethyl dithioacetal hydrochloride. Treatment of the latter with sodium nitrite in dilute hydrochloric acid solution at 0° gave as the major product 2-S-ethyl-2-thio-D-glucose, which was isolated crystalline in 54% yield.[207,208] When the reaction was conducted in dilute acetic acid buffered at pH 5.6, a different major product was formed; this was at first assumed to be 2,5-anhydro-D-glucose diethyl dithioacetal,[209] but the correct structure was subsequently established as ethyl 2-S-ethyl-1,2-dithio-α-D-mannofuranoside by X-ray crystallography[210] and by chemical and spectroscopic methods.[211] The tri-O-acetyl derivative of the furanoside was isolated in 35% yield after chromatography. 2-S-Ethyl-2-thio-D-glucose was also detected chromatographically. Participation of ethylthio groups clearly occurred in these reactions. The product of the reaction at low pH, namely, 2-S-ethyl-2-thio-D-glucose, could conceivably[211] arise by two successive participations, but it appears more probable[211a] that an intermediate *manno* episulfonium ion common to both observed reaction-pathways would give rise in the acidic medium to 2-S-ethyl-2-thio-D-mannose at first; the *manno* product could suffer subsequent epimerization during the deionization procedure used during isolation, as it is exceptionally prone[212] to alkaline epimerization.

In related work,[212a] 2-amino-2-deoxy-D-glucose ethylene dithioacetal was deaminated at different pH values. In a system buffered at pH ~5, three main products were isolated after the mixture

(207) D. Horton, L. G. Magbanua, and J. M. J. Tronchet, *Chem. Ind.* (London), 1718–1719 (1966).

(208) A. E. El Ashmawy, D. Horton, L. G. Magbanua, and J. M. J. Tronchet, *Carbohyd. Res.*, 6, 299–309 (1968).

(209) J. Defaye, *Bull. Soc. Chim. Fr.*, 1101–1102 (1967).

(210) J. Defaye, A. Ducruix, and C. Pascard-Billy, *Bull. Soc. Chim. Fr.*, 4514–4515 (1970).

(211) J. Defaye, T. Nakamura, D. Horton, and K. D. Philips, *Carbohyd. Res.*, 16, 133–144 (1971).

(211a) D. Horton, *Pure Appl. Chem.*, in press.

(212) B. Berrang and D. Horton, *Chem. Commun.*, 1038–1039 (1970).

(212a) P. Angibeaud, C. Bosso, J. Defaye, and D. Horton, *Abstr. Papers Amer. Chem. Soc. Meeting*, 168, CARB 6 (1974).

had been acetylated; one was 3,5,6-tri-*O*-acetyl-2-deoxy-D-*arabino*-hexono-1,4-lactone, the second appeared to be 1,2-*S*-ethylene-1,2-dithio-α-D-mannopyranose 3,4,6-triacetate, and the third was a furanoid analog of the dithioglycoside. It has been suggested[211a] that these products may arise from a common intermediate in the diazotization reaction, with competition between a 1→2 hydride-shift (leading to the deoxy derivative) or 1→2 migration of sulfur leading to the thioglycosides. At strongly acid pH, the main product is the deoxy acid derivative, evidently reflecting decreased nucleophilicity of sulfur in the acidic medium.

The observation[208] that hydroxyl oxygen atoms are less nucleophilic in acidic solution, presumably due to a higher concentration of conjugate acid, suggests that the extent of anhydro-ring formation in the deamination of other acyclic amines might be lessened by conducting the reactions at lower pH.

V. MISCELLANEOUS COMPOUNDS

1. Aziridines

Treatment of methyl 4,6-*O*-benzylidene-2,3-dideoxy-2,3-epimino-α-D-allopyranoside with nitrous acid gave methyl 4,6-*O*-benzylidene-2,3-dideoxy-α-D-*erythro*-hex-2-enopyranoside in high yield (81%). The *manno* isomer likewise gave the same alkene, in 78% yield. The first step in these reactions is the formation of the unstable, yellow *N*-nitrosoepimines, the *allo* isomer being isolated, and crystallized at −20°; it was characterized by u.v. and i.r. measurements and by elemental analysis, and could be stored at −20° for a few days.[213]

2. Amides

Unsubstituted amides can be converted under very mild conditions into the corresponding carboxylic acids by treatment with nitrous acid. This reaction is useful when the usual hydrolysis conditions would affect other functions present, such as hemiacetals,[214] as

(213) R. D. Guthrie and D. King, *Carbohyd. Res.*, **3**, 128–129 (1966).
(214) M. Ishidate and M. Matsui, *Yakugaku Zasshi*, **82**, 662–669 (1962); *Chem. Abstr.*, **58**, 4639g (1963).

in **124**, and esters,[215] as in **125**. The product isolated from compound **124** was the 5,3-lactone.

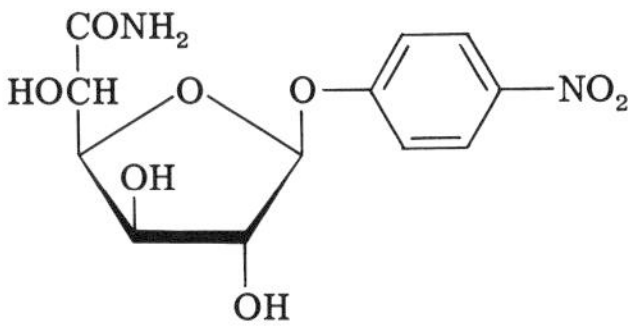

124 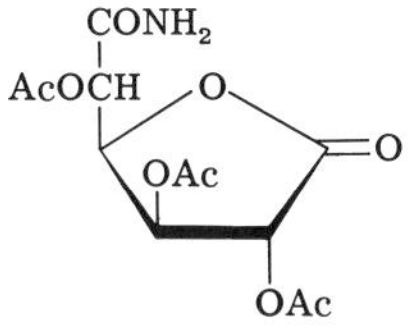   **125**

Monosubstituted amides can be converted into *N*-nitroso derivatives by use of a sufficiently powerful nitrosating agent, such as ni-

$$\text{RNHCOR}' \longrightarrow \text{RN(NO)COR}' \longrightarrow [\text{RN=NOCOR}'] \longrightarrow [\overset{\oplus}{\text{RN}_2}\overset{\ominus}{\text{O}_2\text{CR}'}] \longrightarrow \text{RO}_2\text{CR}' + \text{N}_2$$

trosyl chloride or dinitrogen tetraoxide.[216] The nitrosoamides are thermally labile, the lability increasing in the series $\text{RCH}_2\text{N(NO)COR}' < \text{R}_2\text{CHN(NO)COR}' < \text{R}_3\text{CN(NO)COR}'$. Decomposition of the nitrosoamides in an inert solvent can give esters in high yield (see Scheme 2, p. 14). Homolytic reactions apparently do not occur for aliphatic nitrosoamides at or near room temperature.

The only carbohydrate nitrosoamides thus far studied are derivatives of 2-acetamido-2-deoxy-D-glucose,[217,217a] the D-galactose analog,[217b] and 3-acetamido-3-deoxy-D-mannose.[217b] Nitrosation of methyl 2-acetamido-3,4-6-tri-*O*-acetyl-2-deoxy-β-D-glucopyranoside with nitrosyl chloride in chloroform containing pyridine at 0° gave the crystalline *N*-nitroso derivative (**126**) in 93% yield.[217] Dinitrogen tetraoxide can be used similarly.[217b] The solid was stable at 20° for several days, but, in solution, slow decomposition occurred over several weeks. The decomposition of this nitrosoamide resembled the deamination of the corresponding, unesterified amine, in that ring contraction was the main reaction, apart from denitrosation to give the amide. The amide was isolated in ~40% yield in the reaction in chloroform containing 2% of ethanol. Some substitution also occurred, to give the 2-chloro-2-deoxyglucoside (**129**) as a minor

(215) K. Ochi, Jap. Pat. 19,626 (1971); *Chem. Abstr.*, **75**, 64,218 (1971).

(216) E. H. White, *J. Amer. Chem. Soc.*, **77**, 6008–6010 (1955).

(217) J. W. Llewellyn and J. M. Williams, *Chem. Commun.*, 1386–1387 (1971), and unpublished results.

(217a) D. Horton and W. Loh, *Carbohyd. Res.*, **36**, 121–130 (1974).

(217b) D. Horton and W. Loh, *Carbohyd. Res.*, **38**, 189–204 (1974).

product.[218] The ring-contraction products (**127** and **128**) resulted from reaction of the rearranged cation with counter-ion (acetate) and ethanol. A solution of the nitrosoamide **126** in 50% aqueous acetone at ~20° gave, in seven days, unstable 3,4,6-tri-*O*-acetyl-2,5-anhydro-D-mannose (**10**), which was characterized as its diethyl acetal, diethyl

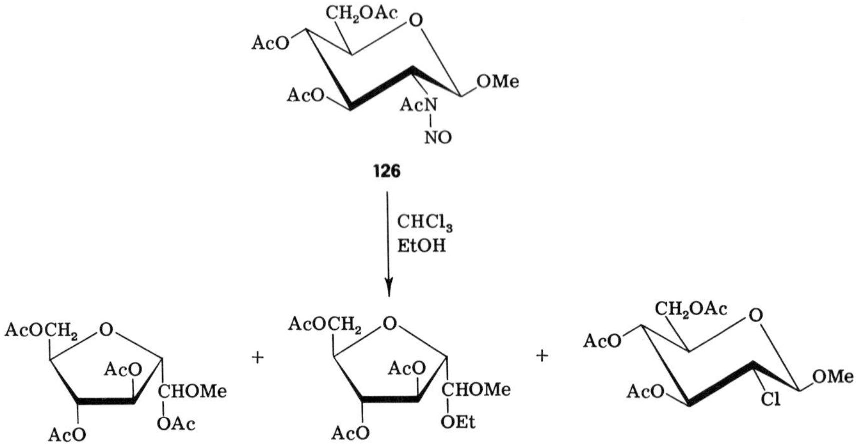

126

$$\downarrow \begin{array}{l} CHCl_3 \\ EtOH \end{array}$$

127   +   **128**   +   **129**

dithioacetal, and 1,1-di-*O*-acetyl aldehydrol derivatives. Substitution also occurred to a minor extent, as methyl β-D-glucopyranoside and its tetraacetate (<5%) were detected chromatographically after deacetylation and acetylation, respectively, of the crude product. The *N*-nitroso derivative of a disaccharide, 6-*O*-(2-acetamido-3,4,6-tri-*O*-acetyl-2-deoxy-β-D-glucopyranosyl)-1,2:3,4-di-*O*-isopropylidene-α-D-galactopyranose, was also prepared, and converted into **10** (see p. 22) and 1,2:3,4-di-*O*-isopropylidene-α-D-galactopyranose in aqueous acetone at room temperature.[217]

The effect of solvent on such reactions was much more marked in the decompositions of 1,3,4,6-tetra-*O*-acetyl-2-deoxy-2-(*N*-nitrosoacetamido)-α-D-glucopyranose[221] (**130**). The extent of ring con-

(218) The formation of chloro compounds in related reactions in chlorine-containing solvents had been observed before, and was attributed to reaction of the carbonium ion with solvent.[219,220]

(219) E. H. White, T. J. Ryan, and K. W. Field, *J. Amer. Chem. Soc.*, **94**, 1360–1361 (1972).

(220) E. H. White, H. P. Tiwari, and M. J. Todd, *J. Amer. Chem. Soc.*, **90**, 4734–4736 (1968).

(221) J. W. Llewellyn and J. M. Williams, *Carbohyd. Res.*, **28**, 339–350 (1973).

traction decreased considerably in the following solvents: 1:1 (v/v) water–acetone, 1:5 (v/v) acetic acid–acetic anhydride (containing sodium acetate), and chloroform containing 2% of ethanol. The major product in the last two media was β-D-glucopyranose pentaacetate (133, R = Ac), and the β-D configuration was accounted for by postulating two successive participations, the first by the ring-oxygen atom to give the bicyclic oxonium ion (131), and the second by the anomeric acetoxyl group to give the acetoxonium ion (132), which is the precursor of β-D-glucopyranose products. The ethyl glycoside (133, R = Et) was the second major product in the reaction in chloroform containing 2% of ethanol. No ring-contraction products were detected. In acetic acid–acetic anhydride, ring contraction occurred to give 1,1,3,4,6-penta-O-acetyl-2,5-anhydro-D-mannose (134) as a minor product. α-D-Glucopyranose pentaacetate was tentatively identified as a minor product by g.l.c.; it could have arisen from the oxonium ion (131) or by cis ring-opening of the acetoxonium ion[222] (132). In aqueous acetone, both the α-nitrosoamide (130) and its β-D anomer gave 3,4,6-tri-O-acetyl-2,5-anhydro-D-mannose (10; see

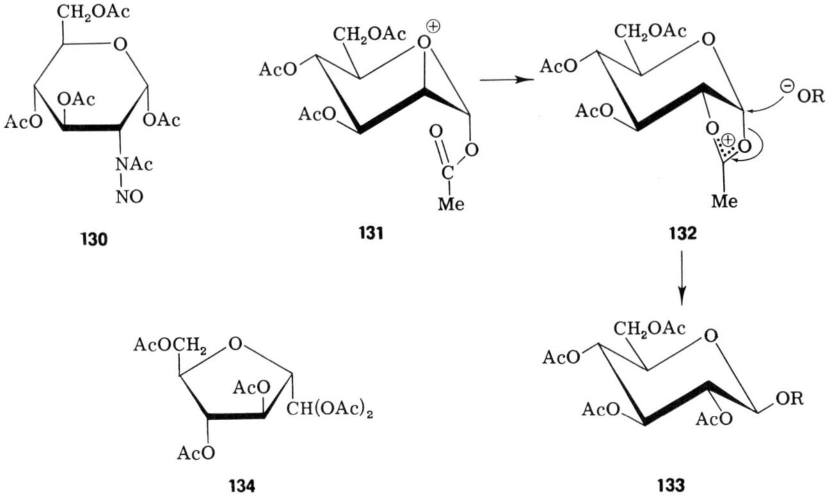

p. 22) as the major product.

The β anomer of nitrosoamide 130 gives back the starting amide (87% yield) when it is photolyzed in acetonitrile.[217a] An unusual reaction takes place[217a] when 130 in ethereal solution is treated with

(222) J. G. Buchanan, J. Conn, A. R. Edgar, and R. Fletcher, J. Chem. Soc. (C), 1515–1521 (1971).

potassium hydroxide in isopropyl alcohol. The C-2 substituent and C-1 are lost, and a C_5 acetylene, 1,2-dideoxy-D-*erythro*-pent-1-ynitol, is formed; it was isolated in 70% yield as its triacetate **135**. The D-galactose analog of **130** reacts in the same way to give the triacetate of 1,2-dideoxy-D-*threo*-pent-1-ynitol.[217b] The N-nitroso derivative (**136**) of methyl 2-acetamido-3,4,5,6-tetra-O-acetyl-2-deoxy-D-gluconate was likewise converted by base into the acetylene **135**,

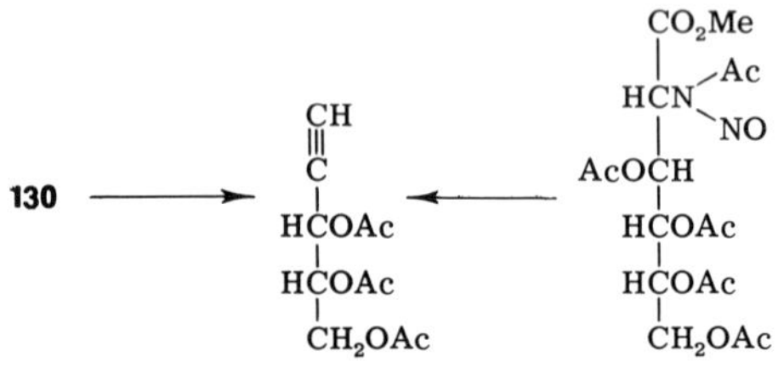

$$
\begin{array}{ccc}
 & & \text{CO}_2\text{Me} \\
 & & | \quad \diagup\text{Ac} \\
 & \text{CH} & \text{HCN}\diagdown \\
 & ||| & | \quad \diagdown\text{NO} \\
 & \text{C} & \text{AcOCH} \\
\textbf{130} \longrightarrow & | & \longleftarrow \quad | \\
 & \text{HCOAc} & \text{HCOAc} \\
 & | & | \\
 & \text{HCOAc} & \text{HCOAc} \\
 & | & | \\
 & \text{CH}_2\text{OAc} & \text{CH}_2\text{OAc} \\
 & & \\
 & \textbf{135} & \textbf{136}
\end{array}
$$

although in lower (30%) net yield.

The relative stabilities of various 2-(N-nitroso)acetamido-2-deoxy-hexoses have been compared. Factors contributing to stability appear to be crystallinity, the presence of electron-withdrawing substituents on the oxygen atoms, and a *trans*-disposition of O-1 to the 2-(N-nitroso)acetamido group.[217b]

VI. APPLICATIONS OF DEAMINATION

1. In Synthesis

The deamination of aliphatic amines is of limited use in synthesis. Although the reactions are notorious for the multitude of products normally produced, one pathway is occasionally much favored, and one product can be isolated in good yield. An example of a high-yielding reaction is provided by semipinacolic deamination; by its use, good yields of ring-contracted products were obtained from pyranose derivatives in which an equatorial amino group was located on C-2 or C-3, and the deamination of methyl 3-amino-3-deoxy-α-D-xylopyranoside was used as the first step in the synthesis of nucleosides containing a branched-chain sugar moiety (see p. 29).

2,5-Anhydroaldonate derivatives are formed in good yield when 2-amino-2-deoxyaldonic acids are deaminated. Thus, deamination of 2-amino-2-deoxy-L-(and D-)gluconic acid was used to form the tetrahydrofuran ring in the synthesis of (+)-muscarine,[223] 2,5-anhydro-1,3,6-trideoxy-1-(trimethylammonium)-L-*ribo*-hexitol chloride (and its enantiomer).[224,225]

1-Amino-1-deoxyketoses are accessible from aldoses by such reactions as Amadori rearrangements, and deamination then affords a route to the ketose. Thus, D-galactose was converted into D-tagatose.[161]

Aminoethyl groups in cyclic compounds in which anhydro-ring formation is not possible can be converted in good yield into hydroxyethyl groups by the action of nitrous acid (see p. 57).

The alkaline decomposition of 2-(*N*-nitroso)acetamido-2-deoxyhexoses provides a preparatively useful route to acetylenic sugar derivatives.[217a,217b]

Occasionally, a per-*O*-acetylated amine can be used in the synthesis of a partially acetylated compound. For example, 2-amino-2-deoxy-*neo*-inositol pentaacetate, which was prepared from hygromycin A, was converted by deamination into 1,2,3,4,6-penta-*O*-acetyl-*myo*-inositol.[226]

2. In Determination of Structure

The micro determination of 2-amino-2-deoxy-D-glucose and -D-galactose by deamination followed by reaction with indole was referred to on p. 25. Deamination has been the basis of several other methods for determination or detection of microgram quantities of 2-amino-2-deoxyaldoses and their derivatives. The 2-amino-2-deoxy derivatives of D-glucose, D-galactose, D-gluconic acid, and D-galactonic acid were differentiated by deamination followed by electrophoresis in borate buffer.[227] Hexosamines that give 2,5-anhydrohexoses on deamination were determined by conversion into a blue chromophore by reaction with 3-methyl-2-benzothiazolinone hydrazone hydrochloride and ferric chloride.[228] The 2-amino-2-deoxy derivatives of D-glucose, D-galactose, D-mannose, 3-*O*-methyl-D-

(223) E. Hardegger and F. Lohse, *Helv. Chim. Acta*, **40**, 2383–2389 (1957).
(224) H. C. Cox, E. Hardegger, F. Kögl, P. Liechti, F. Lohse, and C. A. Salemink, *Helv. Chim. Acta.*, **41**, 229–234 (1958).
(225) It was incorrectly stated in Ref. 6 that the methyl ester of 2-amino-2-deoxy-L-gluconic acid was used, and an incorrect structure for normuscarine was given.
(226) S. J. Angyal and M. E. Tate, *J. Chem. Soc.*, 4122–4128 (1961).
(227) A. R. Williamson and S. Zamenhof, *Anal. Biochem.*, **5**, 47–50 (1963).
(228) S. Tsuji, T. Kinoshita, and M. Hoshino, *Chem. Pharm. Bull.* (Tokyo), **17**, 217–218, 1505–1510 (1969).

glucose, and 3-O-(D-1-carboxyethyl)-D-glucose (muramic acid) were determined by deamination, reduction with sodium borohydride, and g.l.c. analysis of the trimethylsilyl ethers, with pentaerythritol as the internal standard.[229] In a similar method, the reduction products were acetylated prior to g.l.c. analysis.[230]

In a discussion of the use of deamination to obtain structural information, it is convenient to consider two different aspects: (a) applications to monosaccharide and cyclitol amines, and (b) applications to compounds containing amino sugar and inosamine residues. The latter compounds would include oligosaccharides, polysaccharides, glycolipids, glycoproteins, and antibiotics.

With the advent of nuclear magnetic resonance spectroscopy, the application of deamination to the study of monomeric amines is now of less importance. Except for that in 2-amino-2-deoxyaldopyranoses, the amino group can be converted into a hydroxyl group by deaminating the per-O-acetylated amine although this reaction may occur with, or without, inversion of configuration. Deamination can be used as a conformational probe. Thus, 2-amino-2-deoxyaldopyranoses give aldopyranoses with inversion of configuration when the amino group is axial, and 2-5-anhydroaldoses are formed when the amino group is equatorial. Other axial amino groups can be recognized by the presence of ketones, which usually have characteristic n.m.r. spectra, or epoxides among the deamination products. Other equatorial amines give ring-contracted aldehydes as major products, except for 4-amino-4-deoxypyranoses, in the deamination of which the ring-oxygen atom participates.

Applications of deamination of compounds containing an amino sugar residue or an inosamine residue are potentially more useful for several reasons. Hydrolysis of compounds containing an amino sugar is often difficult. For example, the glycosidic linkage in 2-amino-2-deoxyglycosides is resistant to acid hydrolysis,[231] and the conditions used to effect complete hydrolysis can degrade other constituents of the compound (for examples, see later discussion). 4-Amino-4-deoxyaldoses are themselves degraded to pyrrolidine and pyrrole derivatives under acidic conditions, and are thus difficult to isolate from acid hydrolyzates.[232]

(229) S. Hase and Y. Matsushima, *J. Biochem.* (Tokyo), **66**, 57–62 (1969).
(230) W. Niedermeier, *Anal. Biochem.*, 473–474 (1971).
(231a) R. C. G. Moggridge and A. Neuberger, *J. Chem. Soc.*, 745–750 (1938).
(231b) A. B. Foster, D. Horton, and M. Stacey, *J. Chem. Soc.*, 81–86 (1957).
(232) H. Paulsen, *Angew. Chem. Intern. Ed. Engl.*, **5**, 504–506 (1966).

Sometimes, deamination may cleave glycosidic linkages linking amino sugars, and this procedure has two important advantages in structural studies. Firstly, glycosidic linkages can be cleaved under extremely mild conditions (pH 4, at or below room temperature), and secondly, the specific cleavage of certain glycosidic linkages can be accomplished. This is an advantage in the study of oligosaccharide (or other) structures that contain amino sugar residues.

The depolymerization of chitosan (N-deacetylated chitin) by the action of nitrous anhydride to give 2,5-anhydro-D-mannose was reported[233] by Ambrecht in 1919, and, in a mistaken attempt[70] to avoid this depolymerization, dinitrogen tetraoxide was later used at −10° (see also, Ref. 234). The estimation of the nitrogen released in deamination was used to determine the proportion of primary amino functions in chitosan.[235] The difference in the rates of deamination of methyl 2-amino-2-deoxy-α- and -β-D-glucopyranosides was used to assign the configuration of the glycosidic linkages in chitosan, which was deaminated at the same rate as the β-D-glycosides[85] (see p. 21).

Nitrous acid deaminations have been invaluable in the investigation of the structure of heparin.[85,236−253] The potential value of de-

(233) W. Ambrecht, *Biochem. Z.*, **95**, 108–123 (1919).
(234) K. H. Meyer and H. Wehrli, *Helv. Chim. Acta*, **20**, 353–362 (1937).
(235) P. Karrer and S. M. White, *Helv. Chim. Acta*, **13**, 1108 (1930).
(236) A. B. Foster and A. J. Huggard, *Advan. Carbohyd. Chem.*, **10**, 335–368 (1955), especially p. 358.
(237) A. F. Charles and A. R. Todd, *Biochem. J.*, **34**, 114 (1940).
(238) D. Lagunoff and G. Warren, *Arch. Biochem. Biophys.*, **99**, 396–400 (1962).
(239) A. B. Foster, R. Harrison, T. D. Inch, M. Stacey, and J. M. Webber, *J. Chem. Soc.*, 2279–2287 (1963).
(240) M. L. Wolfrom and P. Y. Wang, *Chem. Commun.*, 241–242 (1967).
(241) M. L. Wolfrom, P. Y. Wang, and S. Honda, *Carbohyd. Res.*, **11**, 179–185 (1969).
(242) U. Lindahl and O. Axelsson, *J. Biol. Chem.*, **246**, 74–82 (1971).
(243) A. S. Perlin, N. M. K. Ng Ying Kin, S. S. Bhattacharjee, and L. F. Johnson, *Can. J. Chem.*, **50**, 2437–2441 (1972).
(244) Z. Yosizawa, *Biochem. Biophys. Res. Commun.*, **16**, 336–341 (1964).
(245) U. Lindahl, *Biochim. Biophys. Acta*, **130**, 368–382 (1966).
(246) J. A. Cifonelli and J. King, *Carbohyd. Res.*, **21**, 173–186 (1972).
(247) J. A. Cifonelli, *Carbohyd. Res.*, **8**, 233–242 (1968).
(248) U. Lindahl, *Biochim. Biophys. Acta*, **156**, 203–206 (1968).
(249) T. Helting and U. Lindahl, *J. Biol. Chem.*, **246**, 5442–5447 (1971).
(250) F. Yamauchi, M. Kosaki, and Z. Yosizawa, *Biochem. Biophys. Res. Commun.*, **33**, 721–726 (1968).
(251) A. S. Perlin and G. R. Sanderson, *Carbohyd. Res.*, **12**, 183–192 (1970).
(252) J. Knecht and A. Dorfman, *Biochem. Biophys. Res. Commun.*, **21**, 509–515 (1965).
(253) A. A. Horner, *Can. J. Biochem.*, **45**, 1015–1020 (1967).

amination as a method for selectively cleaving aminohexosidic linkages was pointed out by Foster and Huggard.[236] On the basis of the rate of deamination of N-desulfated heparin,[85] the α-D configuration had been deduced in 1953 for the 2-amino-2-deoxy-D-glucosidic linkages.

The observation that treatment of heparin with nitrous acid gives nitrogen was reported[237] in 1940, and it was subsequently realized that, in deamination, sulfoamino groups behave in the same way as amino groups. Thus, deamination was proposed as the basis for the determination of the 2-deoxy-2-(sulfoamino)hexose content of mucopolysaccharides.[238] The products of deamination were presumed to be oligosaccharides having 2,5-anhydro-D-mannose (reducing) end-groups, which were estimated by color formation with indole.

Deamination studies have aided attempts to locate the positions of the O-sulfate groups in heparin. From the results of deamination and periodate-oxidation studies, it was concluded[239] that half of the uronic acid residues are sulfated at C-2, and sequential periodate oxidation, reduction with sodium borohydride, acid treatment, and deamination gave 2,5-anhydro-D-mannose 6-sulfate.[240,241] Sulfated O-(hexosyluronic acid)-2,5-anhydro-D-mannose and sulfated uronic acids were subsequently isolated, and the isolation (in 65% yield) of the disulfated O-(idosyluronic acid)-2,5-anhydro-D-mannose (137) from heparin, together with the [13]C n.m.r. spectra of 137 and heparin, suggested[243] that at least two thirds of the heparin structure consists of the repeating unit 138.

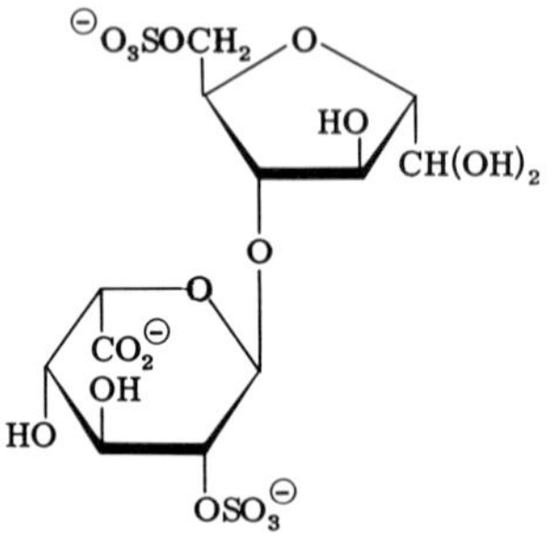

137 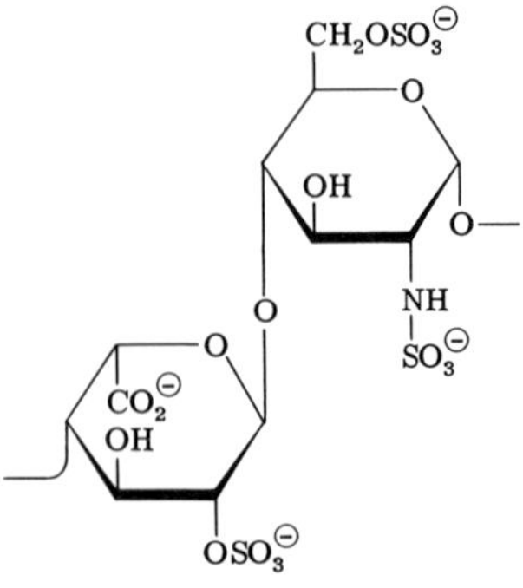   138

As acetamido groups do not react with nitrous acid under the usual conditions, it was possible to identify 2-acetamido-2-deoxy-D-glucose residues in heparin after deamination,[244] and their distribution was deduced from analysis of the oligosaccharides isolated by fractiona-tion of the deamination products.[245,246] Such 2-acetamido-2-deoxy-D-glucose residues were found to be divided approximately equally between the interior region and the protein-linkage region of the chain. In 1968, Cifonelli reported[247] that sulfoamino groups react with nitrous acid at $-20°$ in $3:2$ (v/v) 1,2-dimethoxyethane–water, and that free amines do not react under these conditions. Thus, the L-serine residues in heparin survive such low-temperature deamina-tion, and L-serine-containing oligosaccharides (139) originating from the heparin–protein linkage region have been isolated.[245,246,248,249]

X-GlcNAc-GlcA-Gal-Gal-Xyl-Ser

139

where X = various oligosaccharide residues.

One controversy in the investigation of heparin structure concerns the identity of the uronic acid constituent. This was first thought to be D-glucuronic acid, and then L-iduronic acid was reported as a con-stituent. A report[250] in 1968 that, during deamination, L-iduronic acid is formed from D-glucuronic acid residues by concomitant epimeriza-tion at C-5 has not been substantiated.[242] A degradation sequence that included deamination, and reduction of the uronic acid to hexose, showed that L-iduronic acid is the major uronic acid in heparin,[251] and this conclusion was supported by the isolation of the L-idosiduronic acid derivative (137), and by the ^{13}C n.m.r. spectrum already referred to.[243] Although the detailed structure of heparin is not yet known, the oligosaccharides that can be obtained therefrom by deamination can be useful in studies of metabolic disorders. Thus, the structure of heparitin sulfate, excreted by patients suffer-ing from the Hurler syndrome, was studied by deamination.[252] The use of alkyl nitrites in aqueous 1,2-dimethoxyethane has been recom-mended for deamination of heparin, on the grounds that it is more convenient and more stable than nitrous acid solution.[246]

The depolymerization of N-deacetylated chondroitin sulfate, der-matan sulfate, hyaluronic acid, heparin, and heparan sulfate by nitrous acid has been studied quantitatively by reduction of the products with tritiated sodium borohydride followed by radiochroma-tography[88] (see also, Ref. 254a).

The conditions needed for hydrolysis of the glycosidic linkage in neobiosamine   B   [3-O-(2,6-diamino-2,6-dideoxy-β-L-idopyranosyl)-

(254a) R. L. Taylor and H. E. Conrad, *Biochemistry*, **11**, 1383–1388 (1972).

D-ribose] (16) degraded the D-ribose,[254b] which was therefore identified after cleavage of the glycosidic linkage by the action of nitrous
acid[107] on the methyl glycoside 17 (see p. 26). Similarly, the D-
glucuronic acid component of citrosamine, a diglycosylinositol isolated from citrus leaves, was decomposed during acid hydrolysis.
Deamination of the 2-amino-2-deoxy-D-glucose moiety released an
O-(D-glucosyluronic acid)-myo-inositol, which could be hydrolyzed
in the usual way.[255] Treatment with nitrous acid was found to be a
convenient method, also, for releasing the aglycon in 4-O-(2-amino-2-
deoxy-D-glucopyranosyl)-D-ribitol, isolated from a teichoic acid,[256]
and in 2-O-(2-amino-2-deoxy-D-galactopyranosyl)glycerol,[257] also
from a teichoic acid. The 2,5-anhydro-D-aldoses were identified by
reduction with sodium borohydride.

The work of Baddiley and collaborators[256-259] on teichoic acids
provides excellent examples of the use of deamination in the elucidation of oligosaccharide structure. For example, when treated with
nitrous acid (see Scheme 9), the hexasaccharide 140, the repeating
unit of the Pneumococcus Type XA capsular polysaccharide, gave 2-

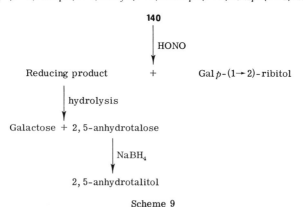

Gal$f$-(1→3)-Gal$p$-(1→4)-Gal$f$-(1→6)-GalNp-(1→3)-Gal$p$-(1→2)-ribitol

140

HONO

Reducing product   +   Gal$p$-(1→2)-ribitol

hydrolysis

Galactose + 2,5-anhydrotalose

NaBH$_4$

2,5-anhydrotalitol

Scheme 9

O-D-galactopyranosylribitol and a reducing product that was hydrolyzed to galactose and 2,5-anhydrotalose.[258] It should be noted that

(254b) T. H. Haskell, J. C. French, and Q. R. Bartz, J. Amer. Chem. Soc., 81, 3481–3483
        (1959).
(255) I. Stewart and T. A. Wheaton, Phytochemistry, 7, 1679–1681 (1968).
(256) J. Baddiley, J. G. Buchanan, U. L. RajBhandary, and A. R. Sanderson, Biochem.
        J., 82, 439–448 (1962).
(257) D. C. Ellwood, M. V. Kelemen, and J. Baddiley, Biochem. J., 86, 213–225 (1963).
(258) E. V. Rao, J. G. Buchanan, and J. Baddiley, Biochem. J., 100, 801–810 (1966).

such degradations complement acid hydrolysis, which, in this instance, gave 3-O-(2-amino-2-deoxy-D-galactopyranosyl)-D-galactose. Similar studies were conducted on the Pneumococcus Type XXIX specific substance,[259] and on a teichoic acid from Staphylococcus albus[257] (see also, Ref. 260). By deamination,[261] an unsubstituted 2-amino-2-deoxy-D-glucopyranosyl group was detected as a nonreducing end-group in two phosphoglycolipids from Bacillus megaterium, and deamination was used in the degradation of phytoglycolipids and a trisaccharide derived therefrom.[262,263] In the study of glycopeptide structure by the action of nitrous acid, the specific cleavage of glycosidic linkages has also been reported.[264]

Almost all of the foregoing examples involved the deamination of derivatives of 2-amino-2-deoxyaldopyranoses in which the amino group is equatorial. The scope of the cleavage of glycosidic linkages by deamination is rather limited, the requirements for such application being the occurrence of a rearrangement to afford a product in good yield. Apart from 2-amino-2-deoxyaldopyranoses having an equatorial amino group, the only other amines that undergo *one* rearrangement in good yield are 3-amino-3-deoxyaldopyranoses having an equatorial amino group. If such amines possessing an alkyl or glycosyl substituent at O-4 were treated with nitrous acid, the O-4 substituent would be cleaved. For example, when methyl 3-amino-4,6-O-benzylidene-3-deoxy-α-D-glucopyranoside was deaminated, non-hydrolytic loss of the O-benzylidene group occurred.[265]

A substituent at C-3 of a 2-amino-1,2-dideoxycyclitol in which the amino group is equatorial would also be removed in deamination. A simple model for this reaction is the conversion of a *cis-trans* mixture of 2-methoxycyclohexylamine into a mixture of products that included cyclopentanecarboxaldehyde.[266] The latter presumably arose from the *trans* amine and from that proportion of the *cis* amine which

(259) E. V. Rao, M. J. Watson, J. G. Buchanan, and J. Baddiley, Biochem. J., 111, 547–556 (1969).

(260) J. L. Mosser and A. Tomasz, J. Biol. Chem., 245, 287–298 (1970).

(261) J. C. MacDougall and P. J. R. Phizackerley, Biochem. J., 114, 361–367 (1969).

(262) H. E. Carter, S. Brooks, R. H. Gigg, D. R. Strobach, and T. Suami, J. Biol. Chem., 239, 743–746 (1964).

(263) H. E. Carter, A. Kisic, J. L. Koob, and J. A. Martin, Biochemistry, 8, 389–393 (1969).

(264) M. Isemura and K. Schmid, Biochem. J., 124, 591–604 (1971).

(265) R. D. Guthrie and G. P. B. Mutter, personal communication; G. P. B. Mutter, M. Sc. Thesis, University of Leicester, England, 1963.

(266) M. Mousseron and M. Canet, Compt. Rend., 233, 484–486 (1951).

existed as the equatorial conformer (compare Scheme 3, p. 23).

When hydride shifts occur in the deamination of axial amines, cleavage of ether or glycosidic functions may result. The small amount of glucose tentatively identified[151] in the product of deamination of 6-O-(2-amino-2-deoxy-β-D-mannopyranosyl)-D-glucose (72; see p. 47) can be attributed to the occurrence of a hydride shift of H-1 to C-2.

The degradation observed[267] when the pneumococcal C-substance was treated with nitrous acid can be attributed[93] to rearrangement of the 2-acetamido-4-amino-2,4,6-trideoxyhexose component. Were the bond to the 4-amino group equatorial, migration of the ring-oxygen atom to give a 4,5-anhydrohexose (see p. 31) would result in cleavage of the glycosidic linkage, and migration of C-2 would result in cleavage of a substituent at C-3. Alternatively, were the bond to the amino group at C-4 axial, hydride shift of H-3 to C-4 would also result in loss of a C-3 substituent.

Many carbohydrate amines occur in Nature as amides, usually N-acetyl derivatives, and N-deacetylation is necessary before deamination with nitrous acid can be effected. Hydrazinolysis has commonly been used to effect N-deacetylation.[268] Hydrazinolysis followed by deamination was reported for chondroitin sulfate,[268-270] and hyaluronate,[269,270] and it was found that more than 60% of the hexosamine content was unaffected by nitrous acid.[269] The chromatographic detection and estimation of glucose and galactose in the products from hyaluronate and chondroitin indicated that substitution was more important than ring contraction.[270] It was suggested that the resistance of the 2-amino-2-deoxyhexose residues to deaminative cleavage (to form 2,5-anhydrohexoses) was due to the presence of a substituent at C-3. Later, it was reported that substitution at C-3 of 2-acetamido-2-deoxy-D-glucopyranose derivatives drastically lowers the yield of N-deacetylated product in hydrazinolysis.[271]

The rather vigorous conditions required for N-deacetylation can cause degradation.[272] A recovery of 43% was reported[273] for the hydrazinolysis product of chondroitin 4-sulfate, the extent of N-

(267) D. E. Brundish and J. Baddiley, Biochem. J., 110, 573–582 (1968).
(268) Y. Matsushima and N. Fujii, Bull. Chem. Soc. Jap., 30, 48–50 (1957).
(269) Z. Yosizawa and T. Sato, J. Biochem. (Tokyo), 51, 155–161 (1962).
(270) Z. Yosizawa, Tohoku J. Exp. Med., 80, 26–31 (1963); Chem. Abstr., 60, 1970d (1964); Nature, 201, 926–927 (1964).
(271) M. Fujinaga and Y. Matsushima, Bull. Chem. Soc. Jap., 39, 185–190 (1966).
(272) Z. Yosizawa and T. Sato, J. Biochem. (Tokyo), 51, 233–241 (1962).
(273) M. L. Wolfrom and B. O. Juliano, J. Amer. Chem. Soc., 82, 2588–2592 (1960).

deacetylation being 60–70%. The recovery was greater when carboxyl-reduced chondroitin sulfate was used, but the extent of
N-deacetylation was much lower (29%). It was subsequently found
that more-efficient N-deacetylation could be accomplished by using
hydrazine sulfate as the catalyst,[274] and the hydrazinolysis–deamination sequence has been used to cleave, selectively, a glycosidic
linkage in disaccharide model-compounds. The 3-O-β-D-galac-
topyranosyl and 6-O-α-D-mannopyranosyl derivatives of benzyl 2-
acetamido-2-deoxy-β-D-glucopyranoside gave the 2,5-anhydro-O-
glycosyl-D-mannoses.[275]

The conversion of the 2-amino-2-deoxy-β-D-mannopyranosyl group
in a disaccharide (72; see p. 46) and of those in a polysaccharide
into β-D-glucopyranosyl groups by deamination facilitated structural
studies, even though the conversions were not quantitative.[151]

The reactions discussed in this Section show that, although applications of deamination in structure-elucidation studies are limited,
those applications that are possible have considerably aided such
studies.

(274) Z. Yosizawa, T. Sato, and K. Schmid, Biochim. Biophys. Acta, 121, 417–420
(1966).
(275) B. A. Dmitriev, Y. A. Knirel, and N. K. Kochetkov, Carbohyd. Res., 29, 451–457
(1973); 30, 45–50 (1973).

THE REACTION OF AMMONIA WITH ACYL ESTERS OF CARBOHYDRATES*

By Maria E. Gelpi and Raúl A. Cadenas

*Departamento de Química, Facultad de Agronomía,
Universidad de Buenos Aires, Argentina*

I. Introduction

Study of the action of ammonia upon carbohydrate acyl esters originated some eighty years ago and, since then, many aspects of this complex reaction have been investigated. The processes that take place in this reaction include migrations, degradations, transesterifications, and deacylations, and their simultaneous occurrence makes the interpretation of the whole scheme very difficult. The present article provides a general description of the facts and a discussion of the different variables that play a role in the yields of products formed and in the mechanisms involved.

II. Acyl Derivatives of Monosaccharides

The first study[1] on the reaction of a carbohydrate ester with ammonia was performed in 1893 by Wohl, who submitted 2,3,4,5,6-

---

* For an early account of the action of ammonia upon acylated nitriles, see V. Deulofeu, *Advan. Carbohyd. Chem.* **4**, 119–151 (1948).

(1) A. Wohl, *Ber.*, **26**, 730–744 (1893).

penta-*O*-acetyl-D-gluconononitrile (1) to the action of aqueous ammonia in the presence of a small proportion of silver oxide, and obtained a nitrogenated substance (2) that was called "arabinose diacetamide." According to modern nomenclature, compound 2 is named 1,1-bis-(acetamido)-1-deoxy-D-arabinitol.

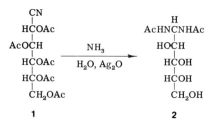

1   2

For a long time, this reaction was conducted exclusively with acetylated nitriles of aldonic acids, and the products obtained were known in general as "aldose-amides." Fischer[2] used this reaction to transform tetra-*O*-acetyl-L-rhamnononononitrile into 1,1-bis(acetamido)-1,5-dideoxy-L-arabinitol, whose subsequent hydrolysis and oxidation allowed him to determine the configuration of *dextro*-tartaric acid (L-threaric acid).

Wohl[3,4] extended this reaction to 2,3,4,5,6-penta-*O*-acetyl-D-galactononitrile (3) and to 2,3,4,5-tetra-*O*-acetyl-L-arabinonononitrile (5), which led to 1,1-bis(acetamido)-1-deoxy-D-lyxitol (4) and 1,1-bis-(acetamido)-1-deoxy-L-erythritol (6), respectively.

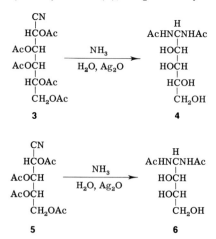

3   4

5   6

(2) E. Fischer, *Ber.*, **29**, 1377–1383 (1896).
(3) A. Wohl and E. List, *Ber.*, **30**, 3101–3108 (1897).
(4) A. Wohl, *Ber.*, **32**, 3666–3672 (1899).

Later,[5,6] the presence of silver oxide was shown to be unnecessary. The reaction was also conducted in methanolic media, and 1,1-bis-(acetamido)-1-deoxy-L-threitol[7] (8) and 1,1-bis(acetamido)-1-deoxy-D-erythritol[8] (10) were obtained in that medium from 2,3,4,5-tetra-O-acetyl-L-xylono- (7) and 2,3,4,5-tetra-O-acetyl-D-arabinono- (9) nitriles, respectively.

$$
\begin{array}{cccc}
\text{CN} & & \text{H} & \\
\text{AcOCH} & & \text{AcHNCNHAc} & \\
| & \xrightarrow[\text{MeOH}]{\text{NH}_3} & | & \\
\text{HCOAc} & & \text{HCOH} & \\
| & & | & \\
\text{AcOCH} & & \text{HOCH} & \\
| & & | & \\
\text{CH}_2\text{OAc} & & \text{CH}_2\text{OH} & \\
\end{array}
$$

7   8   9   10

These studies showed that this degradation always leads to an alditol derivative having one carbon atom less than the acylated nitrile started with. However, Brigl and coworkers[9] obtained a mixture of 1,1-bis(benzamido)-1-deoxy-D-mannitol (12) and N-benzoyl-D-mannosylamine (13) by ammonolysis of hexa-O-benzoyl-D-glycero-D-galacto-heptononitrile (11). The pyranose structure[10] and the β-anomeric configuration[11] of 13 were later ascertained.

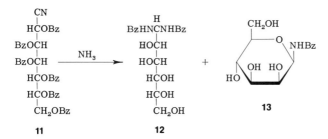

11   12   13

On the other hand, ammonolysis[12] of hexa-O-acetyl-D-glycero-D-gulo-heptononitrile (14) gave only an N-acetyl-D-glucofuranosylamine

(5) L. Maquenne, *Compt. Rend.*, **130**, 1402–1404 (1900); *Ann. Chim. Paris*, **24**, 399–412 (1901).
(6) R. C. Hockett, *J. Amer. Chem. Soc.*, **57**, 2265–2268 (1935).
(7) V. Deulofeu, *J. Chem. Soc.*, 2458–2460 (1929).
(8) V. Deulofeu, *J. Chem. Soc.*, 2602–2607 (1930).
(9) P. Brigl, H. Mühlschlegel, and R. Schinle, *Ber.*, **64**, 2921–2934 (1931).
(10) J. O. Deferrari and V. Deulofeu, *J. Org. Chem.*, **22**, 802–805 (1957); *Anales Asoc. Quím. Argentina*, **46**, 126–136 (1958).
(11) A. S. Cerezo and V. Deulofeu, *Carbohyd. Res.*, **2**, 35–41 (1966).
(12) R. C. Hockett and L. B. Chandler, *J. Amer. Chem. Soc.*, **66**, 957–960 (1944).

(**15**), whose furanose structure was determined[12] by oxidation with lead tetraacetate, and its α-anomeric configuration through periodate oxidation and borohydride reduction.[11]

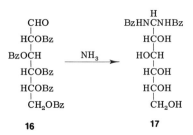

The formation of the free sugar in the ammonolysis reaction of the acylated nitriles was demonstrated chromatographically.[13] Thus, the 2,3,4,5,6,7-hexabenzoates of the D-*glycero*-D-*gulo*-, D-*glycero*-L-*manno*-, and D-*glycero*-D-*galacto*-heptononitriles gave, in methanolic ammonia, D-glucose, D-galactose, and D-mannose, respectively, together with the corresponding 1,1-bis(benzamido)-1-deoxy-D-hexitols and *N*-benzoyl-D-glycopyranosylamines.

Brigl and coworkers[9] discovered the first example showing that the formation of "aldose amides" is not restricted to the degradation of acylated nitriles of aldonic acids. By ammonolysis of 2,3,4,5,6-penta-*O*-benzoyl-*aldehydo*-D-glucose (**16**), they obtained 1,1-bis(benzamido)-1-deoxy-D-glucitol (**17**).

Likewise, 2,3,4-tri-*O*-acetyl-*aldehydo*-L-erythrose[14] (**18**) and 2,3,4,5-tetra-*O*-acetyl-*aldehydo*-L-arabinose[15] afforded 1,1-bis(acetamido)-1-deoxy-L-erythritol (**19**) and 1,1-bis(acetamido)-1-deoxy-L-arabinitol (the L form of **2**).

(13) J. O. Deferrari and B. Matsuhiro, *J. Org. Chem.*, **31**, 905–908 (1966).
(14) V. Deulofeu, *J. Chem. Soc.*, 2973–2975 (1932).
(15) H. S. Isbell and H. L. Frush, *J. Amer. Chem. Soc.*, **71**, 1579–1581 (1949).

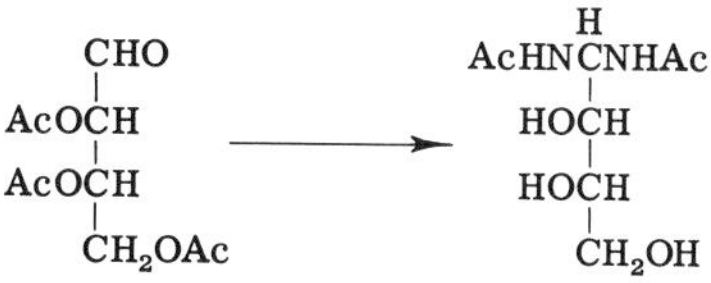

CHO   H
            AcHNCNHAc

AcOCH   →   HOCH

AcOCH   HOCH

CH$_2$OAc   CH$_2$OH

18   **19**

The similarity of the products obtained by ammonolysis, either of acylated nitriles, or of acylated *aldehydo* monosaccharides having one carbon atom less than the corresponding nitriles, was verified[12] by ammonolysis of 2,3,4,5,6-penta-*O*-acetyl-*aldehydo*-D-glucose (**20**), which afforded the same *N*-acetyl-α-D-glucofuranosylamine (**15**) as that obtained from the peracetylated heptononitrile **14**.

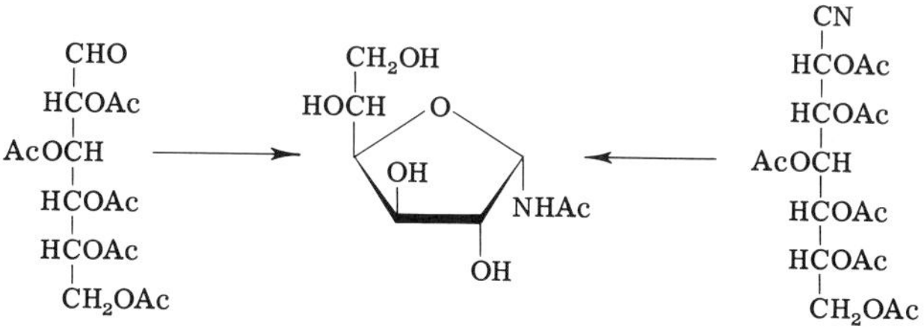

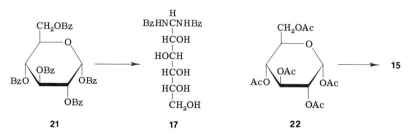

CHO   CH$_2$OH   CN

HCOAc   HOCH   HCOAc

AcOCH   →   OH   ←   HCOAc

HCOAc   NHAc   AcOCH

HCOAc   OH   HCOAc

CH$_2$OAc   HCOAc

                                             CH$_2$OAc

20   **15**   **14**

That starting from an aldehyde structure was not an indispensable requisite was evidenced by the ammonolysis of 3,5,6-tri-*O*-benzoyl-D-glucofuranose[9] and of 1,2,3,4,6-penta-*O*-benzoyl-α-D-glucopyranose[16] (**21**) to give 1,1-bis(benzamido)-1-deoxy-D-glucitol (**17**), showing that the hemiacetal ring does not hinder the formation of the acylamido derivative. Likewise, compound **15** was obtained[17] in 8% yield by treatment of 1,2,3,4,6-penta-*O*-acetyl-α-D-glucopyranose (**22**) with methanolic ammonia.

CH$_2$OBz   H
          BzHNCNHBz   CH$_2$OAc

        O   HCOH   O

            HOCH   **15**

OBz   →   HCOH   OAc

BzO   OBz   HCOH   AcO   OAc

OBz   CH$_2$OH   OAc

21   **17**   **22**

(16) V. Deulofeu and J. O. Deferrari, *J. Org. Chem.*, **17**, 1087–1092 (1952); *Nature*, **167**, 42 (1951).

(17) C. Niemann and J. T. Hays, *J. Amer. Chem. Soc.*, **67**, 1302–1304 (1945).

The reaction was also studied with acetylated and benzoylated pyranose and furanose derivatives of D-galactose,[18] D-mannose,[19] D-lyxose,[20] L-rhamnose,[10] D-ribose, D-xylose, and L-arabinose,[21] and with some partially benzoylated D-glucoses.[22]

The same compounds isolated by Brigl and coworkers,[9] who started from 2,3,4,5,6,7-hexa-O-benzoyl-D-*glycero*-D-*galacto*-heptononitrile, were obtained from the ammonolysis of 1,2,3,4,6-penta-O-benzoyl-α-D-mannose,[19] namely, a mixture of 1,1-bis(benzamido)-1-deoxy-D-mannitol (12) and N-benzoyl-D-mannopyranosylamine (13). Likewise, 1,2,3,4-tetra-O-benzoyl-L-rhamnopyranose,[10] having the same steric relationship at the asymmetric carbon atoms as the perbenzoate of D-mannose, also afforded an N-benzoyl-L-rhamnopyranosylamine directly.

Partially acylated monosaccharides gave variable yields of nitro-genated products. The results and yields obtained in these experi-ments by Deulofeu and coworkers[22] and other authors are given in Table III (see p. 126).

Application of chromatographic techniques has since allowed the isolation of new components from these complex reaction-mixtures. Thus, 1,1-bis(acetamido)-1-deoxy-D-glucitol was isolated as a second-ary product (together with compound 15) by ammonolysis of D-glucose pentaacetate, and the previously known[13] N-benzoyl-β-D-glucopyran-osylamine was found to be formed from 1,2,3,4,6-penta-O-benzoyl-β-D-glucopyranose.[23] Likewise, ammonolysis in aqueous medium[24] of 1,2,3,4,6-penta-O-acetyl-β-D-mannopyranose (23) gave, together with the previously isolated[19] 1,1-bis(acetamido)-1-deoxy-D-mannitol (24), N-acetyl-α- (25) and N-acetyl-β-D-mannofuranosylamine (26) and N-acetyl-α-D-mannopyranosylamine (27).

(18) J. O. Deferrari and V. Deulofeu, *J. Org. Chem.*, **17**, 1097–1101 (1952).

(19) J. O. Deferrari and V. Deulofeu, *J. Org. Chem.*, **17**, 1093–1096 (1952).

(20) V. Deulofeu and J. O. Deferrari, *Anais Acad. Brasil. Cienc.*, **26**, 69–74 (1954); J. O. Deferrari, V. Deulofeu, and E. Recondo, *Anales Asoc. Quim. Argentina*, **46**, 137–142 (1958).

(21) J. O. Deferrari, M. A. Ondetti, and V. Deulofeu, *J. Org. Chem.*, **84**, 183–186 (1959); *Anales Asoc. Quim. Argentina*, **47**, 293–304 (1959).

(22) E. G. Gros, M. A. Ondetti, J. F. Sproviero, V. Deulofeu, and J. O. Deferrari, *J. Org. Chem.*, **27**, 924–929 (1962).

(23) A. S. Cerezo, J. F. Sproviero, V. Deulofeu, and S. Delpy, *Carbohyd. Res.*, **7**, 395–404 (1968).

(24) A. B. Zanlungo, J. O. Deferrari, and R. A. Cadenas, *Carbohyd. Res.*, **14**, 245–254 (1970).

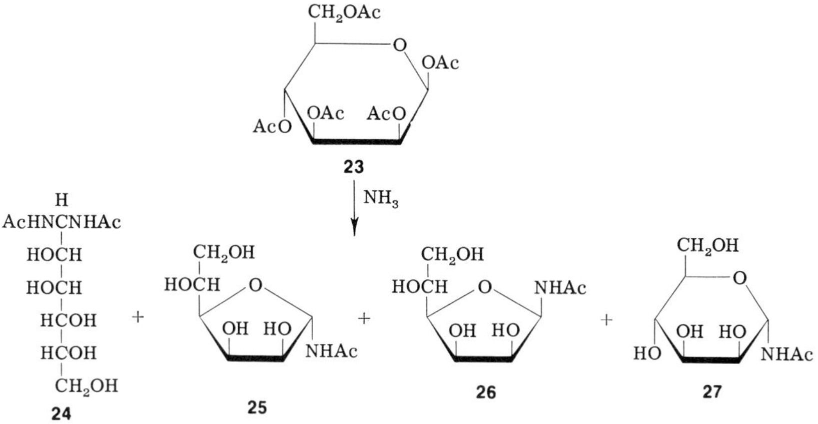

CH$_2$OAc

23

H
AcHNCNHAc
|
HOCH
|
HOCH
|
HCOH   +
|
HCOH
|
CH$_2$OH
24

CH$_2$OH
HOCH
OH HO
NHAc
25

+

CH$_2$OH
HOCH   NHAc
OH HO
26

+

CH$_2$OH
OH HO
HO   NHAc
27

The peracetates of L-rhamnopyranose[24] and D-galactopyranose[25] also afforded N-acylglycosylamines (see Table III, p. 126) and, on the basis of the yields of the different products obtained, the steric factors that could influence the relative formation of these products (see p. 107) were rationalized.[24-26]

The fully acylated heptoses showed similar behavior; from the hexaacetates and hexabenzoates of D-*glycero*-D-*gulo*-, D-*glycero*-L-*manno*-, and D-*glycero*-D-*galacto*-heptoses, the corresponding 1,1-bis(acylamido)-1-deoxy-D-heptitols were obtained[27] in higher yields than with the hexose esters; the hexabenzoate of D-*glycero*-L-*manno*-heptose also afforded an N-benzoyl-D-*glycero*-L-*manno*-heptosyl-amine.

A different result was obtained[28] in the ammonolysis of N-acyl derivatives of 2-amino-2-deoxy-D-glucopyranose. The N-acetyl and N-benzoyl derivatives were found to be stable in methanolic ammonia, leading to recoveries of 90–93% of the starting product. However, 2-acetamido-tetra-O-acetyl-2-deoxy-D-glucose and tetra-O-acetyl-2-benzamido-2-deoxy-D-glucose gave a recovery of only 50% of the respective N-acyl derivative, whereas ammonolysis of 2-benzamido-tetra-O-benzoyl-2-deoxy-D-glucose afforded only 12% of 2-benza-mido-2-deoxy-D-glucose. Extensive degradation, with formation of chromogenic compounds, occurred in these reactions. Likewise, 2-acetamido-tetra-O-acetyl-2-deoxy-D-gluconitrile showed no evi-

(25) A. B. Zanlungo, J. O. Deferrari, and R. A. Cadenas, *Carbohyd. Res.*, **10**, 403–416 (1969).

(26) A. S. Cerezo, A. B. Zanlungo, J. O. Deferrari, and R. A. Cadenas, *Chem. Ind.* (London), 1051–1052 (1970).

(27) J. O. Deferrari and R. M. de Lederkremer, *Carbohyd. Res.*, **4**, 365–370 (1967).

(28) J. F. Sproviero, *Anales Asoc. Quím. Argentina*, **53**, 261–267 (1965).

dence of degradation products,[28] a fact also observed in studies[29] on alkaline transformations of 2-anilino-2-deoxy-D-hexononitriles.

An experiment whose results support certain mechanistic views (see Section VI, p. 110) was the ammonolysis of 2,3,4,6-tetra-O-benzoyl-D-glucopyranosylamine,[30] which afforded a mixture of D-glucose and D-glucosylamine in 95% yield, and only 1.4% of 1,1-bis(benzamido)-1-deoxy-D-glucitol (**17**) with traces of N-benzoyl-D-glucopyranosylamine.

The monosaccharide esters commonly studied were acetates and benzoates. However, a few studies have been reported in which the acyl group was different.[23,31] By ammonolysis of 2,3,4,5,6-penta-O-propionyl-D-glucono- and -D-galactono-nitriles, Deulofeu and Giménez[31] obtained the corresponding 1-deoxy-1,1-bis(propionamido)alditols. Some monosaccharide nicotinates[32] were also studied, and from 1,2,3,4,6-penta-O-nicotinoyl-α-D-glucopyranose[33] (**28**) was isolated (in low yield) 6-(1-deoxy-D-*erythro*-tetritol-1-yl)-2-(D-*arabino*-tetritol-l-yl)pyrazine (**31**), together with 1-deoxy-1,1-bis(nicotinamido)-D-glucitol (**29**) and N-nicotinoyl-D-glucofuranosylamine (**30**).

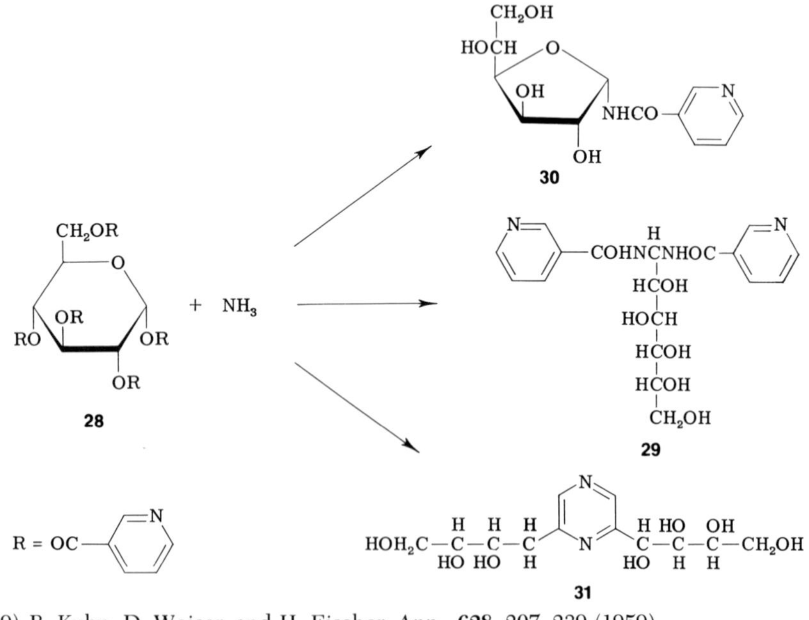

(29) R. Kuhn, D. Weiser, and H. Fischer, *Ann.*, **628**, 207–239 (1959).
(30) J. F. Sproviero, A. Salinas, and E. S. Bertiche, *Carbohyd. Res.*, **19**, 81–86 (1971).
(31) V. Deulofeu and F. Giménez, *J. Org. Chem.*, **15**, 460–465 (1950).
(32) E. A. Forlano, J. O. Deferrari, and R. A. Cadenas, *Carbohyd. Res.*, **21**, 484–486 (1972).
(33) E. A. Forlano, J. O. Deferrari, and R. A. Cadenas, *Carbohyd. Res.*, **23**, 111–119 (1972).

The formation of this pyrazine derivative results from a secondary reaction between the free monosaccharide and ammonia. Its formation involves loss of the asymmetry at C-1 and C-2 of the free sugar, and thus the same heterocycle resulted from the ammonolysis of 1,2,3,4,6-penta-*O*-nicotinoyl-β-D-mannopyranose.[34]

In Table III (see p. 126) are shown the products and yields obtained[34] from the pernicotinates of D-mannose, D-galactose, L-arabinose, and D-xylose upon ammonolysis in aqueous medium. The latter pentose nicotinate also gave[35] 6-(1-deoxy-D-glycerol-1-yl)-2-(D-*threo*-glycerol-1-yl)pyrazine (32). The structure and formation of these heterocyclic compounds will be discussed in Section VI,3 (see p. 124).

$$\text{HOH}_2\text{C}-\overset{\text{H}}{\underset{\text{HO}}{\text{C}}}-\overset{\text{HO}}{\underset{\text{H}}{\text{C}}}\diagdown \text{N} \diagdown \text{CH}_2-\overset{\text{OH}}{\underset{\text{H}}{\text{C}}}-\text{CH}_2\text{OH}$$

32

In summary, it may be stated that the reaction of acyl esters of aldoses and aldobioses (see Section III, p. 92) with ammonia consists of a set of competitive pathways, including intramolecular O → N migrations of acyl groups, deacylations, and transesterifications, with formation of "aldose amides" and variable proportions of the free sugar as the principal products. Significant proportions of basic or insoluble polymeric substances were not observed with aldose acetates or benzoates, although occurrence of extensive browning indicates the probable formation of soluble melanoidins.

On the other hand, the acyl esters of 2-ketoses follow a completely different pathway, as heterocyclic, nitrogenated compounds and melanoidins are the principal products of their reaction with ammonia. The free 2-ketoses show similar behavior, but they present a significant difference in the extent of formation of these compounds, the presence of the esterifying acyl groups in the molecule being decisive in increasing their yields.

Thus, the reaction of 1,3,4,5,6-penta-*O*-acetyl-*keto*-L-sorbose (33) with 24% aqueous ammonia[36] afforded 6-(1-deoxy-L-*threo*-tetritol-1-yl)-2-(L-xylitol-1-yl)pyrazine (34) in 11% yield and 2,5-bis(L-xylitol-1-yl)pyrazine (35) in 0.4% yield. In this reaction mixture, the follow-

(34) E. A. Forlano, J. O. Deferrari, and R. A. Cadenas, *Anales Asoc. Quim. Argentina*, **60**, 323–330 (1972).

(35) E. A. Forlano, J. O. Deferrari, and R. A. Cadenas, *Carbohyd. Res.*, **31**, 405–406 (1973).

(36) M. C. Teglia and R. A. Cadenas, *Carbohyd. Res.*, **26**, 377–384 (1973).

ing imidazole compounds were identified by paper chromatography: 4(5)-(L-xylitol-1-yl)imidazole (**36**), 4(5)-(glycerol-1-yl)imidazole (**37**), 4(5)-(2,3,4-trihydroxybutyl)imidazole (**38**), 4(5)-(2-hydroxyethyl)imidazole (**39**), 4(5)-(hydroxymethyl)imidazole (**40**), and 4(5)-methylimidazole (**41**) (see Scheme 1).

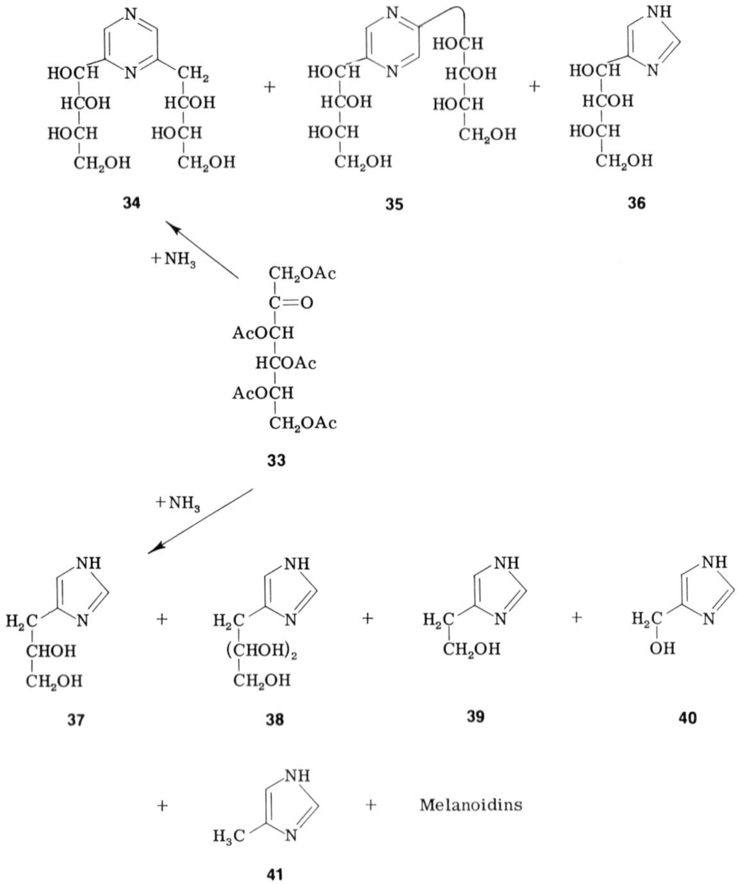

Scheme 1

Imidazole derivatives were not detected in the ammonolysis of aldose esters, and pyrazines only appeared in the case of the pernicotinates of D-glucose and D-xylose, as already described.

The formation of melanoidins in yields exceeding 20% (on the basis of the free sugar involved in the reaction) is a noteworthy feature of this reaction in the 2-ketose field. Direct reaction of L-sorbose with ammonia gave only 2–3% of melanoidins, showing the special influence of the acyl groups attached to the 2-ketose molecule upon the competing mechanisms that intervene in the reaction.

An analogous scheme was revealed in the ammonolysis of 1,3,4,5-tetra-O-benzoyl-β-D-fructopyranose and 1,3,4,5,6-penta-O-acetyl-*keto*-D-fructose,[37] with formation of the imidazole derivatives **37, 38, 40, 41,** and 4(5)-(D-*arabino*-tetritol-1-yl)imidazole (**42**), which were identified by paper chromatography.

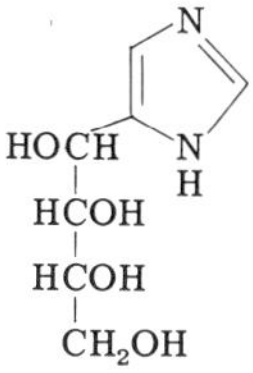

HOCH
|
HCOH
|
HCOH
|
CH₂OH

42

2,5-Bis(D-*arabino*-tetritol-1-yl)pyrazine (**43**) and 5-(1-deoxy-D-*erythro*-tetritol-1-yl)-2-(D-*arabino*-tetritol-1-yl)pyrazine (**44**) were also isolated, in 3.6 and 2.3% yields, respectively; they had previously been synthesized by Kuhn and coworkers[38] from 1-amino-1-deoxy-D-fructose and 2-amino-2-deoxy-D-glucose, respectively.

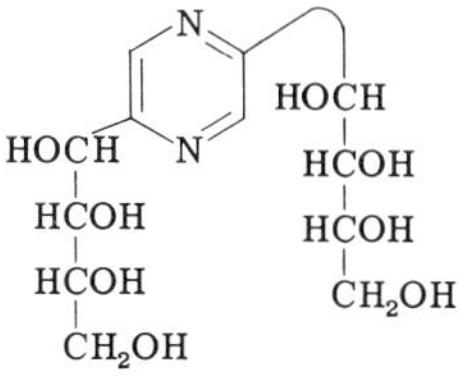

HOCH   HOCH
| |
HCOH   HCOH
| |
HCOH   HCOH
| |
CH₂OH   CH₂OH

43

HOCH   HCOH
| |
HCOH   HCOH
| |
HCOH   CH₂OH
|
CH₂OH

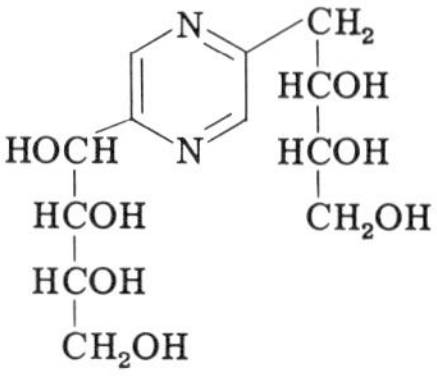

44

The formation of melanoidins amounted to a 20–25% yield (based on free sugar), whereas prolonged reaction (eight days) of ammonia with D-fructose gave only a 2.5% yield of these polymers.

In the reaction of ammonia with 1,3,4,5-tetra-O-benzoyl-β-D-fructopyranose, a significant proportion of benzoic acid was formed, indicating that the ammonolysis of the sugar ester occurred to a certain extent through acyloxy-group rupture. As neither the formation of imidazole nor of pyrazine derivatives (see Section VI,3, p. 124) re-

(37) M. C. Teglia and R. A. Cadenas, *Anales Asoc. Quím. Argentina,* **61,** 153–160 (1973).

(38) R. Kuhn, G. Krüger, H. J. Haas, and A. Seeliger, *Ann.,* **644,** 122–127 (1961).

quires[39] this type of fragmentation, it may be connected with the formation of melanoidins, although this connection through a specific mechanism is not apparent at present.

III. Acyl Derivatives of Disaccharides

In 1935, Zemplén applied the Wohl reaction to octa-O-acetylcellobiononitrile,[40] and obtained amorphous, nitrogenated substances that he supposed to be "acetamido derivatives." His aim being the demonstration of the structure of the reducing disaccharides through successive degradation of their acylated nitriles to the lower aldobiose, he discontinued these studies.

The first work with fully acetylated aldobioses was published in 1936 by Zechmeister and Tóth,[41] who dissolved octa-O-acetylcellobiose in liquid ammonia, and maintained this mixture for 48 h at 55°. From the syrupy reaction-products, they isolated an N-acetylcellobiosylamine and, by subsequent acetylation of the residue, the peracetate of 1,1-bis(acetamido)-1-deoxycellobiitol. For this compound, the authors proposed[41] alternative structures that were later[42] rejected.

From the reaction of octa-O-acetylcellobiose with 40% methanolic ammonia for 120 h at 50°, Micheel and coworkers[43] isolated dicellobiosylamine, cellobiosylamine, and the same N-acetylcellobiosylamine previously described by Zechmeister and Tóth.[41]

In 1963, a series of systematic studies was begun on the reaction of fully acylated disaccharides (acetates and benzoates) with methanolic or aqueous solutions of ammonia at room temperature. Octa-O-acetyl-α-cellobiose in 16% methanolic ammonia afforded[44] 1,1-bis-(acetamido)-1-deoxycellobiitol (45) and N-acetyl-α-cellobiosylamine (46), which is the anomer of the cellobiosylamine previously described by Zechmeister and Tóth[41] and Micheel and coworkers.[43]

Likewise, octa-O-acetyl-β-lactose in 16% methanolic ammonia[45] afforded 1,1-bis(acetamido)-1-deoxylactitol (47) and N-acetyl-α-lactosylamine (48). The anomer of the latter compound had previously been obtained by Kuhn and Krüger,[46] either by reaction of ketene with

(39) M. R. Grimmett, Rev. Pure Appl. Chem., 15, 101–108 (1965); M. J. Kort, Advan. Carbohyd. Chem. Biochem., 25, 311–349 (1970).
(40) G. Zemplén, Ber., 59, 1254–1266 (1926).
(41) L. Zechmeister and G. Tóth, Ann., 525, 14–24 (1936).
(42) J. O. Deferrari and R. A. Cadenas, J. Org. Chem., 30, 2007–2009 (1965).
(43) F. Micheel, R. Frier, E. Plate, and A. Hiller, Chem. Ber., 85, 1092–1096 (1952).
(44) J. O. Deferrari and R. A. Cadenas, J. Org. Chem., 28, 1070–1072 (1963).
(45) R. A. Cadenas and J. O. Deferrari, J. Org. Chem., 28, 1072–1075 (1963).
(46) R. Kuhn and G. Krüger, Chem. Ber., 87, 1544–1547 (1954).

lactosylamine or by acetylation of lactosylamine and subsequent O-deacetylation of this acetate.

Upon acetylation and subsequent O-deacetylation in alkaline media, both the mono- and bis-(amido) derivatives of mono- and disaccharides always showed great stability, with high recoveries of the original amido sugar.

The disaccharide acetates afforded, in methanolic media, very low yields (3–5%) of these nitrogenated derivatives, compared with the yields (20–40%) given by the acetates of monosaccharides. This behavior was attributed to the steric hindrance caused by the bulky monosaccharide portion attached to C-4 of the residue on which the acyl migration takes place; but the absence of an important contributing group (such as the 4-O-acyl group) to the migration undoubtedly plays a major part in lowering the yields. The principal product in these reactions was the free disaccharide.

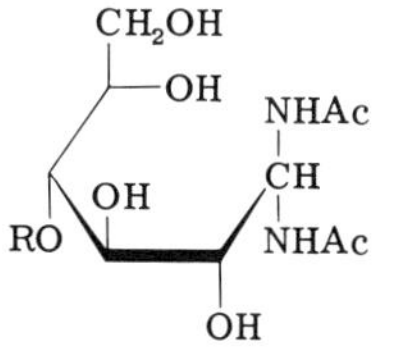

45   R = β-D-glucopyranosyl
47   R = β-D-galactopyranosyl
49   R = α-D-glucopyranosyl

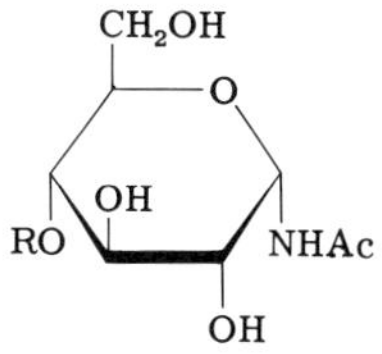

46   R = β-D-glucopyranosyl
48   R = β-D-galactopyranosyl

1,1-Bis(acetamido)-1-deoxymaltitol (49) was isolated with great difficulty, and only in 0.8% yield, from the ammonolysis of octa-O-acetyl-β-maltose with methanolic ammonia.[47] To overcome these difficulties, a parallel study was conducted with maltose octaacetate in which water replaced methanol as the solvent and, in this way, the yield of 49 was increased to 27%. These results will be discussed in Section IV (see p. 99). Subsequent studies showed water to be the better medium for the reaction. The only limitation was the solubility of the starting ester, which prevented use of the benzoates (which are always ammonolyzed in alcoholic media).

The structure of compounds 45, 47, and 49 was ascertained[42] by methylation with methyl iodide in N,N-dimethylformamide in the presence of barium oxide.[48] In each experiment, the corresponding 1,1-bis(acetamido)-1-deoxy-octa-O-methylaldobiitol was obtained; on acid hydrolysis, this afforded two methylated monosaccharides, whose structures confirmed the acyclic formulas proposed.

(47) R. A. Cadenas and J. O. Deferrari, J. Org. Chem., 28, 2613–2616 (1963).
(48) R. Kuhn, H. H. Baer, and A. Seeliger, Ann., 611, 236–241 (1958).

Compound **45**, for example, gave 1,1-bis(acetamido)-1-deoxy-octa-*O*-methylcellobiitol (**50**), which, by mild hydrolysis, gave 2,3,4,6-tetra-*O*-methyl-D-glucopyranose (**51**) and 2,3,5,6-tetra-*O*-methyl-D-glucofuranose (**52**).

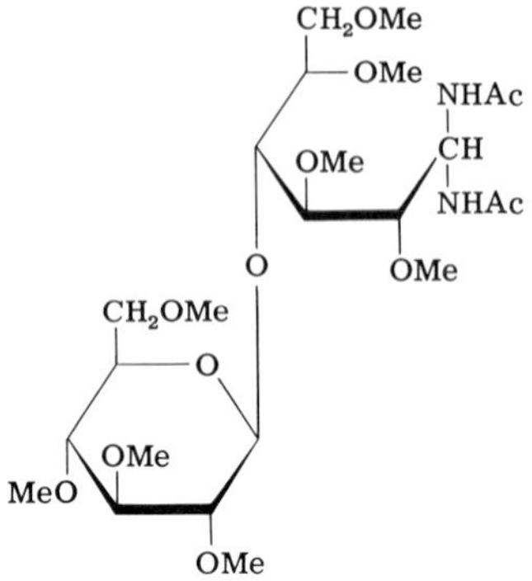

50

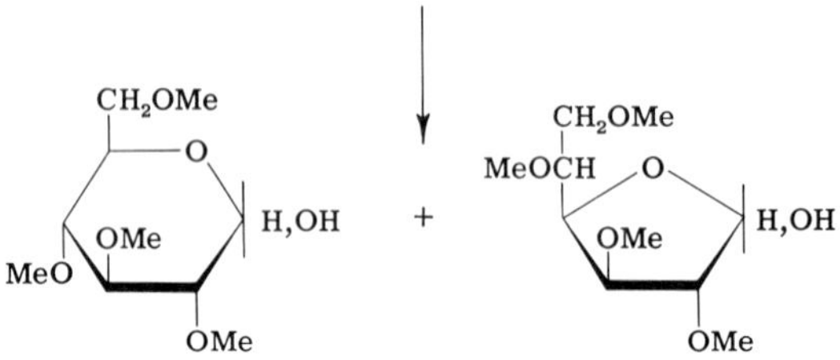

51   **52**

Octa-*O*-acetylmelibiose, possessing a $(1 \rightarrow 6)$-glycosidic linkage, could afford an *N*-acetylmelibiosylamine furanoid in its D-glucose portion. In both aqueous and methanolic media, the same products were obtained,[49] namely, 1,1-bis(acetamido)-1-deoxymelibiitol (**53**) and *N*-acetyl-6-*O*-α-D-galactopyranosyl-β-D-glucofuranosylamine (**54**), but the yields changed significantly according to the medium employed. The yield of the nitrogenated compounds **53** and **54** in

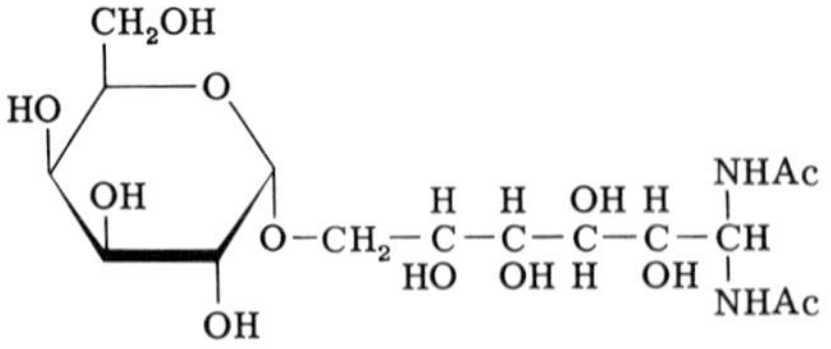

53 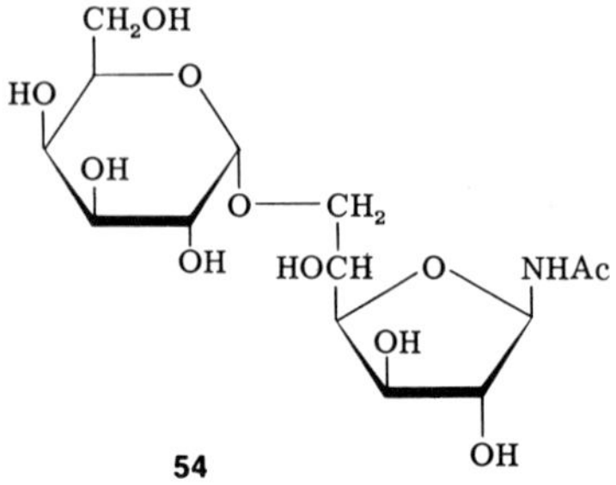   **54**

(49) A. B. Zanlungo, J. O. Deferrari, and R. A. Cadenas, *J. Chem. Soc., C*, 1908–1911 (1970).

aqueous medium was ~27.2%, whereas, in methanolic ammonia, it was only ~5.2%.

Parallel studies with benzoates of these disaccharides also showed the formation of bis- and mono-benzamido derivatives (see Table IV, p. 132), as well as the free disaccharide.

The benzoyl group on O-6 showed unusual resistance to hydrolysis, and 1.5% of 6-O-benzoylcellobiose and 31.6% of 6-O-benzoylmaltose were isolated from ammonolysis of the perbenzoates of cellobiose[50] and maltose,[51] respectively. For the latter, only after 15 days of reaction with 16% methanolic ammonia was the 6-O-benzoyl group completely eliminated.

The persistence of one benzoyl group was attributed to the mutual steric hindrance of both portions of the disaccharide molecule, which would hinder the nucleophilic attack by ammonia. As regards 6-O-benzoylmaltose, obtained in unexpectedly high yield, its stability to ammonia was explained by a particular stereoelectronic relationship of the benzoyl group to the two pyranose ring-oxygen atoms and to the oxygen atom that glycosidically bonds the two portions of the disaccharide, as depicted in formula **55**.

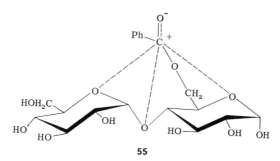

55

The importance of the 3-O-benzoyl group for the formation of nitrogenated migration-products (see also, Section VI, p. 110) is clearly demonstrated in the ammonolysis of 1,2,6,2',3',4',6'-hepta-O-benzoyl-β-maltose,[52] which lacks that group in the residue wherein acyl migrations can occur. Here, a 40% yield of 6-O-benzoylmaltose (and a 56% yield of maltose) was obtained, whereas nitrogenated carbohydrate derivatives were neither isolated nor detected chromatographically.

(50) J. O. Deferrari, I. M. E. Thiel, and R. A. Cadenas, *J. Org. Chem.*, **30**, 3053–3055 (1965).

(51) I. M. E. Thiel, J. O. Deferrari, and R. A. Cadenas, *J. Org. Chem.*, **31**, 3704–3707 (1966).

(52) I. M. E. Thiel, J. O. Deferrari, and R. A. Cadenas, *Ann.*, **723**, 192–197 (1969).

Similar results were obtained in the ammonolysis of 1,2,6,2',3',4',6'-hepta-O-benzoyl-β-lactose,[53] whereas lactose octabenzoate[54] afforded 7.4% of nitrogenated derivatives, together with 82% of lactose and no 6-O-benzoyllactose. In this respect, this benzoate showed behavior analogous to that of cellobiose octabenzoate, with practically the same yields of nitrogenated carbohydrates. On the other hand, these yields for maltose octabenzoate amounted to 20.8%. These differences are noteworthy, and were attributed[54] to the different stereochemistry of the glycosidic linkage between the two portions of the disaccharide molecule that puts maltose octabenzoate in one category and the two other benzoates in the other.

Molecular models showed the situation depicted in formulas 56 and 57, in which both benzoates have their benzoyl groups peripherally disposed, allowing ready *exo* attack by ammonia that would favor elimination of most of these groups as benzamide and, consequently, giving the migration reaction only a small chance of occurrence.

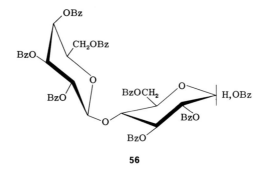

56

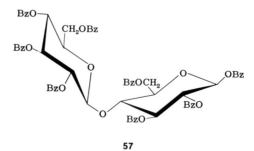

57

(53) I. M. Vazquez, I. M. E. Thiel, and J. O. Deferrari, *Carbohyd. Res.*, **26**, 351–356 (1973).

(54) J. O. Deferrari, I. M. E. Thiel, and R. A. Cadenas, *Carbohyd. Res*, **29**, 141–146 (1973).

Conversely, the conformational features of octa-O-benzoyl-β-maltose (**58**) show a more *endo* distribution of the benzoyl groups, precluding the ability of ammonia to eliminate these groups as benzamide. The three perbenzoates split off their 1-O-benzoyl group first, to allow the migration to C-1 to operate. But it would be in compound **58** where the crowding of the bulky groups would most favor the ini-

58

tial opening of the pyranose ring of the reducing moiety, followed by attack by ammonia on the free carbonyl group and migration, to C-1, of the benzoyl groups still attached to a major extent.

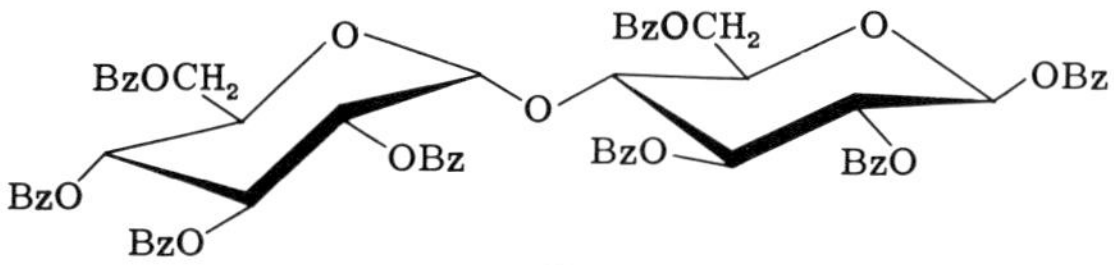

Simultaneously with the studies just described, acylated nitriles of aldobionic acids were also investigated, and several hexosyl-pentose amides were isolated. Ammonolysis of octa-O-acetylcellobiononitrile (**59**) afforded[55] 1,1-bis(acetamido)-1-deoxy-3-O-β-D-glucopyranosyl-D-arabinitol (**60**), N-acetyl-3-O-β-D-glucopyranosyl-D-arabinofuranosylamine (**61**), and a 3-O-β-D-glucopyranosyl-D-arabinose not previously described in the literature.

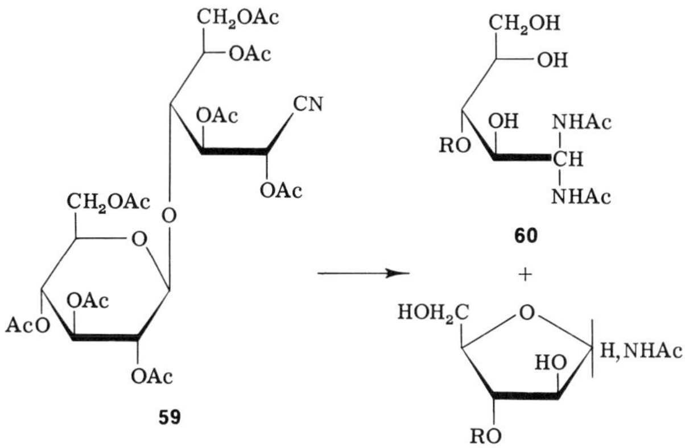

R = β-D-glucopyranosyl group

(55) J. O. Deferrari, M. E. Gelpi, and R. A. Cadenas, *J. Org. Chem.*, **30**, 2328–2330 (1965).

Similar products (but having R = β-D-galactopyranosyl) were obtained from octa-*O*-acetyllactobiononitrile[56] in 25% aqueous ammonia. The furanose structure of the monoamido derivative **61** (and that of its analog *N*-acetyl-3-*O*-β-D-galactopyranosyl-D-arabinofuranosylamine[56]) was ascertained by methylation, hydrolysis, and identification of 2,5-di-*O*-methyl-D-arabinose.

The reaction of octa-*O*-acetylmelibiononitrile (**62**) with aqueous ammonia[57] gave, together with the expected 1,1-bis(acetamido)-1-deoxy-5-*O*-α-D-galactopyranosyl-D-arabinitol (**63**) and *N*-acetyl-5-*O*-α-D-galactopyranosyl-α-D-arabinofuranosylamine (**64**), the two anomers of the free hexosyl-pentofuranose, namely, 5-*O*-α-D-galactopyranosyl-α-D-arabinofuranose (**65a**) and 5-*O*-α-D-galactopyranosyl-β-D-arabinofuranose (**65b**).

62

63

64

65a R_1 = OH, R_2 = H

65b R_1 = H, R_2 = OH

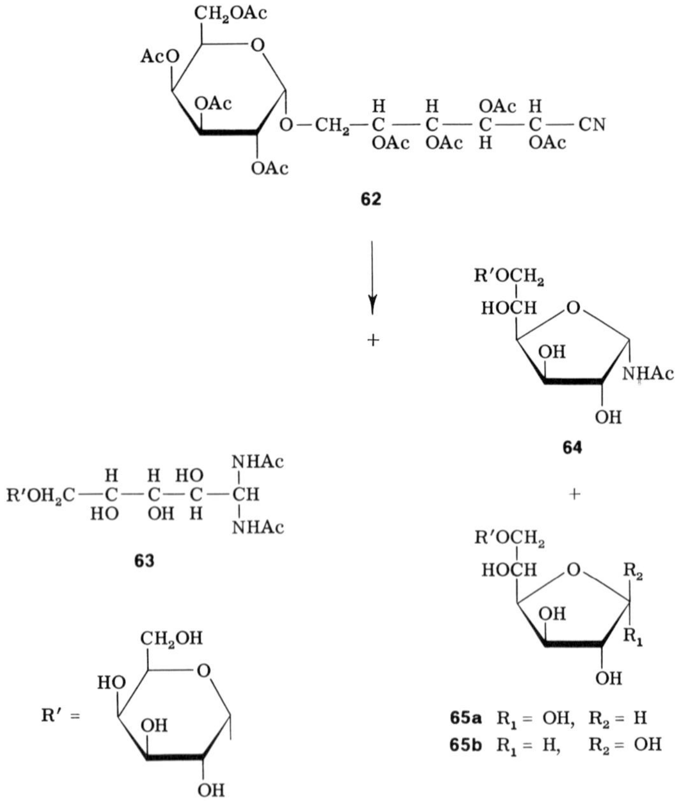

(56) M. E. Gelpi, J. O. Deferrari, and R. A. Cadenas, *J. Org. Chem.*, **30**, 4064–4066 (1965).

(57) J. O. Deferrari, B. N. Zuazo, and M. E. Gelpi, *Carbohyd. Res.*, **30**, 313–318 (1973).

Table IV (see p. 132) shows the products and yields obtained from the ammonolysis of acylated disaccharides.

IV. INFLUENCE OF THE SOLVENT

The solvent in which the ammonolysis of acyl derivatives of carbohydrates is conducted has a great influence on the yields of 1,1-bis-(acylamido)-1-deoxyalditols and N-acylaldosylamines.

Thus, tetra-O-acetyl-D-xylononitrile,[5,6] tetra-O-acetyl-L-xylononitrile,[7] and penta-O-acetyl-D-galactononitrile[58] gave, through a Wohl degradation in methanolic medium, yields of 30 to 40% of bis(amides), whereas, in aqueous medium, the yields rose to 70%. By using liquid ammonia, Zechmeister and Tóth[41] obtained nitrogenated sugars as practically the sole derivatives of carbohydrates, the corresponding free sugars not being isolated.

Treatment of the pentabenzoates of D-glucono-, D-galactono-, and D-mannono-nitriles with 10% ethanolic ammonia[59] led to the corresponding 1,1-bis(benzamido)-1-deoxypentitols having a benzoyl group on O-5. For example, 2,3,4,5,6-penta-O-benzoyl-D-glucononitrile (66) gave 1,1-bis(benzamido)-5-O-benzoyl-1-deoxy-D-arabinitol (67) in 25% yield. Compound 67 readily loses the 5-O-benzoyl group by the action of methanolic ammonia, to give 68.

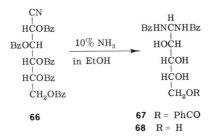

66

67 R = PhCO
68 R = H

If the nitrile 66 is treated with 5.5% methanolic ammonia, 1,1-bis-(benzamido)-1-deoxy-D-arabinitol (68) is obtained, whereas the same concentration of ammonia in isopropyl alcohol gives[60] compound 67. When this reaction was conducted (employing 5.5–8% of ammonia in methanol or isopropyl alcohol) with perbenzoylated D-galactononitrile, and with the perbenzoates of D-glucopyranose, D-galacto-

(58) R. C. Hockett, V. Deulofeu, A. L. Sedoff, and J. R. Mendive, *J. Amer. Chem. Soc.*, **60**, 278–280 (1938).
(59) E. R. de Labriola and V. Deulofeu, *J. Org. Chem.*, **12**, 726–730 (1947).
(60) E. G. Gros, A. Lezerovich, E. F. Recondo, V. Deulofeu, and J. O. Deferrari, *Anales Asoc. Quím. Argentina*, **50**, 185–197 (1962).

pyranose, and *aldehydo*-D-glucose[61] and its bisulfite derivative, results parallel with those just described were obtained. Also, an 8% solution of ammonia in isobutyl alcohol afforded these O-benzoyl nitrogenated derivatives, whereas, in ethanol, a mixture of the debenzoylated with the mono-O-benzoyl bis(amido) derivatives was obtained.[60] The same result, that is, formation of 1,1-bis(benzamido)-ω-O-benzoyl-1-deoxyalditol, is obtained if the benzoates of these nitrogenated alditols are ammonolyzed with 8% ammonia in isopropyl alcohol.[62]

These studies showed the particular resistance of the benzoyl group attached to a primary hydroxyl group. The only exception[62] is the ammonolysis of 2,3,4,5-tetra-O-benzoyl-L-arabinononitrile, which, in isopropyl alcohol containing 5.5% of ammonia, affords 1,1-bis(benzamido)-1-deoxy-L-erythritol (the L enantiomer of **10**; see p. 83) and not its 4-O-benzoylated derivative.

The improvement in the yields of bis(amido) sugars, previously observed[6] when water was used instead of methanol, was re-investigated in the ammonolysis of maltose octaacetate.[47] As already mentioned, the yield of 1,1-bis(acetamido)-1-deoxymaltitol (**49**, see p. 93) changed from 0.8% in methanol to 27% in water. Analogous results were obtained (see p. 94) for octa-O-acetylmelibiose.[49] The interpretation of these differences was based on previous studies[63] on the ammonolysis of simple esters.

In methanolic ammonia, the following equilibrium is established.

$$NH_3 + MeOH \rightleftharpoons NH_4^+ + MeO^- \tag{1}$$

The presence of alkoxide ion would enhance the rate of ammonolysis, and the formation of bis(amido) derivatives by an ortho-ester mechanism (see Section VI, p. 110) would be partially suppressed in the competitive set of reactions. Thus, ammonolysis of penta-O-benzoyl-D-glucose in the presence of 5 mmolar proportions of sodium methoxide showed a decrease of 11% in the yield of the bis(benzamido)-glucitol derivative as compared with the same reaction conducted without added methoxide ion.[47]

The influence of alkoxide ion in causing the pathway of simple ammonolysis to predominate can be interpreted in the following way.

(61) A. Lezerovich, E. G. Gros, J. F. Sproviero, and V. Deulofeu, *Carbohyd. Res.*, **4**, 1–6 (1967).

(62) M. A. Ondetti, Thesis, Facultad de Ciencias Exactas y Naturales, Buenos Aires, Argentina (1957).

(63) H. L. Betts and L. J. Hammett, *J. Amer. Chem. Soc.*, **59**, 1568–1572 (1937); J. F. Bunnett and G. T. Davis, *ibid.*, **82**, 665–674 (1960); R. Baltzly, I. M. Berger, and A. A. Rothstein, *ibid.*, **72**, 4149–4152 (1950).

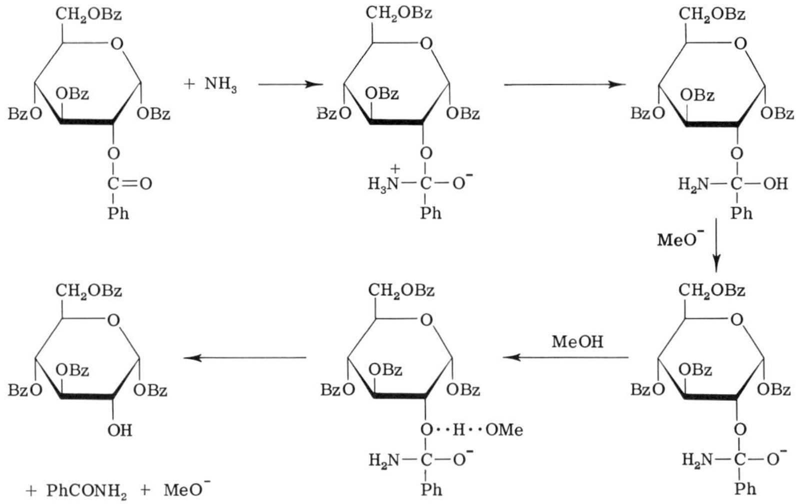

On the other hand, formation of methyl benzoate was also found to occur in methanol, indicating that, together with a general base-catalysis to produce benzamide and with the intramolecular, ortho-ester mechanism (see p. 110) to give the nitrogenated sugars, a transesterification reaction takes place in which the alkoxide ions play an important role. This can be exemplified by the following sequence.

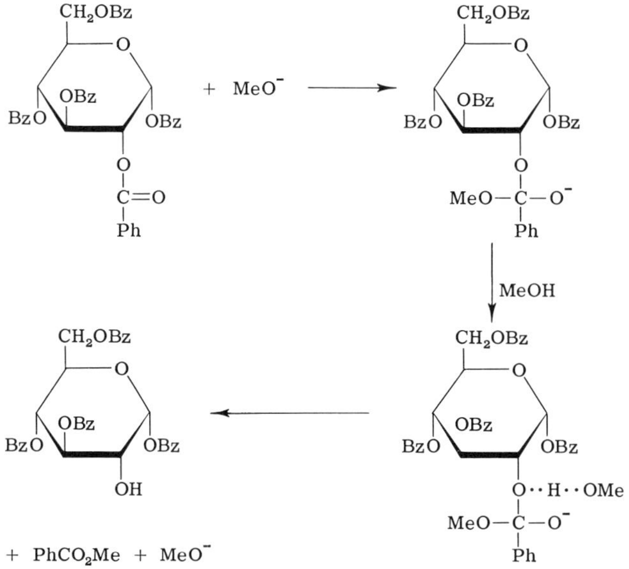

From these results, it follows that the predominance of these path-ways over that of the ortho-ester route to afford nitrogenated sugars is related to the acidity of the solvent employed. Deulofeu and coworkers[60] observed that the degree of ammonolysis of benzoyl groups attached to different sugars depends upon the alcohol employed as the solvent, in the following decreasing order: methanol > ethanol > isopropyl alcohol > isobutyl alcohol.

The Hines[64] showed that, as acids, ethanol, isopropyl alcohol, and tert-butyl alcohol are weaker than water, whereas methanol is stronger. The influence of the solvent could thus be interpreted in terms of equation 1. In methanol, the equilibrium would be more displaced to the right, and the rate of simple ammonolysis and transesterification would be enhanced, with concomitant decrease in the yields of amido sugars. In water (for the ammonolysis of sugar acetates) and in alcohols other than methanol, the equilibrium would be displaced to the left and this would allow operation of the ortho-ester mechanism a better chance. The isolation, from the reaction in isopropyl alcohol, of mono-O-benzoylated bis(benzamido)alditols, could also be explained on this basis.

V. STRUCTURE AND CONFIGURATION

Determination of the structure of an amide of a carbohydrate is not difficult. Examination of acyl derivatives indicates whether a cyclic or an acyclic structure is present. Moreover, the results of oxidations with periodate and lead tetraacetate allow unequivocal determination of these structures; alternatively, methylation studies have been used.

The first oxidative studies on N-acylglycosylamines were conducted on a preparative scale by Niemann and Hays.[65] The N-acetyl-D-glucopyranosylamine (69) obtained by acetylation of D-glucopyranosylamine with ketene (or by acetylation of this amine, followed by ammonolysis) was oxidized by way of the dialdehyde 70 to the acid 71, isolated as the strontium salt. Hydrolysis of 71 led to D(+)-glyceric acid (72), thus showing the pyranose structure of 69.

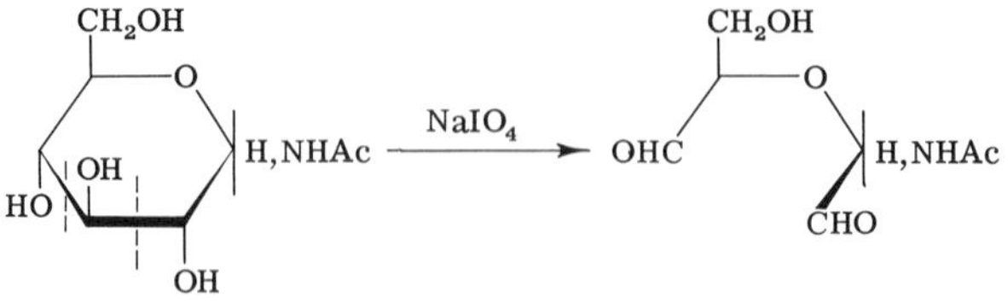

69   70

(64) J. Hine and M. Hine, J. Amer. Chem. Soc., 74, 5266–5271 (1952).
(65) C. Niemann and J. T. Hays, J. Amer. Chem. Soc., 62, 2960–2961 (1940).

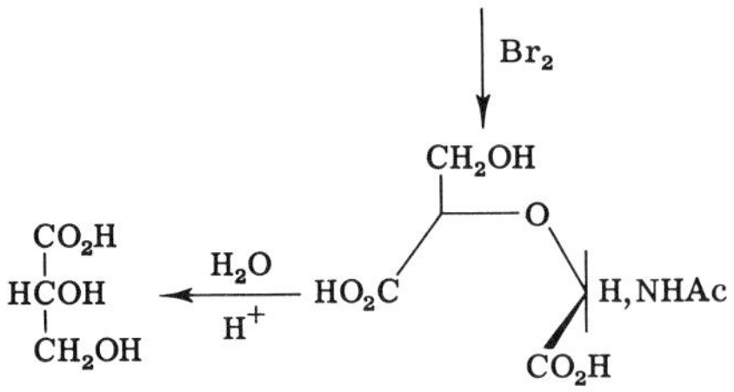

The oxidation of N-acylglycopyranosylamines on a semimicro scale is straightforward and overoxidation does not occur; this was evident in the oxidation of an N-benzoyl-L-rhamnopyranosylamine[10] and of the anomeric pair of N-acetyl-D-galactopyranosylamines.[66] These anomers were obtained by deacetylation of the pentaacetates of α- and β-D-galactopyranosylamine. On periodate oxidation, the "dialdehydes" obtained (73 and 74) showed specific rotations of $[\alpha]_D$ +60° and $[\alpha]_D$ −96°, respectively.

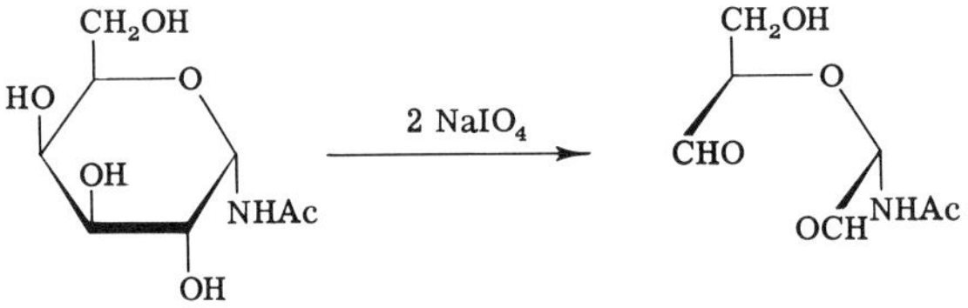

As the only difference between the two oxidation products is the configuration at the carbon atom that had been the anomeric center, it was possible to determine the configuration of C-1 of other N-acetyl-hexopyranosylamines by comparison of the specific rotations of the "dialdehydes" formed by their oxidation with periodate. In this way, the anomeric configuration of N-acetyl-β-D-glucopyranosylamine

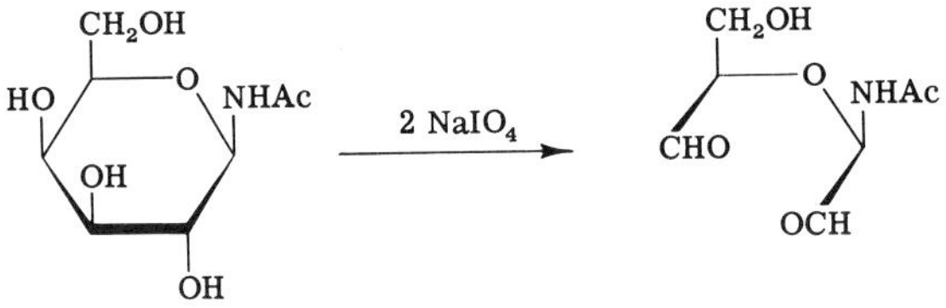

(66) H. L. Frush and H. S. Isbell, *J. Res. Nat. Bur. Stand.*, **47**, 239–247 (1951).

and N-acetyl-β-D-mannopyranosylamine was ascertained. Isbell and Frush[67] also established that the N-acetylpentopyranosylamines that possess the same absolute configuration at C-1 as N-acetyl-α-L-arabinopyranosylamine (75) give, by periodate oxidation, a dialdehyde (76) having $[\alpha]_D -49°$, a value that is exclusively contributed by

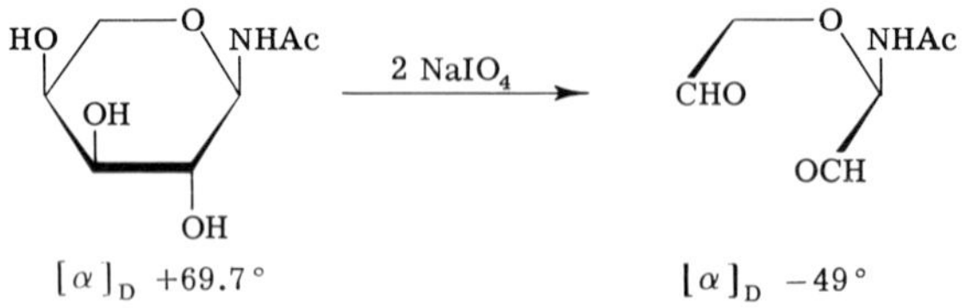

$[\alpha]_D +69.7°$ $[\alpha]_D -49°$

75   76

the original anomeric center. On this basis, the β configuration was assigned[68] to the N-acetyl-D-xylopyranosylamine having $[\alpha]_D -0.7°$ that, on periodate oxidation, gave a dialdehyde having $[\alpha]_D -39.3°$.

When applied to N-acylglycofuranosylamines, the periodate oxidation showed an abnormal uptake of oxidant ("overoxidation"). For example, when oxidized with lead tetraacetate[12] and with periodate,[10,65] N-acetyl-α-D-glucofuranosylamine (15) afforded formaldehyde (indicating a furanose structure), and it consumed more than 5 moles of oxidant per mole. This result can be attributed to subsequent oxidation of the formic acid produced,[69] or to the formation,[10] by hydrolysis, of the intermediate 2-hydroxypropanedial (tartronaldehyde) (77) that would then be oxidized. This tendency to undergo overoxidation has been found common for the furanoid N-acylglycosylamines.[24,25]

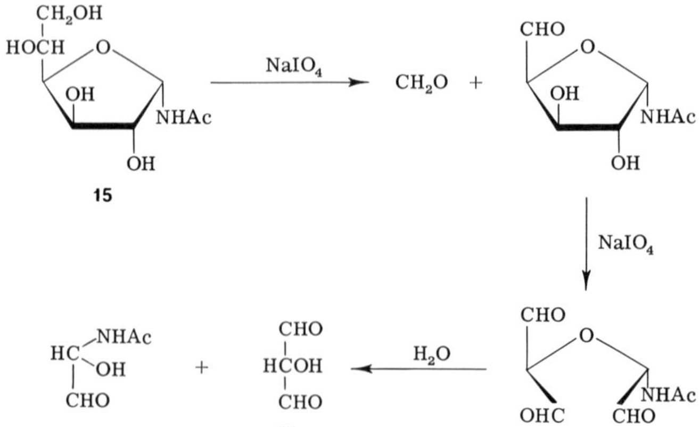

15   77

(67) H. S. Isbell and H. L. Frush, *J. Res. Nat. Bur. Stand.*, **46**, 132–144 (1951).
(68) H. S. Isbell and H. L. Frush, *J. Org. Chem.*, **23**, 1309–1319 (1958).
(69) R. C. Hockett, M. T. Dienes, H. G. Fletcher, Jr., and H. E. Ramsden, *J. Amer. Chem. Soc.*, **66**, 467–468 (1944).

The method of Frush and Isbell[66] just described was modified[11] to allow its use for ascertaining the anomeric configuration of N-acetyl-glycofuranosylamines. After oxidation with periodate, the dialdehydes produced are reduced with borohydride, and the only asymmetric center that remains is the former anomeric carbon atom. In this way, N-acetylhexopyranosylamines, N-acetylhexofuranosylamines, and N-acetylpentofuranosylamines of the same anomeric configuration give the same final product. This method was applied[11] to N-acetyl-α-D-galactopyranosylamine[66] (**78**, of known anomeric configuration) and to the N-acetyl-D-glucofuranosylamine (**15**) and N-acetyl-D-xylo-furanosylamine[11] (**79**) whose anomeric configurations were unknown. These sugars led to a final product (**80**) common to all three, showing the α-D configuration of **15** and **79**. Usually, the N-acetylated α anomers give rotatory values for the final product of oxidation and reduction (**80**) of ~ +7 to +9°, and the β anomers, of −8 to −11°.

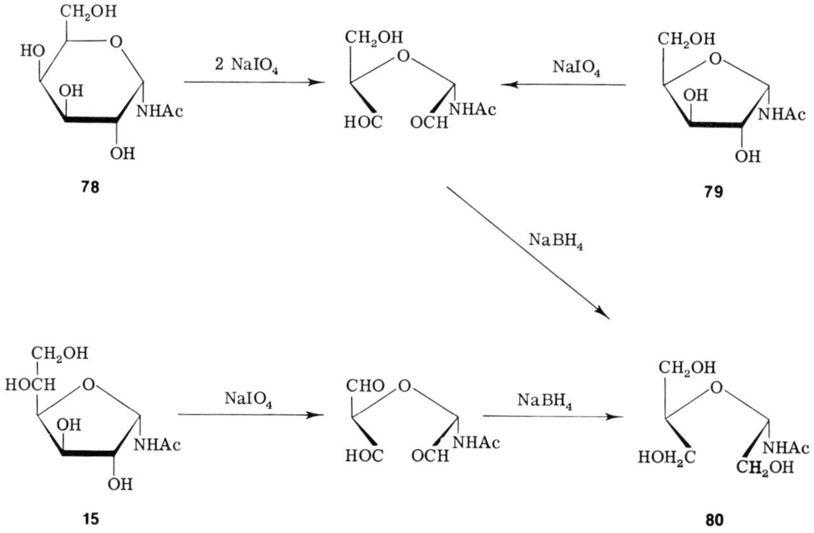

For the N-acylaldobiosylamines, this method requires a second oxidation and further reduction, as shown for N-acetyl-6-O-α-D-galactopyranosyl-β-D-glucosylamine (N-acetyl-β-melibiosylamine)[49] (**54**) in the following sequence.

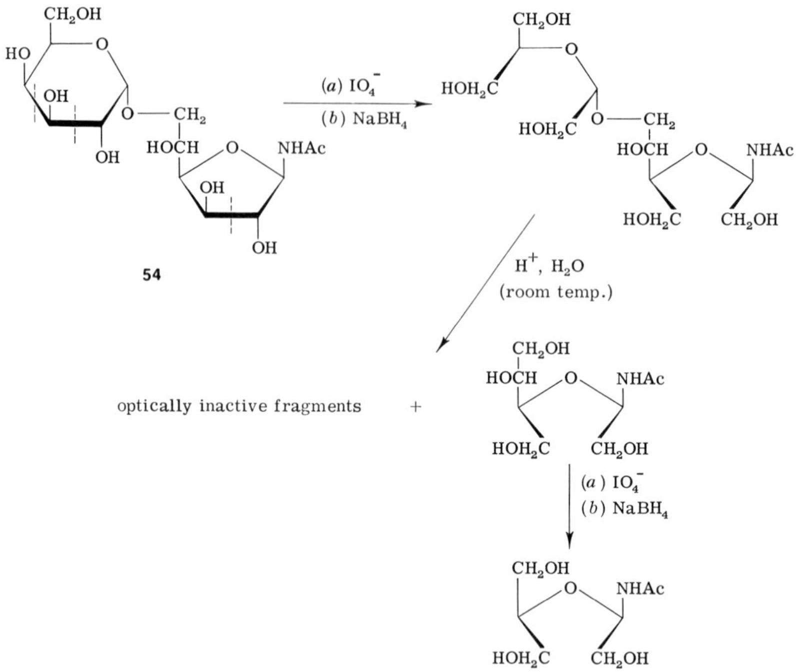

This procedure has been applied[70] to the determination of the ring structure and anomeric configuration of several N-benzoylglycosyl-amines obtained by N-benzoylation of free glycosylamines, and the results showed that the original size of the ring and the anomeric configuration were retained during the benzoylation. Obviously, N-acyl derivatives of rhamnosylamine will give ambiguous results, owing to retention of the asymmetry of C-5; this asymmetry is not removed by the oxidation and reduction sequence, but application of the principle of optical superposition allowed, in this particular case, determination of the anomeric configuration.[70]

These N-acylglycosylamines follow the Hudson Rules of Isorotation, which were also previously used[11] to confirm the α configuration of N-acetyl-D-glucofuranosylamine (15) and of N-acetyl-D-xylofuranosylamine (79). A modification was proposed[70] to the B values usually employed for these N-acyl derivatives, on the basis of a better agreement between the theoretical values and the experimental data; the differences probably originate from the presence of a highly polarizable group (the acyl group) in the aglycon.

(70) S. Delpy and A. S. Cerezo, Anales Asoc. Quím. Argentina, 61, 59–65 (1973).

The infrared spectra of many 1-acylamido sugars and 1,1-bis-(acylamido)-1-deoxyalditols and their acyl esters were studied[71] in a search for correlations analogous to those described in the literature[72] for other classes of sugars. However, no general criteria were discovered, although, depending on the particular nitrogenated sugar under consideration, the presence of bands of some diagnostic value was noted. Also, it was observed that the bis(acylamido) derivatives showed two, non-equivalent, Amide II bands, which suggested that the two amido groups were differently involved in a hydrogen-bonded structure.

A comparative study[24] of the yields of cyclic and acyclic products from the ammonolysis of several peracetylated glucopyranoses showed that, when the acyl substituent in the starting peracetylated sugar was aliphatic, the yields of cyclic nitrogenated products were higher than if the acyl group was aromatic (for example, benzoyl). The low yields in the latter case were attributed[25] to stereo-electronic factors: the slow ammonolysis of these bulky groups having relatively low reactivity would make cyclization difficult, and would allow the favored formation of acyclic 1,1-bis(benzamido) derivatives. The formation of furanose rings in the ammonolysis of benzoates was not observed, N-benzoylglucopyranosylamines in the $^4C_1(D)$ conformation being favored as they are less strained.

On the other hand, the favored formation (and the yields) of N-acetylglucofuranosylamines were explained on the basis of steric considerations.[24,26] The formation of N-acylglycosylamines is a kinetically controlled, irreversible reaction, and the prevalence of the furanose over the pyranose structure has been shown in many irreversible, cyclization reactions of carbohydrates.[73] This preponderance, as well as the favoring of a determined anomeric configuration, would be the result of a balance of relative free-energies of the transition states in the cyclization step, which would have virtually the same geometry as the final, cyclic structure.[74] The five-membered

(71) R. S. Tipson, A. S. Cerezo, V. Deulofeu, and A. Cohen, *J. Res. Nat. Bur. Stand.*, **71**, 53–79 (1967).

(72) H. Spedding, *Advan. Carbohyd. Chem.*, **19**, 23–49 (1964); L. M. J. Verstraeten, *Anal. Chem.*, **36**, 1040–1044 (1964); *Carbohyd. Res.*, **1**, 481–484 (1966); P. Nanasi and P. Cerletti, *Gazz. Chim. Ital.*, **92**, 576–582 (1962); P. Nanasi, E. Nemes-Nanasi, and P. Cerletti, *ibid.*, **95**, 966–974, 975–982 (1965).

(73) J. W. Green and E. Pacsu, *J. Amer. Chem. Soc.*, **59**, 1205–1210 (1937); V. G. Bashford and L. F. Wiggins, *J. Chem. Soc.*, 299–303 (1948); B. G. Hudson and R. Barker, *J. Org. Chem.*, **32**, 3650–3658 (1967).

(74) N. L. Allinger and V. Zalkow, *J. Org. Chem.*, **25**, 701–704 (1960).

ring has lower entropy requirements for its formation from the acyclic form; the latter shows the highest degree of freedom and has a great tendency to cyclize.[75]

It was observed that the configuration of the starting sugars has a strong influence on the cyclization. For example, from acetates having the D-*gluco* or D-*galacto* configuration, only furanose structures are formed as cyclic products, whereas, from acetates having the D-*manno* configuration, a mixture of furanose and pyranose N-acetylglycosyl-amines is obtained. These results can be explained on a steric basis; in the final products, in the D-*galacto* and D-*gluco* configurations (**81** and **15**, respectively), there are two substituents above and below the plane of the ring, whereas, in the D-*manno* configuration (**25, 26**) there are at least three. The greater steric interactions in **25** and **26**

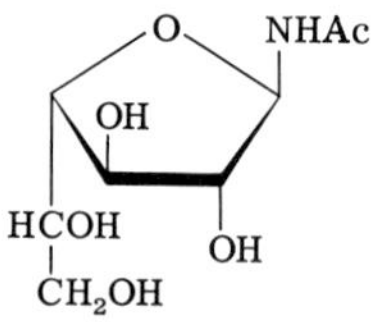

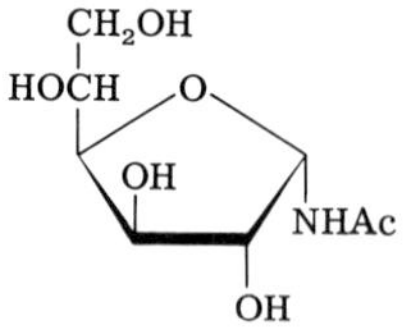

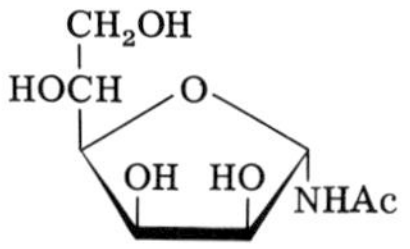

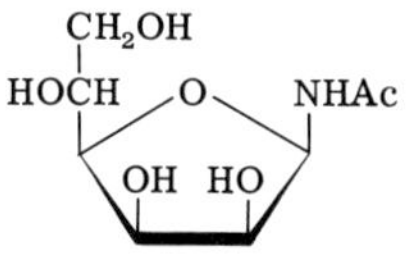

would enhance the activation energy for the cyclization stage, resulting in a lower rate of furanoid ring-closure, and thus making cyclization to the pyranoid form more competitive.

On the other hand, the β configuration in **81**, and the α configuration in **15** are attributable to inversion of configuration at C-4 in passing from the D-*galacto* to the D-*gluco* series, which makes a *trans* relationship between the bulky substituents on C-1 and C-4 the favored orientation with both sugars. The same reasoning could be applied to explain the preponderance of the α anomer **25** in the mixture of cyclic

(75) J. A. Mills, *Advan. Carbohyd. Chem.*, **10**, 1–53 (1955); E. S. Gould, "Mechanism and Structure in Organic Chemistry," Holt, Rinehart and Winston, New York, N. Y., 1960, p. 200.

products obtained by ammonolysis of penta-O-acetyl-β-D-manno-pyranose.

Study[24,25] of N-acetyl-α-D-glucofuranosylamine (15), N-acetyl-β-D-galactofuranosylamine (81), and N-acetyl-α-L-rhamnofuranosylamine by n.m.r. spectroscopy at 100 and 220 MHz indicated that, in solution, these sugars adopt non-planar ring shapes, with preponderance of the $T_2{}^3$ conformation[76] for the former compound (15) and the E_2 and E_3 conformations, respectively, for the other two. These assignments also present the fewest steric interactions, with staggering of bulky substituents, and agree with the general conformational behavior shown by furanoid sugar derivatives.[77] However, the limitations imposed by the facile conformational interconversions of five-membered ring systems, and the questionable applicability of the Karplus equation to the calculation of exact dihedral angles, imply that n.m.r. data should be considered more in terms of conformational equilibria than for revealing precise conformation.[78]

As regards the 1,1-bis(acylamido)-1-deoxyalditols, the first structural study was that of Fletcher and coworkers,[69] who conducted oxidations with lead tetraacetate. These authors postulated an empirical relationship between the position of the oxidation curve in a graph and the number of hydroxyl groups present. They compared the curves afforded by a series of 1,1-bis(acetamido)-1-deoxyalditols with those of alditols having the same number of free hydroxyl groups, and observed a close correspondence. At present, periodate oxidation is widely applied to these mono- or di-saccharide derivatives. Under controlled conditions, the methylation technique can also be applied.[42]

Consideration of rotatory powers of 1,1-bis(acylamido)-1-deoxy-alditols showed[79] certain correlations between optical rotation (D line of sodium) and configuration that can be summarized as follows. (a) When the acyl group is aliphatic (for example, acetyl or propionyl), the alditol will be dextrorotatory in water if the configuration of C-2 is S. (The rule does not apply to peracetylated derivatives.) (b) When the two substituents on C-1 are benzamido and the configuration at C-2 is S, a solution of the alditol in pyridine will be levorotatory. Peracetylation does not alter applicability of this rule. (c) The abso-

(76) L. D. Hall, *Chem. Ind.* (London), 950–951 (1963).

(77) R. U. Lemieux and R. Nagarajan, *Can. J. Chem.*, **42**, 1270–1278 (1964).

(78) P. L. Durette and D. Horton, *Advan. Carbohyd. Chem. Biochem.*, **26**, 49–125 (1971); P. L. Durette, D. Horton, and J. D. Wander, *Ann. N. Y. Acad. Sci.*, **222**, 884–914 (1973).

(79) A. S. Cerezo, *Carbohyd. Res.*, **15**, 315–318 (1970).

lute configuration of the asymmetric centers other than C-2 in the alditol is without effect on the sign of the rotatory power.

The conformations in solution of various acylated 1,1-bis(acylamido)-1-deoxypentitols have been examined by p.m.r. spectroscopy.[79a] In line with general behavior observed[78] with other acyclic-sugar derivatives, it was found that the *arabino* and *lyxo* derivatives adopt extended, zigzag conformations, whereas the *ribo* and *xylo* derivatives favor sickle conformations that result by rotation about one carbon–carbon bond of the backbone chain to alleviate a destabilizing 1,3-interaction of acyloxy substituents that would be present in the extended, zigzag conformation.

VI. MECHANISMS OF FORMATION OF THE NITROGENATED DERIVATIVES

1. 1,1-Bis(acylamido)-1-deoxyalditols

Wohl[1,3,4] supposed that the reaction of acetylated nitriles with ammonia occurs in two stages: (1) elimination of acetyl and cyano groups with formation of an aldehydo sugar, and (2) formation of the bis-(acetamido) derivative by condensation of the aldehyde group with two molecules of acetamide from the medium. This hypothesis was supported by known condensation reactions of simple aldehydes with acetamide and benzamide.[80] However, this type of condensation was not confirmed for carbohydrates. Hockett and Chandler[12] unsuccessfully attempted the condensation of penta-*O*-acetyl-*aldehydo*-D-glucose with acetamide. Previously, by treatment of D-*glycero*-D-*galacto*-heptononitrile with ammonia and silver oxide in the presence of benzamide, Brigl and coworkers[9] were unable to obtain the corresponding 1,1-bis(benzamido)-1-deoxymannitol.

Isbell and Frush[15] proposed an intramolecular mechanism for the acetylated aldehydo forms of aldoses. On the basis of the ease of condensation of aldehydes with ammonia, the reaction would start by addition of ammonia to the aldehyde group of the acylated monosaccharide (82). The nitrogen atom thus introduced would form a cyclic ortho ester (84) with a neighboring acetyl group, through the intermediate attack of the amino group, as shown in 83. Subsequent rearrangement would lead to the N-acetyl derivative 85. The reaction

(79a) B. Coxon, R. S. Tipson, M. Alexander, and J. O. Deferrari, *Carbohyd. Res.*, **35**, 15–31 (1974).

(80) E. Roth, *Ann.*, **154**, 72–80 (1870); A. Schuster, *ibid.*, **154**, 80–83 (1870); V. von Richter, *Ber.*, **5**, 477–478 (1872); M. Nencki, *ibid.*, **7**, 158–163 (1874).

would stop at this point should a hemiacetal ring be capable of formation, but most of the molecules would react with another molecule
of ammonia, to give a nitrogenated adduct having the acetoxyl group
on C-3 (**86**) which, through the cyclic ortho ester (**87**), would rearrange to the 1,1-bis(acetamido)-1-deoxyalditol (**88**). In the ammonoly-

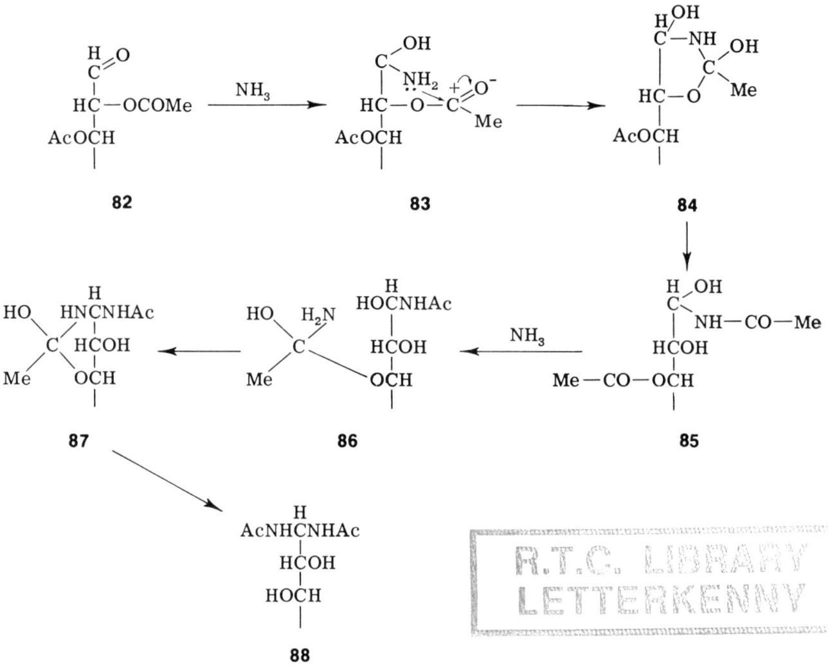

sis of acetylated nitriles, the free aldehyde group would result from
the departure of the cyano group as cyanide ion, and elimination
of the acetyl group on O-2 as acetamide. This mechanism was supported by the formation of 2-acetamido-2-deoxy-D-glucose by ammonolysis[81]   of   1,3,4,6-tetra-O-acetyl-2-amino-2-deoxy-D-glucopy-
ranose through the migration of an acetyl group to the amino group
already present in the molecule.

The formation[10–12] of *N*-acetyl-α-D-glucofuranosylamine (**15**) from
1,2,3,4,6-penta-O-acetyl-D-glucopyranose, and, by extension, of other
N-acyl-D-glycosylamines, would start with the rapid elimination of the
acetyl group on O-1 to give **89**. In this way, C-1 is ready to combine
with ammonia to give the intermediate **90** and, subsequently, the

---

(81) T. White, *J. Chem. Soc.*, 1498–1500 (1938).

cyclic ortho ester **91**. By ammonolysis of the remaining acetyl groups, a subsequent (or simultaneous) rearrangement to **92** and **93**, and cyclization to **15**, would take place.

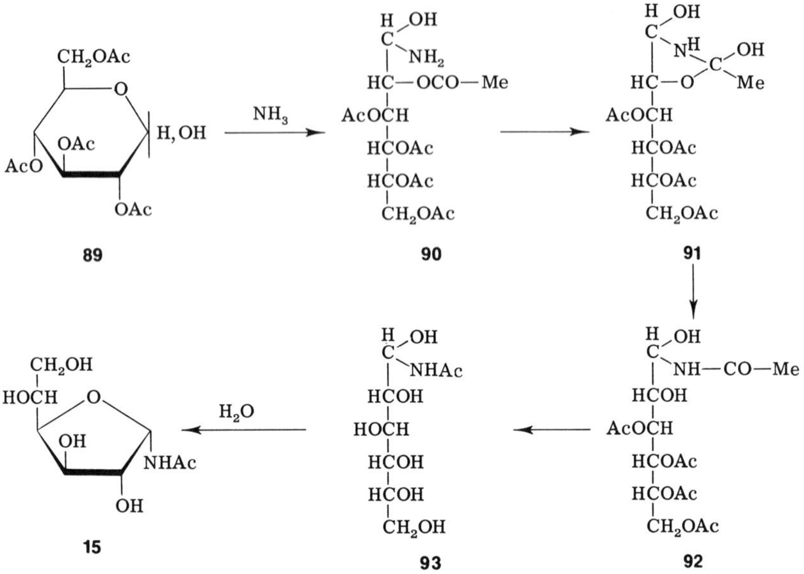

Further experiments disproved Wohl's theories[1,3,4] as to a direct condensation mechanism. By degradation of 2,3,4,5-tetra-O-acetyl-L-arabinononitrile[82] (**5**) in the presence of propionamide or benzamide, 1,1-bis(acetamido)-1-deoxy-L-erythritol (**6**) was obtained in yields comparable to those obtained when the reaction was conducted in the absence of these extraneous amides. Also, an excess of acetamide in the medium did not increase the formation of nitrogenated carbohydrate derivatives.

Compound **5** (see p. 82) was allowed to react[83] with ethanolic ammonia-^{15}N in the presence of an excess of acetamide-^{14}N, and compound **6** containing a percentage of ^{15}N similar to that of the ammonia employed was obtained, showing that, actually, the reaction occurs by an intramolecular mechanism. Were the mechanism intermolecular, in accordance with Wohl's views, the percentage of ^{15}N in **6** would have been lower. Likewise, ammonolysis of D-glucose pentabenzo-

(82) V. Deulofeu and J. O. Deferrari, *Anales Asoc. Quím. Argentina*, **38**, 241–251 (1950).

(83) R. C. Hockett, V. Deulofeu, and J. O. Deferrari, *J. Amer. Chem. Soc.*, **72**, 1840–1841 (1950).

ate[22] in the presence of benzamide containing carbonyl-^{14}C did not show incorporation of radioactivity into the 1,1-bis(benzamido)-1-deoxy-D-glucitol (17) isolated.

An alternative to the mechanism of Isbell and Frush was proposed by Deulofeu and coworkers.[22] The initial stage of the reaction would be similar in the formation of the aminol precursor 94. This intermediate would be in resonance with an immonium ion 95 that, after acetyl migration, would afford 97. Nucleophilic attack by ammonia on C-1 of this intermediate would give 98, and, finally, the 1,1-bis-(acylamido)-1-deoxyalditol (100), after migration of another acyl group as depicted in 99.

The carbonium ion 97 would lead to the N-acylglycosylamine (102), usually produced in low yield, by cyclization through attack (101) by a lone electron-pair of a hydroxyl group conveniently located in the chain (see Scheme 2). N-Acylglycosylamines (102) are not precursors

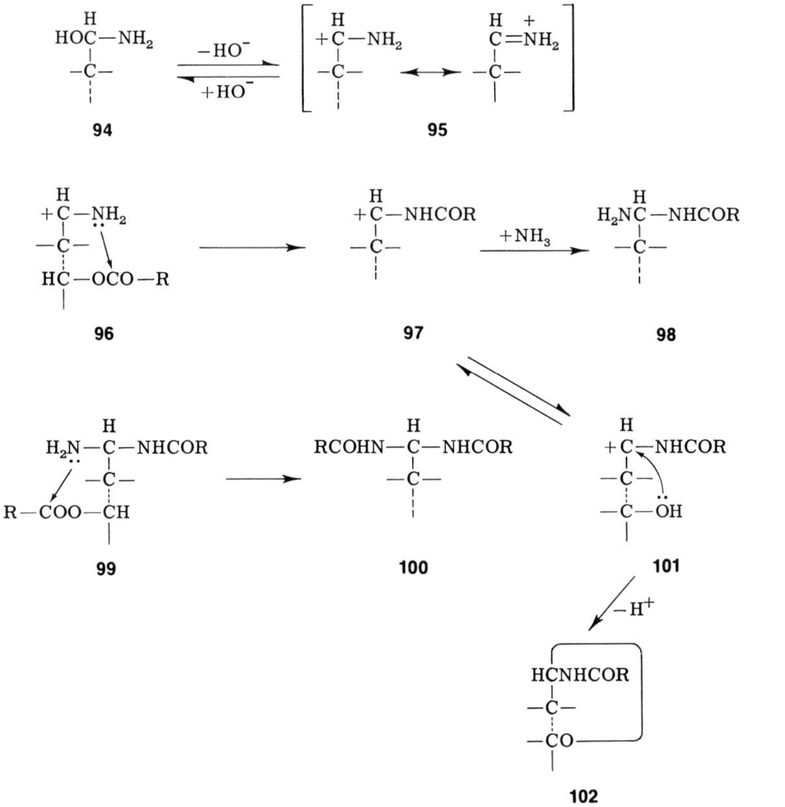

Scheme 2

of aldose bis(amides), as they are recovered in high yield on ammonolysis of their peracetates.[23]

Some attempts have been made to ascertain which acyl groups
contribute to the formation of the bis(amides). Thus, by ammonolysis
of 3,4,5,6-tetra-O-benzoyl-2-S-ethyl-2-thio-*aldehydo*-D-mannose,[83a]
Brigl and coworkers[9] obtained an N-benzoyl-2-S-ethyl-2-thio-D-
glycopyranosylamine, suggesting that the absence of the 2-O-benzoyl
group precluded the formation of the bis(amide). Likewise, ammonolysis of 4,5,6-tri-O-acetyl-2,3-di-O-methyl-*aldehydo*-D-glucose afforded[84] only 2,3-di-O-methyl-D-glucose. However, from these experiments, it could not be conclusively decided that the acyloxy groups
on C-2 and C-3 are the only contributing ones, because the ethylthio
or methyl substituents could hinder the transposition, either by steric
hindrance or by blocking carbon atoms necessary as intermediate
stages in the migration to C-1. Militating against Brigl's conclusion is
the fact that ammonolysis of 3,4,5,6-tetra-O-benzoyl-*aldehydo*-D-
glucose affords 1,1-bis(benzamido)-1-deoxy-D-glucitol.

A demonstration that an acyloxy group at any position of the carbon
chain contributes to the reaction was achieved[22] by ammonolysis of
D-glucose pentabenzoates successively substituted at different
oxygen atoms by benzoyl groups containing carbonyl-[14]C. The measure of the radioactivity of the 1,1-bis(benzamido)-1-deoxy-D-glucitols
obtained in each individual experiment gave a measure of the "apparent" (as a prior migration to another position cannot be excluded)
contribution of each benzoyl group (at a different position) to the
formation of the bis(amido) sugar.

The results, shown in Table I, provided evidence that, for cyclic
benzoates, there is a great contribution of the benzoyl groups on O-3
and O-4, an intermediate contribution by that on O-6, and a small
contribution by the benzoyl group on O-2, whereas that on O-1 does
not contribute at all to the formation of the 1,1-bis(benzamido)-1-
deoxyalditol.

The increased yields observed when acylated aldehydo sugars are
ammonolyzed was explained on the basis of a greater concentration
of the intermediate **96,** because the previous stage of ammonolysis of
the acyloxy group on C-1 is obviated, and, meanwhile, the elimination
of the remaining acyl groups as the corresponding amide would be
lessened.

(83a) A. Ducruix, C. Pascard-Billy, D. Horton, and J. D. Wander, *Carbohyd. Res.*, **29,**
        276–279 (1973).
(84) R. Allerton and W. G. Overend, *J. Chem. Soc.*, 35–36 (1952).

TABLE I

Contribution (Moles/mole) of Each Benzoyl Group to the Migration,
With Formation of 1,1-Bis(benzamido)-1-deoxy-D-glucitol[85]

Benzoate	C-1	C-2	C-3	C-4	C-6	Yield (%)
1,2,3,4,6-Penta-O-benzoyl-D-glucose	0	0.12 ± 0.03	0.76 ± 0.02	0.82 ± 0.02	0.31 ± 0.02	21
2,3,4,6-Tetra-O-benzoyl-D-glucose		0.12 ± 0.03	0.80 ± 0.03	0.81 ± 0.02	0.27 ± 0.02	28
1,2,3,5,6-Penta-O-benzoyl-D-glucose	0	0.10 ± 0.01				53
2,3,5,6-Tetra-O-benzoyl-D-glucose		0.11 ± 0.02				62
3,5,6-Tri-O-benzoyl-D-glucose						16
2,3,4,5,6-Penta-O-benzoyl-*aldehydo*-D-glucose		0.81 ± 0.02				61
2,3,4,5,6-Penta-O-benzoyl-D-glucose (bisulfite derivative)		0.83 ± 0.02				62
1,2,3,4,6-Penta-O-benzoyl-D-galactopyranose[86]		0.13 ± 0.01	0.62 ± 0.01	1.02 ± 0.02	0.18 ± 0.01	

The conformation of the intermediate **95** favored for a direct
O-2 → O-1 or O-3 → O-1 migration would be a zigzag planar one
(**103**) (but see Refs. 78 and 79a) whose C-2–C-3 rotamer is represented
by **104**; but, should the intermediate **95** adopt conformation **105**, the
direct migration would be hindered, although there would be facile
O-4 → O-1 and O-2 → O-1 migrations.

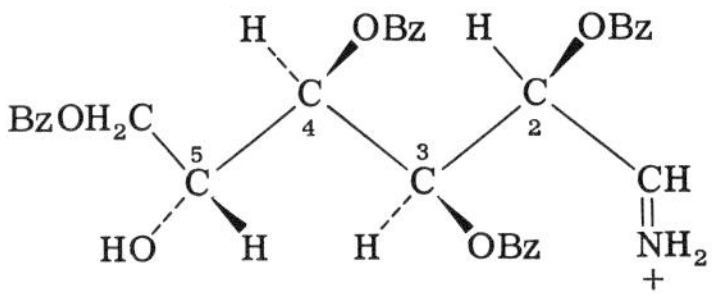

103 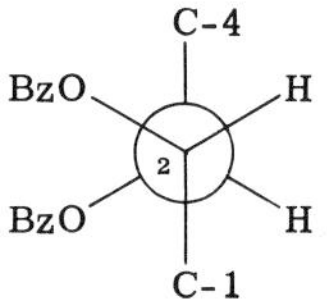   104 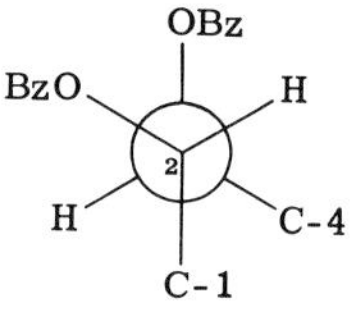   105

It was also observed[85] that the contribution of the benzoyl group on O-2 in furanose benzoates evidently does not change from the values shown in Table I for pyranose benzoates, but in esters of the acyclic compounds discussed, there is a major contribution by that benzoyl group. This difference in behavior was interpreted by assuming the formation, for the cyclic derivatives, of an ortho ester (**106**) that would stabilize the positive charge on C-1 and would thus facilitate the capture of ammonia to give **107** and **108**. The participation of the 2-O-benzoyl group in these intermediates would preclude its participation in the O → N migration.

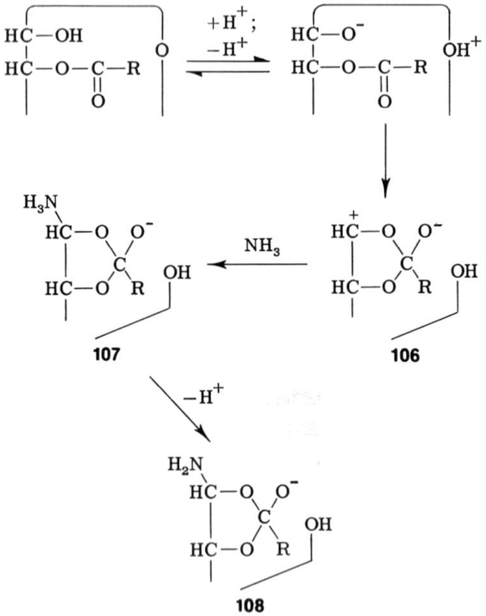

In the acyclic derivatives, the possibility of ortho-ester formation between C-1 and C-2 would be lessened by a more facile reaction of the carbonyl group with ammonia.

The hypothesis that, in cyclic derivatives, the low contribution of the benzoyl group on O-2 to the O → N migration might be attributable to its ready elimination as benzamide was rejected on the basis of the results of the ammonolysis of 2,3,5,6-tetra-O-benzoyl-D-glucofuranose and 3,5,6-tri-O-benzoyl-D-glucofuranose. The former com-

(85) V. Deulofeu, E. G. Gros, and A. Lezerovich, *Anales Real Soc. Españ. Fís. Quím. Ser. B*, **60**, 157–166 (1964).

pound gave 62% of 1,1-bis(benzamido)-1-deoxy-D-glucitol (**17**, p. 84), and the latter gave only 16%. As the only structural difference between the two benzoates is the benzoyl group on O-2, the difference in the yields (62% compared to 16%) does not accord with the low contribution to migration (0.11 mole/mole) of that benzoyl group. From these results, it follows that, despite its low contribution, the presence of this group is necessary to formation of high yields of **17**; this conclusion reinforces the hypothesis that the ortho-ester formation (as shown in **106**) allows a fast opening of the hemiacetal ring and attack by ammonia.

Parallel studies were conducted[86] with penta-O-benzoyl-β-D-galactopyranose (**109**) containing, at different positions, benzoyl

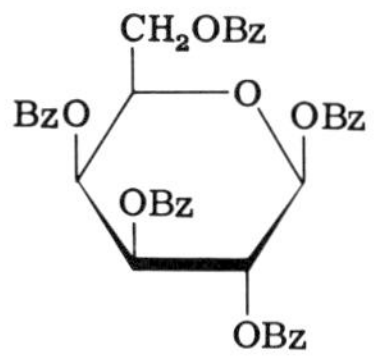

109

groups labelled with carbonyl-^{14}C. Again, the contribution of each benzoyl group was ascertained from the specific activity shown by the 1,1-bis(benzamido)-1-deoxy-D-galactitol obtained (see Table I).

The value of 1.02 moles/mole for the contribution of the 4-O-benzoyl group in **110** implies a rise of 0.18 mole/mole with regard to the 4-O-benzoyl group of penta-O-benzoyl-D-glucopyranose (**21**, see p. 85). This enhanced contribution in the galactose benzoate is concomitant with a lower contribution of the benzoyl groups on O-3 and O-6, whereas the value for that located on O-2 is similar. The total migration of the 4-O-benzoyl group in **109** was attributed[86] to a better location for migration in the zigzag conformation of the acyclic intermediate **110**.

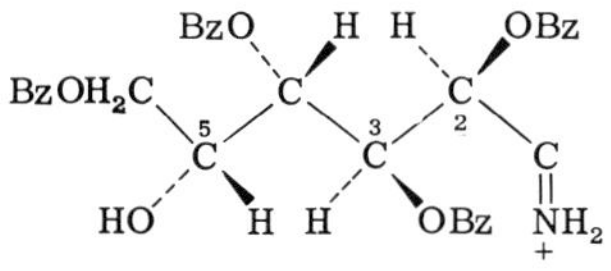

110

(86) E. G. Gros and V. Deulofeu, *J. Org. Chem.*, **29**, 3647–3654 (1964).

A conformational analysis[86] through the Böeseken (Newman) projections of the intermediate 110 for the D-*galacto* derivative, compared
with the D-*gluco* analog 103, suggested that the most stable conformation for 110 would also be the most favorable for direct migration from
O-4 to O-1, the distances between these oxygen atoms being larger
in 103 for migration. However, an alternative interpretation, based
more on configurational and steric approaches than on conformational
considerations, was advanced[25] in which it was considered, that,
according to the Curtin–Hammett principle,[87] the energy of reaction
probably overcomes the energy barriers among the conformers,
making less relevant the consideration of any particular ground-state
conformation for the intermediates of the reaction. The differences
observed in the contributions of similarly situated benzoyl groups,
depending on the cyclic or acyclic structure of the starting sugar,
suggested that the ammonolysis reaction is controlled by steric approach,[88] and that the magnitudes of the contributions are related to
the direction of attack by ammonia.

For the acyclic derivatives prior to reaction, the favored conformation of the C-1–C-2 rotamers would have the carbonyl group eclipsed
with the hydrogen atom on C-2 (111), and, consequently, the favored
side of attack for ammonia would be the same side as that of the acyl
group on O-2, to give 112 as the preponderant orientation for the
initial addition-compound. Thereafter, migrations from the chain take
place. This migration would be limited by the necessity of an ade-

quate distance between the amino group of the intermediate 112 and
the acyl groups in the chain, but the configurations of the intermediate, cyclization adducts would also play a significant role.

Considering that 112 would be the favored orientation for the
initial addition-compound (when the starting sugar is acyclic), the
intermediate cyclic structure for migration would be 113, in which a

(87) D. Y. Curtin, *Record Chem. Progr. Kresge-Hooker Sci. Lib.*, **15**, 111–128 (1954).
(88) W. G. Dauben, G. J. Fonker, and D. S. Noyce, *J. Amer. Chem. Soc.*, **78**,
    2579–2582 (1956); J. C. Richer, *J. Org. Chem.*, **30**, 324–325 (1965).

trans relationship is observed between the hydroxyl group on C-1 and the bulky substituent on C-2. This favored migration-interme-

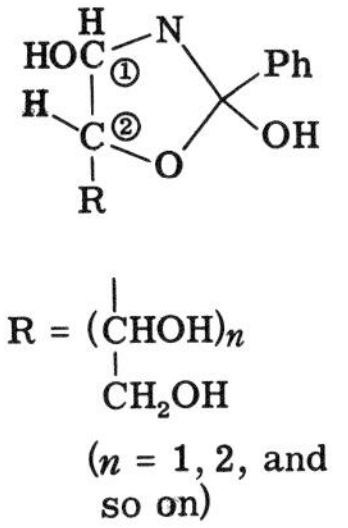

R = (CHOH)$_n$
CH$_2$OH

(n = 1, 2, and
so on)

113

diate would explain the high contribution (0.81 mole/mole) of the benzoyl group on O-2 in acyclic sugar benzoates. Other orientational alternatives,[25] both for the initial, ammonia addition-compound and for the migration intermediate, lead to structures that would resist further change into migration products.

For the migration of the acyl group on O-3, the alternative, six-membered, cyclic intermediates (for example, **114**) would offer less

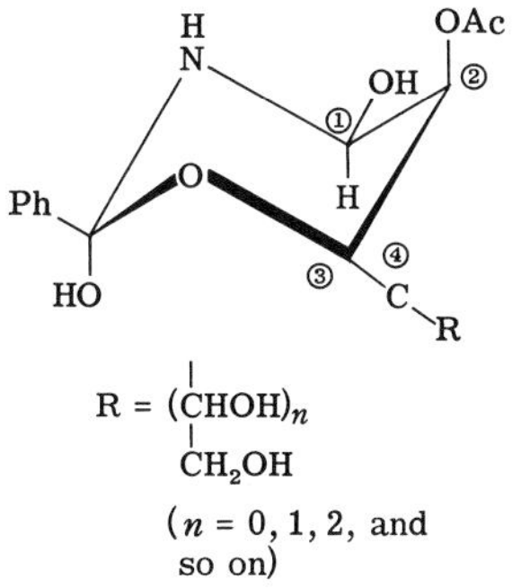

R = (CHOH)$_n$
CH$_2$OH

(n = 0, 1, 2, and
so on)

114

steric hindrance, the orientation of the starting aldehyde–ammonia adduct being less decisive.

When the sugar benzoates have a furanose or pyranose structure, the scheme of contributions evidently changes (see Table I, p. 115). The low contribution of the O-2 (0.11 mole/mole) and the high contribution of the O-4 (0.84 mole/mole) acyl groups would be opposed to that expected if a planar, zigzag, acyclic structure is assumed from the start.[22] For D-galactose benzoates,[86] the contribution of the 4-O-benzoyl group (1 mole/mole) is exceedingly high for a poorly favored position (assuming the intermediate **110**). The explanation for these incongruencies may be arrived at by considering the molecule of the cyclic benzoate, in its totality, as a puckered reacting-system. After a first step in which the ring opens and the carbonyl group is regenerated, the molecule would not immediately adopt a planar, zigzag structure, but would react with ammonia as a puckered one. For example, for penta-*O*-benzoyl-β-D-galactopyranose (**109**), the more accessible side for attack by ammonia, after the 1-*O*-benzoyl group has been split off as benzamide to give **115**, would be opposed to the 2-*O*-benzoyl group. This direction of attack (**111**) for the C-1–C-2 rotamer would give rise to an orientation for the ammonia-addition intermediate (**116**) that would be more unfavorable for the migration

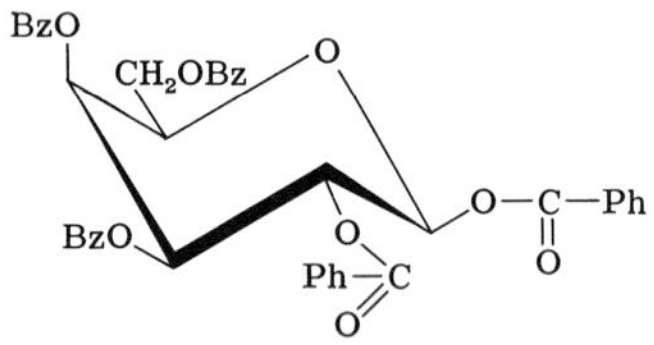

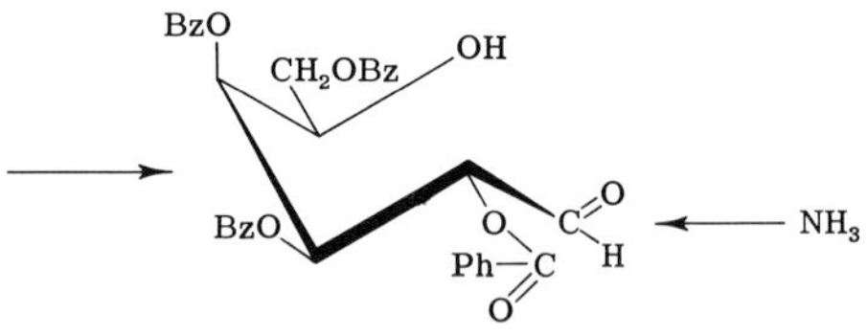

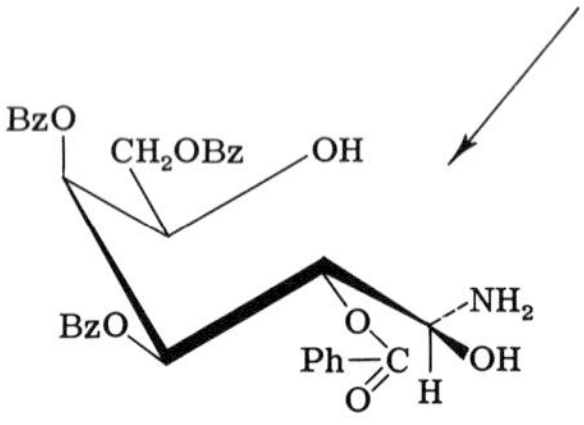

of the 2-*O*-benzoyl group. The C-1–C-2 rotamer (**117**) for that orientation is opposed to that favored (**112**) for the acyclic sugar benzoates.

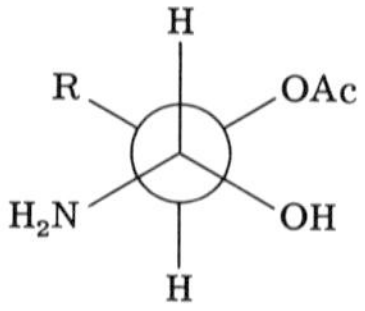

The high contribution of the 4-*O*-benzoyl group could be explained by the proximity of that group to the 1-amino group in the initial adduct **116**. Both groups are brought closer together during the rotation of the bonds to form a zigzag, stretched structure through a transient, helical conformation. By inspection of a molecular model of the D-*galacto* intermediate **116**, it may be seen that that is the most favored orientation, whereas, for the glucose adduct **118**, the situation would not be so favored, a lower contribution of the 4-*O*-benzoyl group being predictable, according to the results shown in Table I (see p. 115).

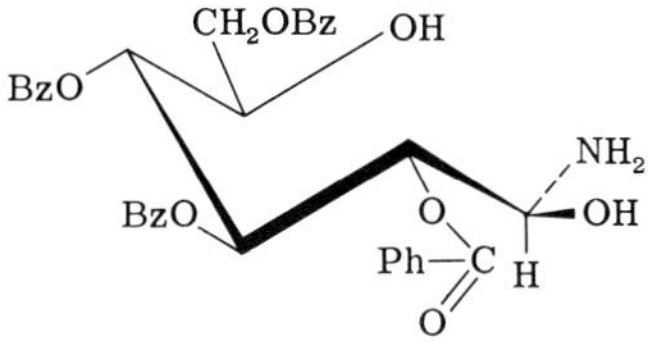

118

With regard to the variation of yields of 1,1-bis(amido)-1-deoxy-alditols being dependent on the structure of the starting acyl esters, it was observed that, in general, high yields (50–78%) of these compounds were obtained in the ammonolysis of acylated aldehydo and furanose sugars, and lower yields (8–48%) from the corresponding pyranoid esters. This difference in behavior was attributed[25] to a different rate of ring opening, which is low for pyranoid esters because of their greater stability, which allows mechanisms other than migration to compete.

As the steric interactions on the ring are relieved through the formation of an acyclic structure, the greater initial free-energy in the ground state of the furanose esters would enhance the rate of ring opening. As was previously mentioned, the high yield (62%) of 1,1-bis(benzamido)-1-deoxy-D-glucitol obtained by ammonolysis of 2,3,5,6-tetra-*O*-benzoyl-D-glucofuranose, compared with the 16% yield produced from 3,5,6-tri-*O*-benzoyl-D-glucofuranose, can, from this point of view (considering the low contribution of the 2-*O*-benzoyl group), be ascribed to a greater instability of the former benzoate, which is a more crowded molecule and, consequently, has a higher rate of ring opening.

2. Formation of *N*-Acylaldosylamines

Some molecules having the intermediate, aminol structure **94** can cyclize to give *N*-acylaldosylamines of the general structure **102** (see Scheme 2, p. 113). Postulation of the acyclic intermediate **94** is supported by the observation[30] that the ammonolysis of 2,3,4,6-tetra-*O*-benzoyl-D-glucopyranosylamine (**119**) gave a very low yield (1.4%)

of the 1,1-bis(benzamido) derivative (**17**) and only traces of the N-benzoyl-β-D-glucopyranosylamine (**120**).

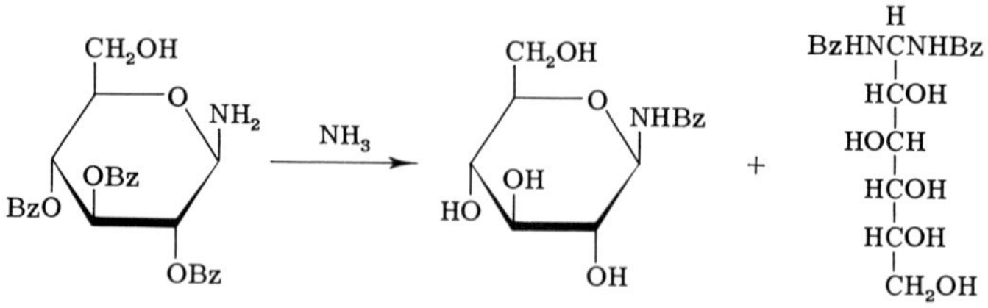

This result demonstrates that a free amino group on C-1 of a cyclic structure would not be an adequate intermediate for the formation of bis(amides), although an $O \rightarrow N$ migration without ring opening would be feasible on steric grounds. The latter possibility has strong steric requirements, as shown by the ammonolysis of 2,3,4,6-tetra-*O*-benzoyl-β-D-mannopyranosylamine[89] (**121**), which gave N-benzoyl-β-D-mannopyranosylamine (**13**) in good yield (60%). In this reaction,

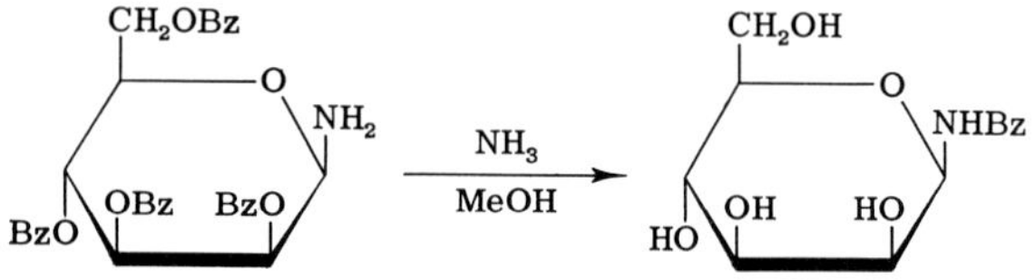

there was no evidence for the formation of the bis(amido) derivative. On this basis, an acylated glycosylamine having at least a 2-*O*-acyl group would be an important intermediate, following (in sequence) the aminol **94** in the formation of N-acylmannosylamines. The effective migration of the 2-*O*-acyl group needs confirmation by the use of isotopic tracers.[89]

The N-acylglycosylamines (in the β-pyranose form) were isolated in low yield from sugar benzoates. On the other hand, in the ammonolysis of the acetates of D-glucose and D-galactose, the formation of these compounds plays an important role. Thus, the ammonolysis of penta-*O*-acetyl-β-D-glucopyranose gave[90] a high yield (53%) of N-acetyl-α-D-glucofuranosylamine (**15**), whereas penta-*O*-acetyl-β-D-galactopyranose gave a 21% yield of N-acetyl-β-D-galactofuranosyl-

(89) J. F. Sproviero, *Carbohyd. Res.*, **26**, 357–363 (1973).
(90) M. E. Gelpi and R. A. Cadenas, *Carbohyd. Res.*, **28**, 147–149 (1973).

amine (**81**) and 28% of the bis(acetamido) derivative. This difference in the yields of N-acetylaldosylamines (as between the two sugar acetates) can be related to the difference in accessibility of the acyl groups on O-4 for migration (see p. 120), which, for the pentaacetate of D-galactose, favors the double migration to give the bis(acetamido) derivative.

The observation that, by ammonolysis, acetates having the D-*xylo* configuration afforded N-acetyl-D-glycofuranosylamines as the principal products, whereas other acetylated aldoses mainly gave other products, namely, bis(acetamido) derivatives and the free sugar, indicates that the configuration of the sugar exerts a strong influence in the cyclization. From Table II, it may be seen that the yields of cyclic N-acetylglycosylamines, as related to those of acyclic compounds (ratio A:B), decrease in the order D-*gluco*, D-*galacto*, and D-*manno*.[26] The factors that determine the favored formation of furanoid rings in the acylglycosylamines[24] were discussed on a configurational basis in Section V (see p. 102).

TABLE II

Relation Between Yields of Cyclic and Acyclic Nitrogenated Products
in Some Ammonolysis Reactions[26]

Peracetate of	N-Acetylglycopy-ranosylamine (%)	Anomer	N-Acetylglycofur-anosylamine (%)	Anomer	Total yield (%) of cyclic products (A)	1,1-Bis(acetamido)-1-deoxyalditols (%) (B)	Ratio, A:B
β-D-Glucopyranose[23,a]			12.1	α	12.1	8.0	1.51
β-D-Galactopyranose[25]			20.7	β	20.7	28.0	0.73
β-D-Mannopyranose[24]	3.9	α	{8.4 / 3.9	α / β	16.2	32.1	0.50
α,β-L-Rhamnopyranose[24]	14.7	α	4.9	α	19.6	37.5	0.52

a This reaction was conducted in methanolic ammonia.

The higher yield of the pyranose derivatives (see Table II) in the ammonolysis of α,β-L-rhamnopyranose tetraacetate shows the influence of polar effects, because the methyl substituent on C-5 would enhance the nucleophilic character of its hydroxyl group in relation to that of the one on C-4.

Usually, ammonolysis of O-acylated N-acylglycosylamines produces the corresponding free N-acylglycosylamines in high yield. However, when a strong electron-attracting group is present on C-1 as an N-acylamino group, it increases the electrophilic character of that carbon atom, and migrations can occur by ammonolysis of the O-acyl derivative. For example, on ammonolysis of 2,3,4,6-tetra-

O-acetyl-N-p-tolylsulfonyl-β-D-glucopyranosylamine,[23,91] a 36% yield of 1-acetamido-1-deoxy-1-p-toluenesulfonamido-D-glucitol (**122**) was obtained, together with 6% of N-p-tolylsulfonyl-β-D-glucopyranosyl-amine (**123**). Splitting of the N-p-tolylsulfonyl group also occurred, as both glucose and p-toluenesulfonamide were detected[23] in the product.

$$AcHN\overset{\displaystyle H}{\underset{\displaystyle |}{C}}NHSO_2C_6H_4Me\text{-}p$$
HCOH
HOCH
HCOH
HCOH
CH$_2$OH

122

CH$_2$OH
NHSO$_2$C$_6$H$_4$Me-p
OH
HO
OH

123

3. Formation of Heterocyclic Compounds

As previously indicated (see pp. 88–91), formation of heterocyclic compounds, mainly pyrazines, was found only in the ammonolysis of some aldose nicotinates[33,35] and acetates and benzoates of ketoses.[36,37] For ketose esters, whose behavior differed from that of the aldose esters, the formation of imidazole derivatives was also observed; these heterocyclic compounds also result from the direct action of ammonia upon the corresponding free sugars, but the presence of the esterifying acyl groups evidently increases their ease of formation and raises their yields.

The formation of pyrazine derivatives can be interpreted by taking as examples[36] 6-(1-deoxy-L-*threo*-tetritol-1-yl)-2-(L-xylitol-1-yl)pyrazine (**34**) and 2,5-bis(L-xylitol-1-yl)pyrazine (**35**) (see p. 90).

Compound **34** would be formed through the following sequence: L-sorbose, produced by ammonolysis of all of the acyl substituents (see Scheme 1, p. 90) would be in equilibrium with L-gulose (or L-idose) by a Lobry de Bruyn–Alberda van Ekenstein rearrangement. The two sugars in the ammoniacal medium would give L-sorbosyl-amine and L-gulosylamine (or L-idosylamine), respectively. The L-gulosylamine, through an Amadori rearrangement, would afford 1-amino-1-deoxy-L-sorbose (**124**), whereas the L-sorbosylamine, by a Heyns rearrangement, would produce 2-amino-2-deoxy-L-gulose (**125**) (or 2-amino-2-deoxy-L-idose). Condensation of **124** with **125** would give **34**, through the intermediate formation of a dihydropyraz-ine (**126**) and its dehydration product (**127**). Formation of compound **35** can be rationalized on the basis that two molecules of **124** would interact to give **128**, which, through oxidation by atmospheric oxygen,

(91) B. Helferich and A. Mitrowsky, *Chem. Ber.*, **85**, 1–8 (1952).

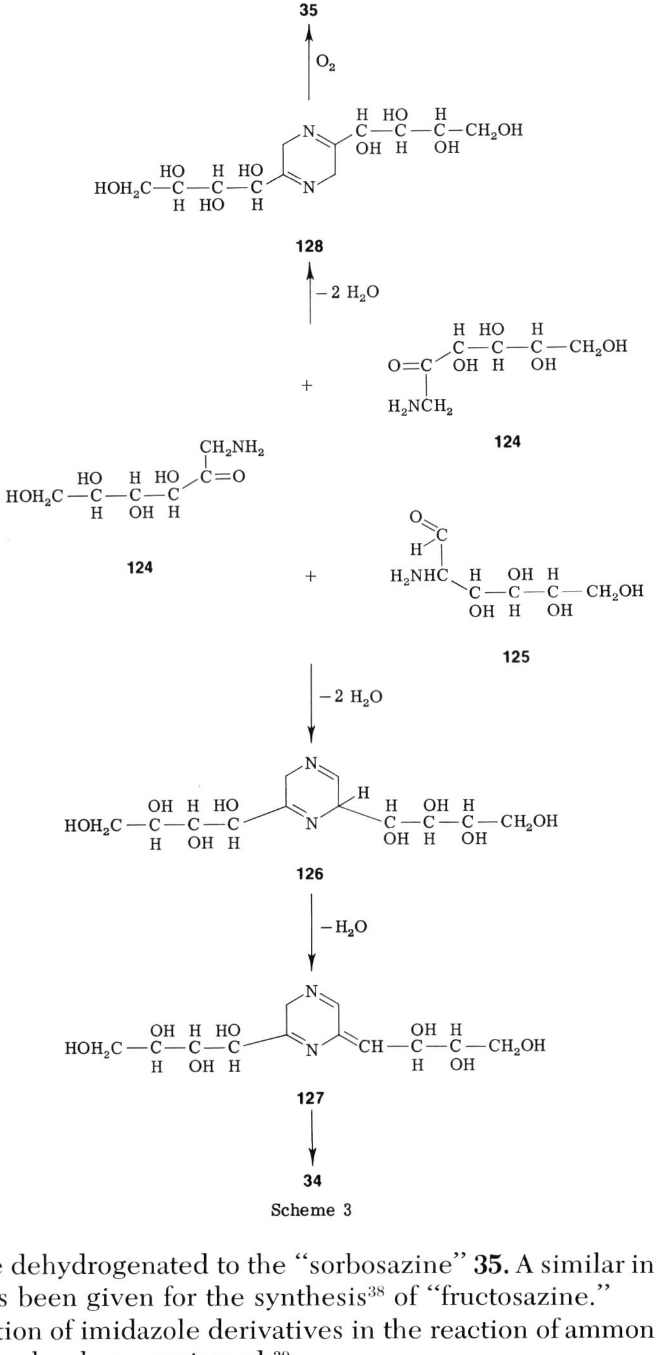

35

$\uparrow$ O_2

$$
\begin{array}{c}
\text{H HO H} \\
\text{C—C—C—CH}_2\text{OH} \\
\text{OH H OH}
\end{array}
$$

$$
\begin{array}{c}
\text{HO H HO} \\
\text{HOH}_2\text{C—C—C—C} \\
\text{H HO H}
\end{array}
$$

128

$\uparrow$ $-2\ H_2O$

$$
\begin{array}{c}
\text{H HO H} \\
\text{C—C—C—CH}_2\text{OH} \\
\text{O=C OH H OH}
\end{array}
$$
+
$$\text{H}_2\text{NCH}_2$$

124

$$
\begin{array}{c}
\text{CH}_2\text{NH}_2 \\
\text{HO H HO C=O} \\
\text{HOH}_2\text{C—C—C—C} \\
\text{H OH H}
\end{array}
$$

124

+

$$
\begin{array}{c}
\text{O=C} \\
\text{H} \\
\text{H}_2\text{NHC H OH H} \\
\text{C—C—C—CH}_2\text{OH} \\
\text{OH H OH}
\end{array}
$$

125

$\downarrow$ $-2\ H_2O$

$$
\begin{array}{c}
\text{OH H HO} \\
\text{HOH}_2\text{C—C—C—C} \\
\text{H OH H}
\end{array}
\quad
\begin{array}{c}
\text{H OH H} \\
\text{C—C—C—CH}_2\text{OH} \\
\text{OH H OH}
\end{array}
$$

126

$\downarrow$ $-H_2O$

$$
\begin{array}{c}
\text{OH H HO} \\
\text{HOH}_2\text{C—C—C—C} \\
\text{H OH H}
\end{array}
\quad
\begin{array}{c}
\text{OH H} \\
\text{CH—C—C—CH}_2\text{OH} \\
\text{H OH}
\end{array}
$$

127

$\downarrow$

34

Scheme 3

would be dehydrogenated to the "sorbosazine" **35.** A similar interpretation has been given for the synthesis[38] of "fructosazine."

Formation of imidazole derivatives in the reaction of ammonia with free sugars has been reviewed.[39]

VII. Tables of Derivatives

In Tables III and IV, the yields and properties of the products isolated from the reaction of ammonia with acylated mono- and disaccharides are listed.

Table III

Derivatives Isolated from the Reaction of Ammonia with Acylated Monosaccharides

Compounds isolated	Yield (%)	M.p. (degrees)	$[\alpha]_D$ (degrees)[a]	Starting material	Solvent	References
A. Tetroses						
1,1-Bis(acetamido)-1-deoxy-D-erythritol (10)	40	210	–	2,3,4,5-tetra-O-acetyl-D-arabinononitrile (9)	EtOH–H$_2$O	8
1,1-Bis(acetamido)-1-deoxy-L-erythritol (6)	45	210	−7.9(w)	2,3,4,5-tetra-O-acetyl-L-arabinononitrile (5)	EtOH–H$_2$O	4
				2,3,4-tri-O-acetyl-aldehydo-L-erythrose (18)	EtOH	14
2,3,4-tri-O-acetyl-		147	+31.7(w)			14,93
1,1-Bis(benzamido)-1-deoxy-L-erythritol	19	220	+13.1(Py)	2,3,4,5-tetra-O-benzoyl-L-arabinononitrile	MeOH	21
2,3,4-tri-O-acetyl-		183	+8.9(Chlf)			21
1,1-Bis(acetamido)-1-deoxy-L-threitol (8)	30	166	+10.2(w)	2,3,4,5-tetra-O-acetyl-L-xylononitrile (7)	H$_2$O;EtOH	7
	70	165			H$_2$O	58
2,3,4-tri-O-acetyl-		178	−74.0(Chlf)			58
2,3,4-tri-O-benzoyl-		155	−110.1(Chlf)			58
2,4-O-benzylidene-		265	–			58
1,1-Bis(acetamido)-1-deoxy-D-threitol	78	165	−10.8(w)	2,3,4,5-tetra-O-acetyl-D-xylononitrile	H$_2$O	5,6
2,3,4-tri-O-acetyl-		179	+74.2(Chlf)			6
2,3,4-tri-O-benzoyl-		155–156	+109.7(Chlf)			58
2,4-O-benzylidene-		265	–			58
1,1-Bis(benzamido)-1-deoxy-D-threitol	18.3	189	+1.7(Py)	2,3,4,5-tetra-O-benzoyl-D-xylononitrile	MeOH	21
2,3,4-tri-O-acetyl-		183	+79.3(Chlf)			21
B. Pentoses						
1,1-Bis(acetamido)-1-deoxy-D-arabinitol (2)	47	187	−9.5(w)	2,3,4,5,6-penta-O-acetyl-D-gluconononitrile (1)	H$_2$O	1
	65	–	–	2,3,4,5,6-penta-O-acetyl-D-mannononitrile	H$_2$O	58
	33.8	–	–		EtOH–H$_2$O	8
2,3,4,5-tetra-O-acetyl-		218	+72.5(Chlf)			58
1,1-Bis(acetamido)-1-deoxy-L-arabinitol	53	189	+9.8(w)	2,3,4,5-tetra-O-acetyl-aldehydo-L-arabinose	MeOH	15
	7.9	193	+8.8(w)	1,2,3,4-tetra-O-acetyl-α-L-arabinopyranose	MeOH	21
1-Deoxy-1,1-bis(propionamido)-D-arabinitol	33	177	−9.8(w)	2,3,4,5,6-penta-O-propionyl-D-gluconononitrile	EtOH–H$_2$O	31
2,3,4,5-tetra-O-acetyl-		171	+67.7(Chlf)			31
1,1-Bis(benzamido)-1-deoxy-D-arabinitol	41.7	198	+5.1(Py)	2,3,4,5,6-penta-O-benzoyl-D-glucononitrile	MeOH	21
2,3,4,5-tetra-O-acetyl-		144–145	+73.4(Chlf)			21
2,3,4,5-tetra-O-benzoyl-		134–135	+62.5(Chlf)			21

(92) E. Votoček, *Ber.*, **50**, 35–42 (1917).

(93) E. Restelli de Labriola, Thesis, Facultad de Ciencias Exactas y Naturales, Buenos Aires, Argentina (1938).

TABLE III (*Continued*)

Compounds isolated	Yield (%)	M.p. (degrees)	$[\alpha]_D$ (degrees)[a]	Starting material	Solvent	References
5-O-benzoyl-	25	206	+8.5(Py)		EtOH	59
	1.4	–	–	2,3,4,5,6-penta-O-benzoyl-D-mannononitrile	EtOH	59
2,3,4-tri-O-acetyl-		193	+69.0(Chlf)			59
1,1-Bis(benzamido)-1-deoxy-L-arabinitol	35.6	197	-5.2(Py)	1,2,3,4-tetra-O-benzoyl-β-L-arabinopyranose	MeOH	21
2,3,4,5-tetra-O-acetyl-		143	-73.3(Chlf)			21
1-Deoxy-1,1-Bis(nicotinamido)-L-arabinitol	28	231	-9.7(w)	1,2,3,4-tetra-O-nicotinoyl-β-L-arabinopyranose	H₂O	34
2,3,4,5-tetra-O-acetyl-N-Nicotinoyl-L-arabinosylamine	3.1	136–137 / 200–201	-60(Chlf) / -7.6(w)			34 / 34
2,3,5-tri-O-acetyl-		syrup	-34(Chlf)			34
1,1-Bis(acetamido)-1,5-dideoxy-L-arabinitol	35	202	+19.8(w)	2,3,4,5-tetra-O-acetyl-L-rhamnononitrile	EtOH	93
1,1-Bis(benzamido)-1,5-dideoxy-L-arabinitol		225	-2.1(Py)	2,3,4,5-tetra-O-benzoyl-L-rhamnononitrile	EtOH	59
2,3,4-tri-O-acetyl-		193	-92.9(Chlf)			59
2,3,4-tri-O-benzoyl-		212	–			59
1,1-Bis(acetamido)-1,5-dideoxy-D-lyxitol	40	233	–	2,3,4,5-tetra-O-acetyl-D-fucononitrile	H₂O	92
1-Deoxy-1,1-bis(propionamido)-D-lyxitol	30	180	-8.5(w)	2,3,4,5,6-penta-O-propionyl-D-galactononitrile	EtOH–H₂O	31
1,1-Bis(acetamido)-1-deoxy-D-lyxitol (4)	40	222	–	2,3,4,5,6-penta-O-acetyl-D-galactononitrile (3)	H₂O	4
	72	230	-9.2(w)		H₂O	58
1,1-Bis(acetamido)-1-deoxy-D-lyxitol	46	232	-10.3(w)	1,2,3,4-tetra-O-acetyl-α-D-lyxopyranose	MeOH	20
1,1-Bis(benzamido)-1-deoxy-D-lyxitol	30	244	–	1,2,3,4-tetra-O-benzoyl-α-D-lyxopyranose	MeOH	20
	30.5	244	+4.6(Py)	2,3,4,5,6-penta-O-benzoyl-D-galactononitrile	MeOH;EtOH	20,60
2,3,4,5-tetra-O-acetyl-		163	+37.7(Chlf)			20
2,3,4,5-tetra-O-benzoyl-		227–229	+39.1(Chlf)			59
5-O-benzoyl-	22	220	+36.1(Py)		EtOH	59
	–	220	–		iPrOH	60
2,3,4-tri-O-acetyl-	16	189	+36.2(Chlf)			59
1,1-Bis(acetamido)-1-deoxy-D-ribitol	25	154	+14.7(w)	1,2,3,4-tetra-O-acetyl-β-D-ribopyranose	MeOH	21
2,3,4,5-tetra-O-acetyl-		184	+18.5(Chlf)			21
1,1-Bis(benzamido)-1-deoxy-D-ribitol	35	190	-7.3(Py)			21
2,3,4,5-tetra-O-acetyl-		172	-10.4(Chlf)			21
N-Acetyl-α-D-xylofuranosylamine	9.2	153	+100(w)	1,2,3,4-tetra-O-acetyl-β-D-xylopyranose	MeOH	11
2,3,5-tri-O-benzoyl-		141	+31.1(Chlf)			11
N-Acetyl-β-D-xylopyranosylamine	3.1	218	-1.2(w)			11
1-Deoxy-1,1-bis(nicotinamido)-D-xylitol	13.2	211	-4(w)	1,2,3,4-tetra-O-nicotinoyl-α-D-xylopyranose	H₂O	34
2,3,4,5-tetra-O-acetyl		syrup	-35(Chlf)			34
6-(D-*glycero*-2,3-Dihydroxypropyl)-2-(D-*threo*-trihydroxypropyl)pyrazine (32)	2.7	120	-128(w)			35
1,1-Bis(benzamido)-1-				1,2,3,4-tetra-O-benzoyl-α-		

(*Continued*)

TABLE III (Continued)

Compounds isolated	Yield (%)	M.p. (degrees)	$[\alpha]_D$ (degrees)[a]	Starting material	Solvent	References
deoxy-D-xylitol	30	184	−2.2(Py)	D-xylopyranose	MeOH	21
2,3,4,5-tetra-O-acetyl-		178	−50.4(Chlf)			21
C. Hexoses						
N-Acetyl-α-D-				2,3,4,5,6,7-hexa-O-acetyl-D-		
glucofuranosylamine				glycero-D-gulo-		
(15)	26	192–194	+86.9(w)	heptononitrile	H_2O	12
				2,3,4,5,6,7-hexa-O-acetyl-D-		
				glycero-D-ido-		
	34.9			heptononitrile	MeOH	16
				2,3,4,5,6-penta-O-acetyl-		
	45			aldehydo-D-glucose (20)	MeOH	17
	56				H_2O	12
				1,2,3,4,6-penta-O-acetyl-β-		
	8			D-glucopyranose (21)	MeOH	17
	12.1				MeOH	23
	53				H_2O	90
2,3,5,6-tetra-O-acetyl-		121	+32.7(Chlf)			17
2,3,5,6-tetra-O-benzoyl-		150	−19.1(Chlf)			23
1,1-Bis(acetamido)-1-						
deoxy-D-glucitol	8	165	+3.7(w)			23
2,3,4,5,6-penta-O-acetyl-		189	+21.1(Chlf)			23
1-Acetamido-1-deoxy-1-p-				2,3,4,6-tetra-O-acetyl-N-p-		
toluenesulfonamido-D-				toluenesulfonamido-β-D-		
glucitol	36	195	−61.9(w)	glucopyranosylamine		23
N-p-Tolylsulfonyl-β-D-						
glucopyranosylamine	6	195	+30(w)			23
1,1-Bis(benzamido)-1-				2,3,4,5,6-penta-O-benzoyl-		
deoxy-D-glucitol (17)	42	202	+1.5(Py)	aldehydo-D-glucose	MeOH	9
				3,4,5,6-tetra-O-benzoyl-		
	78			aldehydo-D-glucose	MeOH	9
				1,2,3,4,6-penta-O-benzoyl-		
	21.2			α-D-glucopyranose (22)	MeOH	16
				1,2,3,4,6-penta-O-benzoyl-		
	18.5			β-D-glucopyranose	MeOH	16
	16.1				MeOH	60
	21.4				MeOH	23
				2,3,4,6-tetra-O-benzoyl-D-		
	28			glucopyranose	MeOH	22
				1,2,3,5,6-penta-O-benzoyl-		
	53			D-glucofuranose	MeOH	22
				2,3,5,6-tetra-O-benzoyl-D-		
	62			glucofuranose	MeOH	22
				3,5,6-tri-O-benzoyl-D-		
	16			glucofuranose	MeOH	9
				1,2,3,6-tetra-O-benzoyl-D-		
	traces			glucose	MeOH	22
				2,3,4,5,6,7-hexa-O-benzoyl-		
				D-glycero-D-gulo-		
	37			heptononitrile	MeOH	13
				2,3,4,6-tetra-O-benzoyl-β-		
	1.4		−	D-glucopyranosylamine		30
2,3,4,5,6-penta-O-acetyl-		196	−39.7(Chlf)			16
				1,2,3,4,6-penta-O-benzoyl-		
6-O-benzoyl-	27.6	208	−	β-D-glucopyranose	2-BuOH	60
	25	208	+6.3(Py)		iPrOH	60
				1-O-acetyl-2,3,4,6-tetra-O-		
	32	−	−	benzoyl-D-glucopyranose	MeOH	22
				2,3,4,5,6,7-hexa-O-benzoyl-		

Tᴀʙʟᴇ III (*Continued*)

Compounds isolated	Yield (%)	M.p. (degrees)	$[\alpha]_D$ (degrees)[a]	Starting material	Solvent	References
N-Benzoyl-β-ᴅ-glucopyranosylamine	8	236	−12.5(w)	ᴅ-*glycero*-ᴅ-*gulo*-heptononitrile	MeOH	13
	traces			2,3,4,6-tetra-O-benzoyl-β-ᴅ-glucopyranosylamine	MeOH	30
	1	230	−11.6(w)	1,2,3,4,6-penta-O-benzoyl-β-ᴅ-glucopyranose	MeOH	23
2,3,4,6-tetra-O-acetyl-		196	−7.9(Chlf)			23
N-Benzoyl-2-S-ethyl-2-thio-ᴅ-mannosylamine	96.3	186–190		3,4,5,6-tetra-O-benzoyl-2-S-ethyl-2-thio-*aldehydo*-ᴅ-mannose	MeOH	9,83a
1-Deoxy-1,1-bis(propion-amido)-ᴅ-glucitol	3.9	155	+6.8(w)	1,2,3,4,6-penta-O-propionyl-ᴅ-glucopyranose	MeOH	23
2,3,4,5,6-penta-O-acetyl-		169	+25.3(Chlf)			23
N-Propionyl-α-ᴅ-glucofuranosylamine	5.6	146	+82.7(w)			23
N-Propionyl-β-ᴅ-glucopyranosylamine	0.9	191	−21.8(w)			23
2,3,4,6-tetra-O-acetyl-	−	151	+16.1(Chlf)			23
1-Deoxy-1,1-bis(nicotin-amido)-ᴅ-glucitol	3.8	188	−3.0(w)	1,2,3,4,6-penta-O-nicotinoyl-α-ᴅ-glucopyranose	H₂O	33
2,3,4,5,6-penta-O-acetyl-		164	−20.0(Chlf)			33
N-Nicotinoyl-ᴅ-glucofuranosylamine	16	176	+38(w)			33
2,3,5,6-tetra-O-acetyl-		129	+39.0(Chlf)			33
2-(ᴅ-*arabino*-Tetrahydroxy-butyl)-6-(ᴅ-*erythro*-2,3,4-trihydroxy-butyl)pyrazine (**31**)	7.6	166	−110(w)			33
hepta-O-acetyl-		141	−8.0(Chlf)			33
1,1-Bis(acetamido)-1-deoxy-ᴅ-galactitol	49	196	+6.9(w)	2,3,4,5,6-penta-O-acetyl-*aldehydo*-ᴅ-galactose	MeOH	18
	44.5			1,2,3,5,6-penta-O-acetyl-α-ᴅ-galactofuranose	MeOH	18
	41.9			1,2,3,5,6-penta-O-acetyl-β-ᴅ-galactofuranose		18
	26.5			1,2,3,4,6-penta-O-acetyl-α-ᴅ-galactopyranose	MeOH	18
	24.4			1,2,3,4,6-penta-O-acetyl-β-ᴅ-galactopyranose	MeOH	18
	34.6			2,3,4,5,6,7-hexa-O-acetyl-ᴅ-*glycero*-ʟ-*manno*-heptononitrile	MeOH	18
	28.2			1,2,3,4,6-penta-O-acetyl-β-ᴅ-galactopyranose	H₂O	25
2,3,4,5,6-penta-O-acetyl-		200	−38.3(Chlf)			18
N-Acetyl-β-ᴅ-galactofuranosylamine	20.7	133	−109.5(w)		H₂O	25
2,3,5,6-tetra-O-acetyl-		111	−15.7(Chlf)			25
1,1-Bis(benzamido)-1-deoxy-ᴅ-galactitol	35	207	−6.2(Py)	1,2,3,4,6-penta-O-benzoyl-α-ᴅ-galactopyranose	MeOH	18
	15.8				MeOH	60
	37			2,3,4,5,6,7-hexa-O-benzoyl-ᴅ-*glycero*-ʟ-*manno*-heptononitrile	MeOH	13
6-O-benzoyl-	32.1	220	−		iPrOH	60

(*Continued*)

TABLE III (Continued)

Compounds isolated	Yield (%)	M.p. (degrees)	$[\alpha]_D$ (degrees)[a]	Starting material	Solvent	References
N-Benzoyl-β-D-galactopyranosylamine	4.4	179	+21.5		MeOH	13
1-Deoxy-1,1-bis(nicotin-amido)-D-galactitol	28	180	+7.4(w)	1,2,3,4,6-penta-O-nicotinoyl-α-D-galactopyranose	H_2O	34
2,3,4,5,6-penta-O-acetyl-		110	−30.0(Chlf)			34
N-Nicotinoyl-D-galactofuranosylamine	1.4	166	+18.0(w)			34
1,1-Bis(acetamido)-1-deoxy-D-mannitol	33	219	−	2,3,4,5,6,7-hexa-O-acetyl-D-glycero-D-galacto-heptononitrile	MeOH	9
	35	218	−13.8(w)	1,2,3,4,6-penta-O-acetyl-β-D-mannopyranose	MeOH	19
	29				H_2O	29
N-Acetyl-α-D-mannofuranosylamine	8.4	162	+120.9(w)			24
2,3,5,6-tetra-O-acetyl-		123	+43.5(Chlf)			24
N-Acetyl-β-D-mannofuranosylamine	3.9	148	−113.2(w)			24
2,3,5,6-tetra-O-acetyl-		131	+29.0(Chlf)			24
N-Acetyl-α-D-mannopyranosylamine	4.2	183	+87.5(w)			24
2,3,4,6-tetra-O-acetyl-		181	+139.6(Chlf)			24
1,1-Bis(benzamido)-1-deoxy-D-mannitol (12)		226	+3.6(Py)	2,3,4,5,6,7-hexa-O-benzoyl-D-glycero-D-galacto-heptononitrile	MeOH	9
	22					13
	20.8			1,2,3,4,6-penta-O-benzoyl-D-mannopyranose	MeOH	19
2,3,4,5,6-penta-O-benzoyl-		140	+25.3(Chlf)			19
N-Benzoyl-β-D-mannopyranosylamine (13)	6	253	+6.4(Py)		MeOH	19
	4.4	−	−	2,3,4,5,6,7-hexa-O-benzoyl-D-glycero-D-galacto-heptononitrile	MeOH	9
					MeOH	13
	60			2,3,4,6-tetra-O-benzoyl-β-D-mannopyranosylamine	MeOH	89
2,3,4,6-tetra-O-acetyl-		135	−28.8(Chlf)			19
1-Deoxy-1,1-bis(nicotin-amido)-D-mannitol	24.2	191	−4.0(w)	1,2,3,4,6-penta-O-nicotinoyl-β-D-mannopyranose	H_2O	34
2,3,4,5,6-penta-O-acetyl-		204	+24.0(Chlf)			34
2-(D-arabino-Tetrahydroxy-butyl)-6-(D-erythro-2,3,4-trihydroxybutyl)-pyrazine (31)	3.8	165	−110.0(w)			34
1,1-Bis(acetamido)-1-deoxy-L-rhamnitol	40.5	239	+23.1(w)	2,3,4,5,6-penta-O-acetyl-L-glycero-L-galacto-heptononitrile	MeOH	10
	38.6			1,2,3,4-tetra-O-acetyl-L-rhamnopyranose	MeOH	10
	37.5			1,2,3,4-tetra-O-acetyl-α-L-rhamnopyranose	H_2O	24

TABLE III (*Continued*)

Compounds isolated	Yield (%)	M.p. (degrees)	$[\alpha]_D$ (degrees)[a]	Starting material	Solvent	References
N-Acetyl-α-L- rhamnopyranosylamine	14.7	206	−99.0(w)		H₂O	24
2,3,4-tri-O-acetyl-		syrup	−39.9(Chlf)			24
N-Acetyl-α-L- rhamnofuranosylamine	4.9	135	−122.1(w)			24
1,1-Bis(benzamido)-1- deoxy-L-rhamnitol	19	222	+14.4(Py)	1,2,3,4-tetra-O-benzoyl-α-L- rhamnopyranose	MeOH	10
2,3,4,5-tetra-O-benzoyl-		213	−32.1(Chlf)			10
N-Benzoyl-L- rhamnopyranosylamine	1.8	240	+10.6(Py)			10
2,3,4-tri-O-acetyl-		208	+25.1(Chlf)			10
D. Heptoses						
1,1-Bis(acetamido)-1- deoxy-D-*glycero*-D- *gulo*-heptitol	49	194	+7.3(w)	hexa-O-acetyl-D-*glycero*-β- D-*gulo*-heptose	MeOH	27
2,3,4,5,6,7-hexa-O-acetyl-		109	+1.1(Chlf)			27
1,1-Bis(benzamido-1-deoxy- D-*glycero*-D-*gulo*-heptitol	60.4	209	−11.2(Py)	hexa-O-benzoyl-D-*glycero*- β-D-*gulo*-heptose	MeOH	27
2,3,4,5,6,7-hexa-O-acetyl-		172–173	−17.0(Chlf)			27
1,1-Bis(acetamido)-1- deoxy-D-*glycero*-L- *manno*-heptitol	49.8	213	+10.2(w)	hexa-O-acetyl-D-*glycero*-α- L-*manno*-heptose	MeOH	27
2,3,4,5,6,7-hexa-O-acetyl-		170	+15.8(Chlf)			27
1,1-Bis(benzamido)-1- deoxy-D-*glycero*-L- *manno*-heptitol	34.5	211	−8.7(Py)	hexa-O-benzoyl-D-*glycero*- L-*manno*-heptose	MeOH	27
2,3,4,5,6,7-hexa-O-acetyl-		117	+4.7(Chlf)			27
N-Benzoyl-D-*glycero*-L- *manno*-heptosylamine	2.5	211	−103(w)			27
penta-O-acetyl-		169	−28.7(Chlf)			27
1,1-Bis(acetamido)-1- deoxy-D-*glycero*-D- *galacto*-heptitol	43.3	226	+11.0(w)	hexa-O-acetyl-D-*glycero*-α- D-*galacto*-heptose	MeOH	27
2,3,4,5,6,7-hexa-O-acetyl-		185	−24.6(Chlf)			27
1,1-Bis(benzamido)-1- deoxy-D-*glycero*-D- *galacto*-heptitol	42.5	225	−1.3(Py)	hexa-O-benzoyl-D-*glycero*- α-D-*galacto*-heptose	MeOH	27
2,3,4,5,6,7-hexa-O-acetyl-		119	−49.0(Chlf)			27
E. Ketoses						
2-(L-*xylo*-Tetrahydroxy- butyl)-6-(L-*threo*-2,3,4- trihydroxybutyl)- pyrazine (**34**)	11	122	−93.0(w)	1,3,4,5,6-penta-O-acetyl- *keto*-L-sorbose		36
hepta-O-acetyl-		syrup	−51.7(Chlf)			36
2,5-Bis(L-*xylo*-tetrahydroxy- butyl)pyrazine (**35**)		175	−27.0(w)			36
2,5-Bis(D-*arabino*-tetrahy- droxybutyl)pyrazine (**43**)	3.6	236	−80.8(w)	1,3,4,5-tetra-O-benzoyl-β- D-fructopyranose or 1,3,4,5,6-penta-O-acetyl- *keto*-D-fructose		37
2-(D-*arabino*-Tetrahydroxy- butyl)-5-(D-*erythro*- 2,3,4-trihydroxybutyl)- pyrazine (**44**)	2.2	162	−80.0(w)			37

[a] Rotation solvent given in parentheses: Chlf, chloroform; Py, pyridine; and w, water.

TABLE IV

Derivatives from the Reaction of Ammonia with Acylated Disaccharides

Compounds isolated	Yield (%)	M.p. (degrees)	$[\alpha]_D$ (degrees)[a]	Starting material	Solvent	References
Cellobiose	89.3	234	+26.7(w)	octa-O-acetyl-α-cellobiose	MeOH	44
1,1-Bis(acetamido)-1-deoxy-4-O-β-D-glucopyranosyl-D-glucitol	3.7	113	−23.3(w)			44
2,3,5,6,2′,3′,4′,6′-octa-O-acetyl-		195	+6.6(Chlf)			44
N-Acetyl-2,3,6,2′,3′,4′,6′-hepta-O-acetyl-4-O-β-D-glucopyranosyl-α-D-glucopyranosylamine		242	+54.0(Chlf)			44
Lactose	71	222	+52.8(w)	octa-O-acetyl-β-lactose	MeOH	45
1,1-Bis(acetamido)-1-deoxy-4-O-β-D-galactopyranosyl-D-glucitol	4.6	114	−14.8(w)			45
2,3,5,6,2′,3′,4′,6′-octa-O-acetyl-		223	−3.9(Chlf)			45
N-Acetyl-4-O-β-D-galactopyranosyl-α-D-glucopyranosylamine		162	+71.5(w)			45
2,3,6,2′,3′,4′,6′-hepta-O-acetyl-		181	+66.6(Chlf)			45
Maltose		128	+129.6(w)	octa-O-acetyl-β-maltose	MeOH	47
1,1-Bis(acetamido)-1-deoxy-4-O-α-D-glucopyranosyl-D-glucitol	0.8	−	−			47
	27	84	+80.9(w)		H₂O	47
2,3,5,6,2′,3′,4′,6′-octa-O-acetyl-		95	+46.6(Chlf)			47
α-Melibiose	7.7	176	+144.3(w)	octa-O-acetyl-β-melibiose	H₂O	49
N-Acetyl-6-O-α-D-galactopyranosyl-β-D-glucofuranosylamine	22.5	219	+66.6(w)			49
2,3,5,2′,3′,4′,6′-hepta-O-acetyl-		159	+32.0(Chlf)			49
1,1-Bis(acetamido)-1-deoxy-6-O-α-D-galactopyranosyl-D-glucitol	4.7	syrup	+52.6(w)			49
2,3,4,5,2′,3′,4′,6′-octa-O-acetyl-		200	+64.6(Chlf)			49
Cellobiose	49.6	230	+32.0(w)	octa-O-benzoyl-β-cellobiose	MeOH	50
6-O-benzoyl-	1.5	syrup	+44.0(w)			50
1,1-Bis(benzamido)-1-deoxy-4-O-β-D-glucopyranosyl-D-glucitol	7.8	148	−33.2(w)			50
2,3,5,6,2′,3′,4′,6′-octa-O-acetyl-		116	−21.1(Chlf)			50

TABLE IV (Continued)

Compounds isolated	Yield (%)	M.p. (degrees)	$[\alpha]_D$ (degrees)[a]	Starting material	Sol-vent	Refer-ences
N-Benzoyl-4-O-β-D-gluco-pyranosyl-D-gluco-pyranosylamine	0.92	syrup	+33.8(Py)			50
2,3,6,2′,3′,4′,6′-hepta-O-acetyl-		104	+37.2(Chlf)			50
Maltose	37.8	124	+131(w)	octa-O-benzoyl-β-maltose	MeOH	51
6-O-benzoyl-	31.6	140	+105.8(w)			51
1,1-Bis(benzamido)-1-deoxy-4-O-α-D-glucopyranosyl-D-glucitol	20.8	225	+42.6(Py)			51
2,3,5,6,2′,3′,4′,6′-octa-O-acetyl-		130	+31.9(Chlf)			51
N-Benzoyl-4-O-α-D-gluco-pyranosyl-D-glucopyranosylamine	0.1	syrup	–			51
2,3,6,2′,3′,4′,6′-hepta-O-acetyl-		syrup	+115.8(Chlf)			51
Lactose	82	220	+58.9(w)	octa-O-benzoyl-lactose	MeOH	54
1,1-Bis(benzamido)-1-deoxy-4-O-β-D-galactopyranosyl-D-glucitol	6.7	196	−20.9(w)			54
2,3,5,6,2′,3′,4′,6′-octa-O-acetyl-		178	−12.5(Chlf)			54
N-Benzoyl-4-O-β-D-galacto-pyranosyl-D-glucopyranosylamine	0.74	syrup	+63.3(w)			54
2,3,6,2′,3′,4′,6′-hepta-O-acetyl-		108	+47(Chlf)			54
Lactose	75.1	219	+55.4(w)	1,2,6,2′,3′,4′,6′-hepta-O-benzoyl-β-lactose	MeOH	53
6-O-benzoyl-	21.5	–	+55.6 (MeOH)			53
1,1-Bis(acetamido)-1-deoxy-3-O-β-D-glucopyranosyl-D-arabinitol	24.2	214	+25.2(w)	octa-O-acetyl-cellobiono-nitrile	H_2O	55
2,4,5,2′,3′,4′,5′-hepta-O-acetyl-		74	+55.3(Chlf)			55
N-Acetyl-3-O-β-D-gluco-pyranosyl-D-arabino-furanosylamine	4.2	130	+72.7(w)			55
2,5,2′,3′,4′,5′-hexa-O-acetyl-		124	−3.9(Chlf)			55
3-O-β-D-Glucopyranosyl-D-arabinose	0.38	143	+37.4(w)			55
1,1-Bis(acetamido)-1-deoxy-3-O-β-D-galactopyranosyl-D-arabinitol	17.6	185	+49.7(w)	octa-O-acetyl-lactobiono-nitrile	H_2O	56
2,4,5,2′,3′,4′,5′-hepta-O-acetyl-		70	+42.9(Chlf)			56

(Continued)

TABLE IV (*Continued*)

Compounds isolated	Yield (%)	M.p. (degrees)	$[\alpha]_D$ (degrees)[a]	Starting material	Sol-vent	Refer-ences
N-Acetyl-3-O-β-D-galacto-pyranosyl-D-arabino-furanosylamine	4.4	174	+26(w)			56
2,5,2′,3′,4′,5′-hexa-O-acetyl-		54	−11.6(Chlf)			56
3-O-β-D-Galactopyranosyl-D-arabinose	1.9	155	+5.3(w)			56

[a] Rotation solvent given in parentheses: Chlf, chloroform; Py, pyridine; and w, water.

CHEMISTRY AND BIOCHEMISTRY OF APIOSE*

By Ronald R. Watson† and Neil S. Orenstein‡

Department of Microbiology, University of Mississippi Medical Center, Jackson, Mississippi 39216; Department of Microbiology, Harvard School of Public Health, Boston, Massachusetts 02115

I. Introduction

Chemists have long held intense interest in compounds whose structure appears to differ radically from the norm. Thus, the isolation and partial characterization, 73 years ago, of a branched-chain sugar, apiose, sparked enthusiasm to study this variation of the more common, straight-chain sugars.[1] Subsequently, a second branched-

---

* We thank the Department of Microbiology, Harvard School of Public Health, for post-doctoral support during part of the writing of this article.

† Present address: Department of Microbiology, Indiana University School of Medicine, 1100 West Michigan St., Indianapolis, Indiana 46202.

‡ Present address: Department of Pathology, Massachusetts General Hospital, 2 Hawthorne Place, Boston, Mass. 02114.

(1) C. S. Hudson, *Advan. Carbohyd. Chem.*, 4, 57–74 (1949).

chain sugar, D-hamamelose [2-C-(hydroxymethyl)-D-ribose] was isolated from *Hamamelis virginiana* (witch hazel).[2,3] These two compounds were the only known examples of branched-chain sugars
until the advent of the age of antibiotics; since then, a number of
branched-chain carbohydrates have been isolated from bacterial and
fungal products.

Fifty years after the discovery of apiose, a review of the literature[1]
showed that it was still a curiosity, and had been found only in the
parsley plant. Apart from its structure, and those of the flavonoids
from which it was isolated (primarily, apiin), little else was known
about apiose. In the next two decades, this branched-chain sugar was
shown to have very widespread occurrence in the plant kingdom,
being identified, in most of the plants investigated,[4-6] by paper
chromatography.

Its *in vivo* and *in vitro* biosynthetic pathways have been partly
elucidated, and an increased understanding of its structure has been
obtained.[7] The chemical synthesis of D-apiose and some of its derivatives has been more completely defined and investigated.[7-9] However, many areas of knowledge concerning it lie untapped and unin-
vestigated; these include (*a*) the degradation and metabolism of
apiose, (*b*) the control of the biosynthesis of apiose, (*c*) the role of
apiose-containing glycosides, (*d*) the confirmation of the special advantages conferred on plant cell-wall polysaccharides by the presence of residues of a branched-chain sugar, and (*e*) the possible role
of apiose in promoting winter hardiness, suggested by its occurrence
in winter-hardy plants, namely, duckweed, celery, and parsley.[9a]

It is the purpose of this article to bring together the current knowledge concerning apiose, and to stimulate further investigation of this
widely distributed and unique sugar, as much by mentioning what is
not known as by recording what is known.

(2) H. Grisebach, "Biosynthetic Patterns in Microorganisms and Higher Plants,"
    Wiley, New York, 1967, pp. 84–101.
(3) F. Shafizadeh, *Advan. Carbohyd. Chem.* 11, 263–283 (1956).
(4) R. B. Duff, *Biochem. J.,* 94, 768–772 (1965).
(5) R. B. Duff and A. H. Knight, *Biochem. J.,* 83, 33P–34P (1963).
(6) R. B. Duff, *Biochem. J.,* unpublished observations.
(7) P. K. Kindel and R. R. Watson, *Biochem. J.,* 133, 227–241 (1973).
(8) J. M. J. Tronchet and J. Tronchet, *Helv. Chim. Acta,* 53, 1174–1180 (1970).
(9) D. H. Ball, F. H. Bissett, I. L. Klundt, and L. Long, Jr., *Carbohyd. Res.,* 17,
    165–174 (1971).
(9a) R. D. Watson, unpublished observations.

II. Nomenclature and Structure of D-Apiose

At the turn of the century, the first branched-chain sugar, apiose, was isolated.[10–12] Its structure was largely established by Vongerichten.[10–14] From the analysis of its substituted osazones, he concluded that a pentose of an uncommon type was present, as treatment with acids yielded no 2-furaldehyde.[11–14] The elucidation of the structure of apiose was primarily achieved by oxidation with bromine water, leading to the formation of apionic acid.[12] Subsequent reduction with hydriodic acid yielded an acid considered to be isovaleric acid;[12] this conclusion was later fully confirmed, and the D configuration at C-2 was clearly established.[15,16] The structure and configuration of D-apiose have been determined[17] to be as depicted in formula 1, and have been confirmed by the synthesis of D-apiose[18] and L-apiose.[19] Additional proof of the configuration was derived from the synthesis of D-apiose from *keto*-D-fructose pentaacetate; this study provided corollary evidence as to the D configuration of the naturally occurring form of apiose in the apiin isolated from parsley.[20] Extensive characterization has shown that acyclic D-apiose is 3-C-(hydroxymethyl)-D-*glycero*-aldotetrose[21] (1).

In the furanose forms, the four carbon atoms involved in the ring are numbered 1 to 4, starting with the anomeric carbon atom.[3,22] The symbol D or L now distinguishes between the two furanoses derived from one apiose, the respective symbol referring to the reference (highest-numbered), asymmetric carbon atom (C-3) of the cyclic forms, as depicted in formulas 2 and 3. After cyclization, the ring-

(10) E. Vongerichten, *Ber.*, **33**, 2334–2342 (1900).

(11) E. Vongerichten, *Ann.*, **318**, 121–136 (1901).

(12) E. Vongerichten, *Ann.*, **321**, 71–83 (1902).

(13) E. Vongerichten, *Ber.* **33**, 2904–2909 (1900).

(14) E. Vongerichten and F. Müller, *Ber.*, **39**, 235–240 (1906).

(15) S. K. Chakraborti, *Trans. Bose Res. Inst.* (Calcutta), **21**, 61–66 (1956–58).

(16) R. K. Hulyalkar, J. K. N. Jones, and M. B. Perry, *Can. J. Chem.*, **43**, 2085–2091 (1965).

(17) R. A. Raphael and C. M. Roxburgh, *J. Chem. Soc.*, 3405–3408 (1955).

(18) P. A. J. Gorin and A. S. Perlin, *Can. J. Chem.*, **36**, 480–485 (1958).

(19) F. Weygand and R. Schmiechen, *Chem. Ber.*, **92**, 535–540 (1959).

(20) A. Khalique, *J. Chem. Soc.*, 2515–2516 (1962).

(21) R. Hemming and W. D. Ollis, *Chem. Ind.* (London), 85–86 (1953).

(22) L. Hough and A. C. Richardson, in "Chemistry of Carbon Compounds," E. H. Rodd, ed., American Elsevier Co., New York, 2nd Edition, 1968, Vol. 1F, pp. 529–530.

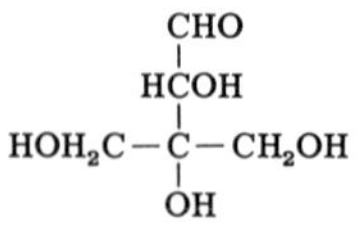
CHO
HCOH
HOH$_2$C—C—CH$_2$OH
OH

D-Apiose
[3-*C*-(Hydroxymethyl)-
D-*glycero*-aldotetrose]
1

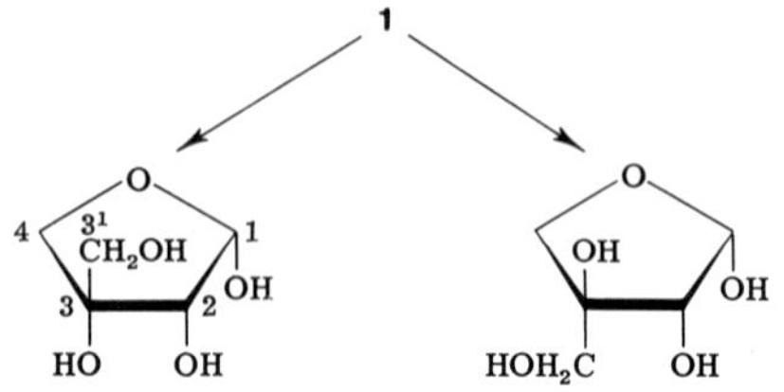
O
4
3^1
CH$_2$OH
1
OH
3
2
HO
OH
O
OH
OH
HOH$_2$C
OH

3-*C*-(Hydroxymethyl)-
α-D-erythrofuranose
2

3-*C*-(Hydroxymethyl)-
β-L-threofuranose
3

carbon atom in the hydroxymethyl group attached to C-3 of the tetrose is numbered [10–21] 3^1 (see formula **2**).

III. Occurrence of Apiose in Nature

The chemistry of D-apiose (**1**) and apiin (**4**), from which this

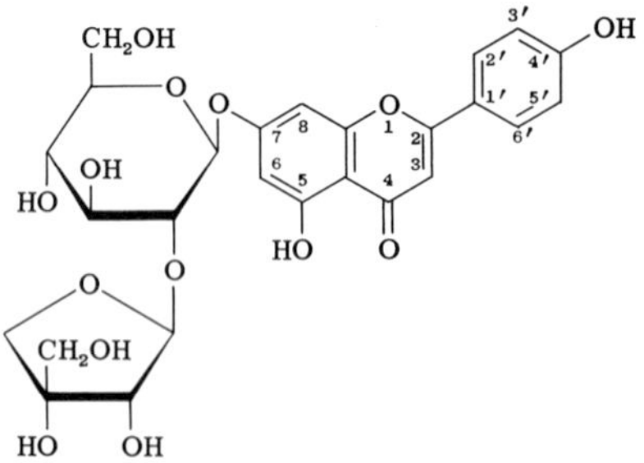
CH$_2$OH
O
O
OH
HO
O
O
CH$_2$OH
HO
OH
7
8
6
5
O
1
2
3
4
HO
O
3′
OH
2′
4′
1′
5′
6′

Apiin
4

branched-chain sugar was originally isolated, was reviewed half a century after its discovery.[1] During those early years, apiose was considered to be a very rare sugar and a curiosity. However, research published in the last 20 years indicates that apiose is not rare; it occurs in a wide variety of plants. More than 50 years after D-apiose had been isolated from parsley, a second plant source of apiose was discovered.[23] The sugar was isolated from extracts of the bark of *Hevea brasiliensis* (rubber tree) and was identified as an apiose by paper chromatography and the properties of its phenylosazone.[23] Whether apiose was a cell-wall polysaccharide, a simple glycoside, a constituent of some other compound of low molecular weight, or a free monomer was not clear. Apiose has been also found in cell-wall polysaccharides, and here its concentration seems more stable than when it is a constituent of a flavonoid glycoside, as evidenced by the uniformly high levels of it recorded for a group of aquatic monocotyledons.[24] Duff[4-6] clearly showed the widespread occurrence of apiose in the plant kingdom. He investigated hydrolyzates from 175 plants belonging to many families, and found apiose in about 60% of the plants. Unfortunately, the identity of these plants was not published. Apiose was detected by hydrolysis of the dried plant-material, with subsequent paper chromatography of the neutralized hydrolyzate.[4-6] No taxonomic pattern was discerned in the plants giving a negative test for apiose,[4] and so, a random sample of these plants was re-examined more carefully; all were found to contain traces of the sugar.[4] This result encouraged the suggestion that all of the plants that did not originally give a positive test for apiose did, in fact, actually contain the branched-chain sugar.

As a starting point for future investigations, most of the plants that Duff[6] tentatively identified as containing apiose are listed from his ongoing research at the time of his death (with the permission of the Macaulay Institute and J. S. D. Bacon).

The following plants were found to contain traces, or marginal amounts, of apiose: *Acer pseudo-plantanus* L. (false plane-tree, sycamore maple), *Angelica sylvestris* L. (angelica), *Aponogeton distachyos*, *Arctostaphylos* (berryberry), *Arrhenatherum elatius* L. (oat grass), *Aucuba japonica* (dogwood), *Berberis vulgaris* L. (common barberry), *Buddleia procumbens* L. (buddleia, butterfly bush), *Campanula* L. (bellflower), *Carpinus betulus* L. (ironwood or hornbean), *Castanea sativa* (chestnut), *Cerastium tomentosum* L. (snow-in-summer), *Convolvulus minor* (morning glory, bindweed), *Cucumis savitus* L. (cucumber), *Eriophorum vaginatum* (bog cotton), *Eschscholzia* (California poppy), *Fagus sylvatica* (European beech), *Fraxinus excelsior* (European blue ash), *Fraxinus* L. (ash), *Galium* L. (bedstraw, cleavers), *Garrya elliptica* (tasseltree), *Gladiolus* sp., *Hippuris vulgaris* L. (mare's tail), *Hydrangea* L. (hydrangea), *Impatiens* L. (balsam, jewelweed, snapweed, touch-me-not), *Juglans nigra* L. (black walnut), *Lavandula spica* L. (lavender), *Limnantes* sp., *Listera ovata* (an orchid, twayblade), *Lupinus luteus* (lupine), *Magnolia soulangeana* (magnolia), *Mahonia aquifolium*, *Melilotus alba* Desr. (white sweet clover), *Myosotic* L. (scorpiongrass, forget-me-not), *Myrica gale* (bayberry), *Nymphaea alba* L. (waterlily, waternymph), *Orchis ustulata* (orchid), *Oxalis* L. (wood sorrel, lady's sorrel), *Parrotia persica* (witch-hazel family), *Pentstemon* Mitchell (beard-tongue), *Petunia* Juss. (petunia),

(23) A. D. Patrick, *Nature*, **178**, 216 (1956).

(24) C. F. Van Beusekom, *Phytochemistry*, **6**, 573–576 (1967).

Plantago coronopus L. (plantain, ribwort), *Plantago lanceolata* (narrow-leafed plantain), *Polemonium caeruleum* L. (greek valerian, jacob's ladder), *Polygonum bistorta* (knotweed), *Populus* L. (poplar, aspen), *Poterium obtusum, Quercus* L. (oak), *Rhus continus* (snije-tree), *Ribes nigrum* L. (black currant), *Salix daphnoides* (willow), *Succisa pratensis* Moench (devil's bit), *Symphoricarpos albus* (snowberry), *Symphytum officinale* L. (common comfrey), *Syringa reflexa* (lilac; not the common one), *Taraxacum officinale* Weber (officinale dandelion, blowballs), *Tradescantia* L. (spiderwort), *Trichophorum* sp., *Trifolium* (subterranean clover), *Typha latifolia* L. (common cattail), and *Verbena* L. (vervain).

The following plants contained *moderate* proportions of apiose, compared to the foregoing: *Amaranthus* L. (amaranth, red root, pigweed), *Astrantia helleborifolia* (masterwort), *Atriplex rubrum* (shadscale or greasewood), *Callitriche* L. (water starwort, water chickweed), *Cistus laurifolius* (rockrose), *Epilobium pedunculare* (willowherb), *Geranium molle* L. (canesbill), *Helianthemum chamaecistus* (sunrose), *Heuchera* sp. (alumroot), *Hippophae rhamnoides* (sea buckthorn), *Laurus rotundifolia, Lonicea nitida* (honeysuckle), *Myriophyllum* L. (water-milfoil), *Philadelphus* L. (mockorange or syringa), *Saxifraga umbrosa* (saxifrage), *Stachyslanata* (woolly woundwort, hedgenettle), *Tilia vulgaris* (lindess, basswood), *Trifolium pratinse* (red clover), *Trifolium repens* L. (white clover), and *Zara lanceolatia.*

The following plants contained *substantial* proportions of apiose, about the same as found in parsley: *Aegopodium pologaria* (bishop's weed), *Apium graveolens* L. (celery), *Conopodium majus* (earthnut), *Datus aucaparia* (rowan), *Desfontania spinosa, Ilex aquifolium* (holly), *Lavatera annual, Lemna wolffia* (duckweed), *Menyanthes trifoliata* (bob bean), and *Vinca minor* (periwinkle).

This widespread occurrence of apiose[5] has not been substantiated by Beck.[24a] In lower plants (1 lichen, 3 mosses, 8 ferns, and 5 Gymnosperms), he detected no apiose. Beck also found no apiose in any of the plants representing families tested by Duff and Knight,[5] except the *Lemnaceae* and *Zosteraceae.* It should be noted that Duff and Knight[5] (also, unpublished data of Duff) did not originally find apiose in all species of any one family in which a member contained apiose, except when some species were re-examined more closely.

1. Characterization of Apiose-containing Compounds of Low Molecular Weight

Other investigators have found apiose in a wide variety of plants (see Table I). Apiose has usually been isolated as glycosides in which it is linked to compounds of low molecular weight. These include only derivatives of flavone, isoflavone, phenol, and anthraquinone (see Table I).

The parsley plant, *Petroselinum crispum,* for example, contains at least two of the D-apiose flavonoids, namely, apiin {4′,5-dihydroxyflavon-7-yl   2-O-[3-C(hydroxymethyl)-β-D-erythrofuranosyl]-β-D-glucopyranoside} (4) and petroselinin {3′,4′,5-trihydroxyflavon-7-yl 2-O-[3-C-(hydroxymethyl)-β-D-erythrofuranosyl]-β-D-glucopyranoside}.[1] The principal compound containing D-apiose in parsley

(24a) E. Beck, personal communication.

TABLE I

D-Apiose-containing Compounds of Low Molecular Weight

Compound	Aglycon		Source (plant)
	(Trivial name)	(Systematic[a] name)	
Apiin	apigenin	4',5,7-trihy-droxyflavone	*Petroselinum crispum* (parsley),[25-27] *Apium graveolens* (celery),[26,27] *Chrysanthemum leucanthemum,*[28] *Chrysanthemum uliginosum,*[28] *Chrysanthemum maximum,*[28] *Cuminum cyminum* (zira),[15] *Bellis perennis* (English daisy),[28] *Anthemis nobilis,*[28] *Chrysanthemum frustescens,*[28] *Matricaria chamomilla,*[28] *Matricaria inodora,*[28] *Centaurea scabiosa,*[28] *Centaurea cyanus,*[28] *Serratula coronata,*[28] *Echinops gmelini,*[28] *Vicia hirsuta,*[29] *Digitalis purpurea,*[30,31] *Crotalaria anagyroides,*[32] *Petroselinum sativum,*[33] *Petroselinum crispum* (parsley)[34]
Petroselinin (Graveolbioside A)	luteolin	3',4',5,7-tetra-hydroxy-flavone	*Petroselinum crispum* (parsley)[25] and *Apium graveolens* (celery)[27,35]
Graveolbioside B	chrysoeriol	4',5,7-trihy-droxy-3'-methoxy-flavone	*Apium graveolens,*[27] *Petroselinum crispum* (parsley),[34] *Luffa echinata*[36]
Kaempferol-3 β-glucoapioside	kaempferol	4',3,5,7-tetra-hydroxy-flavone	*Cicer arietinum* (chick pea)[37]
Lanceolarin	biochanin-A	5,7-dihydroxy-4'-methoxy-isoflavone	*Dalbergia lanceolaria*[38]
2-O-D-Apiosyl-D-glucose dibenzoate	none	none	*Daviesia latifolia*[39]
Furcatin	none	*p*-vinylphenol	*Viburnum furcatum*[40]
Frangulin B	none	1,6,8-trihy-droxy-3-methyl-anthraqui-none	*Rhamnus frangula*[41]

[a] Flavone = 2-phenylbenzopyrone; isoflavone = 3-phenylbenzopyrone.

(25) C. G. Norström, T. Swain, and A. J. Hamblin, *Chem. Ind.* (London), 85 (1953).
(26) M. O. Farooq, I. P. Varshney, and W. Rahman, *Naturwissenschaften,* **44,** 444 (1957).
(27) M. O. Farooq, I. P. Varshney, and W. Rahman, *Naturwissenschaften,* **45,** 265 (1958).

(*Continued on page 142*)

seeds, stems, and leaves is apiin.[25,42] The results obtained by several workers led to determination of the structure of apiin (4). After methylation of apiin, two methylated sugars were isolated by hydrolysis: a tri-O-methyl derivative of D-apiose and 3,4,6-tri-O-methyl-D-glucose.[21] The linkage between the two sugars is, therefore, from C-1 of D-apiose to O-2 of the D-glucosyl residue. Then, the stereochemistry of the carbon atoms in the D-apiose moiety was investigated by methylation of apiin, hydrolysis, and isolation of the resulting tri-O-methyl-D-apiose;[16] this was shown to be identical, by optical rotation, paper, thin-layer, and gas–liquid chromatography, and infrared spectroscopy, to the chemically synthesized 2,3,3^1-tri-O-methyl-3-C-(hydroxymethyl)-D-erythrofuranose.[16] Therefore, 3-C-(hydroxymethyl)-D-erythrofuranose is the naturally occurring form of apiose, at least in parsley. Periodate oxidation indicated that the hydroxyl groups on C-2 and C-3 are in the *cis* relationship. From a knowledge of the molecular rotation of 4′,5-dihydroxyflavon-7-yl β-D-glucopyranoside and of apiin, and by application of the molecular rotation theory of Klyne,[43] it was concluded[16] that the sterochemistry of the linkage between C-1 of D-apiose and O-2 of D-glucose is β-D.

(28) H. Wagner and W. Kirmayer, *Naturwissenschaften*, **44**, 307–308 (1957).

(29) T. Nakaoki, N. Morita, H. Motosune, A. Hiraki, and T. Takeuchi, *Yakugaku Zasshi*, **75**, 172–177 (1955); *Chem. Abstr.*, **49**, 8562 (1958).

(30) R. R. Paris, M. Paris, and J. Fries, *Ann. Pharm. Fr.*, **24**, 245–249 (1966).

(31) R. R. Watson, Ph.D. Thesis, Michigan State University, East Lansing, Michigan (1971) (available from University Microfilms, Inc., Ann Arbor, Michigan; Order Number 71-31,330).

(32) S. S. Subramanian and S. Nagarajan, *Phytochemistry*, **9**, 2581–2584 (1970).

(33) F. Tomas, J. J. Mataix, and O. Carpena, *Rev. Agroquim. Tecnol. Alimentos*, **12**, 263–268 (1972); *Chem. Abstr.*, **77**, 149,671 (1972).

(34) H. Grisebach and W. Bilhuber, *Z. Naturforsch.*, **22B**, 746–751 (1967).

(35) M. O. Farooq, S. R. Gupta, M. Kiamuddin, W. Rahman, and T. R. Seshadri, *J. Sci. Ind. Res.* (India), **12B**, 400–407 (1953).

(36) T. R. Seshadri and S. Vydeeswaran, *Phytochemistry*, **10**, 667–669 (1971).

(37) W. Hösel and W. Barz, *Phytochemistry*, **9**, 2053–2056 (1970).

(38) A. Malhotra, V. V. S. Marti, and T. R. Seshadri, *Tetrahedron Lett.*, 3191–3196 (1965).

(39) S. Hattori and H. Imaseki, *J. Amer. Chem. Soc.*, **81**, 4424–4427 (1959).

(40) B. Hansson, I. Johansson, and B. Lindberg, *Acta Chem. Scand.*, **20**, 2358–2362 (1966).

(41) H. Wagner and G. Demuth, *Tetrahedron Lett.*, 5013–5014 (1972).

(42) A. U. Rahman, *Z. Naturforsch.*, **13b**, 201–202 (1958).

(43) W. Klyne, *Biochem. J.*, **47**, xli–xlii (1950).

Apiin is, therefore, 4',5-dihydroxyflavon-7-yl 2-O-[3-C-(hydroxy-methyl)-β-D-erythrofuranosyl]-β-D-glucopyranoside.[16,44]

In addition to apiin, there is in parsley one other flavonoid glyco-side (and, perhaps, two) that contains apiose.[25,26,34] The identity of the other glycoside(s) is somewhat unclear. Vongerichten[10-12] ob-tained, from *Petroselinum crispum* leaves, a crude mixture of two glycosides; he demonstrated this by hydrolyzing the mixture and obtaining two aglycons, partially characterized as apigenin (4',5,7-trihydroxyflavone) and chrysoeriol (4',5,7-trihydroxy-3'-methoxy-flavone). Whether or not each was linked to apiose prior to hydrolysis was not conclusively demonstrated.[1] Later, Gupta and Seshadri[45] isolated crystalline apiin from *P. crispum* (moss-curled variety), and obtained only one glycoside containing apiose; it was care-fully characterized, and found to be apiin. It was not purified by paper or other chromatography, which might have shown it to be a mixture.[25] Some of the apiin isolated had a melting point 10° lower than that of authentic apiin, although it was identical in every other respect. Shortly thereafter, Nordström and coworkers[25] isolated, from parsley, flavonoid glycosides containing apiose. The two glycosides were extensively characterized after separation by paper chroma-tography. The aglycon moiety of one of these compounds was iden-tified as apigenin by its rate of migration in paper chromatography in two solvent systems, its color reactions, and its ultraviolet spectrum in alcohol.[25] Methylation of the partially hydrolyzed glycoside, with subsequent complete hydrolysis, yielded a compound spectrally identical to authentic 4',5-di-O-methylapigenin; therefore, the origi-nal compound isolated from parsley was a 7-O-(apiosylglucosyl)api-genin.

The aglycon of the second compound was identified [25] as luteolin (3',4',5,7-tetrahydroxyflavone) in the same way. The product from the partial hydrolysis of the second glycoside was methylated and hydro-lyzed; the aglycon appeared to be luteolin, as it gave a spectrum (in 2 mM sodium ethoxide) indistinguishable from that of 3',4',5-tri-O-methylluteolin. Thus, extensive characterization of both glycosides before and after hydrolysis showed that they were apiin and an apiosyl-D-glucoside of luteolin.[25] In light of these observations, the work of Gupta and Seshadri,[45] who isolated only apiin from parsley,

(44) N. Narasimhachari and T. R. Seshadri, *Proc. Indian Acad. Sci.*, Sect. A, **30**, 151–162 (1949).

(45) S. R. Gupta and T. R. Seshadri, *Proc. Indian Acad. Sci.*, Sect. A, **35**, 242–248 (1952).

was subjected to a preliminary re-investigation.[26] Apiin from Indian parsley was purified and found to have a melting point slightly higher than that reported previously. Paper chromatography suggested that this "apiin" was actually a mixture of three components: the major component appeared to be apiin, and the others did not migrate like graveolbioside A, the 7-(apiosylglucoside) of luteolin, identified by Nordström and coworkers.[25] No evidence was presented as to the nature of the sugars or the aglycon moiety of these compounds, and they may not even have contained apiose.[26]

Grisebach and Bilhuber[34] also partially characterized two glycosides isolated from parsley (P. hortense Hoffman, synonymous with P. crispum). The major component was found to be apiin. The minor component was identified by its hydrolytic and spectral properties as the apiosyl-D-glucoside of chrysoeriol; this result agreed with that of Vongerichten,[10-14] obtained many years before. The apiose-containing flavonoid glycosides in P. crispum (moss-curled variety) are apiin,[25,42,45] a small proportion of an apiosyl-D-glucoside of luteolin,[25] and, perhaps, small proportions of an apiosyl-D-glucoside of chrysoeriol.[34] Both of the two minor constituents may exist in the same sample of parsley, and variations in growth and in isolation techniques may explain the apparent discrepancies discussed.[25,34]

It appears likely that apiin and, perhaps, other flavonoid glycosides exist in the living leaf or tissue in the form of some kind of soluble precursor. Parsley-leaf extracts prepared with boiling water do not gel, and the solutes remain in solution on cooling.[46] The addition of a thermolabile, protein extract from either parsley or chick pea (Cicer arietinum) causes liberation of malonate from the soluble, crude glycoside fractions, and results in the precipitation of 95% of the total flavonoid, primarily[46] apiin and graviolbioside B. In contrast, Spinacea oleracea (spinach) and Phaseolus vulgaris (kidney or haricot bean) do not possess a similar enzyme, nor do their flavonoid glycoside fractions appear to yield malonate in the presence of the pea or parsley enzyme.[46] Flavonoid glycosides containing apiose have been isolated from both parsley[25] and Cicer arietinum.[37] The importance of malonate in solubilizing the flavonoid glycosides is suggested by (a) its presence in the supernatant liquor (containing organic acid) after hydrolysis of the flavonoid glycosides[46] by parsley enzyme, and (b) the isolation, from Trifolium (subterranean clover),[47] of three flavonoid glycosides malonated on O-6 of the D-glucose moiety.

(46) H. E. Davenport and M. S. Dupont, Biochem. J., 129, 18P–19P (1972).
(47) A. B. Beck and J. R. Knox, Aust. J. Chem., 24, 1509–1518 (1971).

A flavonoid extracted from *Digitalis purpurea* (foxglove) was compared to an unnamed "apioside" (glycoapiosylapigenin) by paper chromatography in four different solvent-systems.[30] The foxglove flavonoid was chromatographically identical to the unidentified apioside (which was probably apiin), but it was not further characterized. Later, Watson[31] isolated from foxglove an enzyme that synthesized a compound having the paper-chromatographic mobility of apiin from uridine 5'-([U-^{14}C]apiosyl pyrophosphate) (UDP-[U-^{14}C]apiose) plus 4'5-dihydroxyflavon-7-yl β-D-glucopyranoside; the compound was identified by paper chromatography in several solvent systems. This result further suggested that the compound isolated[30] from foxglove was apiin.[31]

Apiin has also been isolated from the white flowers of *Chrysanthemum uliginosum* (high daisy), *C. leucanthemum* (ox-eye daisy), *C. maximum* (max daisy), *Bellis perennis* (English daisy), and *Anthemis nobilis* (chamomile).[28] The products of acid hydrolysis of the apiin-like material from each plant were apiose and 4',5-dihydroxyflavon-7-yl β-glucopyranoside, identified by paper chromatography.[28] Apiin was also isolated from the flowers of *Chrysanthemum frutescens* (marguerite daisy), *Marticaria chamomilla* (sweet false-chamomile), *M. inodora* (scentless false-chamomile), *Centaurea scabiosa*, *Centaurea cyanus* (cornflower), *Serratula coronata*, and *Echinops gmelini* (globe thistle).[28] In these instances, the presence of more than one flavonoid glycoside containing apiose was not precluded by the results of paper chromatography in solvent systems[25] that separate flavonoid glycosides.

Apiin has also been reported in *Centaurea* by other workers.[48] Apiin may be present in *Vicia hirsuta* (vetch);[29] the yield was 2.5–3.0% of a crystalline flavonoid (identified as apiin by its melting point only). Chakraborti[15] isolated apiin by precipitation with basic lead acetate from an ethanol extract of one of five Indian spices belonging to the family *Umbelliferae*. The flavonoid glycoside was isolated from seeds of *Cuminum cyminum* (commonly called zira), and clearly identified as having the properties of apiin by (*a*) its melting point, (*b*) its rate of migration in paper chromatography, and (*c*) identification of its acid hydrolysis products as apigenin, glucose, and apiose. No apiin was observed in the four other spices.[15] By very similar procedures, apiin was isolated, and identified, as a constitu-

(48) V. A. Bandyukova, E. T. Oganesyan, L. I. Lisevitaskaya, V. I. Sidel'nikova, and A. L. Shinkarenko, *Fenol'nye Soedin. Ikhe Biol. Funkts., Mater. Vses. Simp., 1st, 1966*, 95–100 (1968); *Chem. Abstr.*, **71**, 19518n (1969).

ent of stem bark of *Crotalaria anagyroides* (rattle-box).[32] Very preliminary evidence suggested that apiin may be present in *Petroselinum sativum* (parsley).[33]

Several apiose-containing flavonoids have been reported to be present in *Apium graveolens* (celery).[27,35,42] Originally, general and very early reports had suggested that apiin might be present in celery seeds and celery plants.[49–51] In continuing investigations of the presence of anthoxanthins in common plants used as human food, different proportions of the two principal glycosides containing apiose, namely graveolbiosides A and B, were isolated from two varieties of celery.[35] These compounds were separated from each other by fractional precipitation with neutral and basic lead acetate. Complete methylation of the glycosides, followed by hydrolysis, as well as graded hydrolysis of the unmethylated glycosides, showed both glycosides to have the structure of a 7-(apiosyl-glycosyl)-aglycon.[35] The presence of apiose and glucose was established by their rate of migration in paper chromatography, and preparation of their osazones.[35] The aglycon portion of graveolbioside A was demonstrated to be luteolin, and that of graveolbioside B proved to be chrysoeriol (4′,5,7-trihydroxy-3′-methoxyflavone), the 3′-methyl ether of luteolin (3′,4′,5,7-tetrahydroxyflavone).[35] Very tentative evidence suggested that apiin is present in celery, and it was isolated[27] along with crystalline graveolbioside B. It had a melting range broader than that originally reported. Paper chromatography indicated that graveolbioside B contains two components, one of which migrates like apiin. The other component was identified as the 7-apiosyl-glucoside of chrysoeriol, as its hydrolytic products migrated at the same rates as authentic apiose, chrysoeriol, and glucose on chromatography.[27]

Graveolbioside B has also been isolated from a common, Indian herbal drug made from the plant *Luffa echinata* (loofah, vegetable sponge, or desk-cloth gourd).[36] The flavonoid was obtained from alcohol extracts of the fruit by extraction with light petroleum, ether, and acetone. Hydrolysis of the flavonoid glycoside with acid yielded apiose, glucose, and chrysoeriol. This flavonoid glycoside was shown to be identical to graveolbioside B isolated from celery. Permethylation followed by acid hydrolysis, and the multitude of tests applied

(49) H. Braconnot, *Ann. Chim. Phys.*, [3] **9**, 250 (1843).
(50) O. T. Schmidt, *Pharm. Chem.*, **2**, 2107 (1923).
(51) C. Wehmer, *Die Pflanzenstoffe*, **2**, 874 (1931).

to identify the graveolbioside B from celery, confirmed that the glycoside is 4',5-dihydroxy-3'-methoxyflavon-7-yl 2-O-D-apiosyl-β-D-glucopyranoside (common name,[36] graveolbioside B).

In addition to the flavonoids of parsley, and the apiin contained in other plants, an apiose-containing flavonoid glucoside has been found in *Dalbergia lanceolaria*.[38] The compound, lanceolarin (an apiosylglucoside of biochanin-A), occurs in the root-bark of this tree. Lanceolarin was isolated as a crystalline compound from the acetone extract of the root-bark by extraction with hot petroleum ether and ether. Quantitative hydrolysis of lanceolarin with acid demonstrated that it contained apiose, glucose, and biochanin-A in equimolar proportions. Glucose and apiose were only tentatively identified, their migration in paper and thin-layer chromatography being the sole criteria for identification.[38] Biochanin-A, on the other hand, was rigorously identified as the aglycon by (*a*) its mixed melting-point, (*b*) its color reactions, (*c*) its distinctive migration in paper and thin-layer chromatography, and (*d*) the preparation of two derivatives. Partial hydrolysis of lanceolarin with acid yielded a glucoside of biochanin-A that had the properties of sissotrin, the 7-glucoside of biochanin-A. Methylation of the glucoside of biochanin-A followed by hydrolysis showed[38] that it, like sissotrin, had a glucosyl group linked to the 7-hydroxyl group of biochanin-A. The structure of lanceolarin is 5-hydroxy-4'-methoxyisoflavon-7-yl O-apiosylglucopyranoside.[38]

Another apiose-containing compound was isolated from the aerial parts of *Cicer arietinum* (chick pea).[37] After extensive purification, the compound was found to afford, on hydrolysis, kaempferol, glucose, and apiose in the ratios of 1:0.87:0.90. Selective hydrolysis with acid resulted in liberation of apiose and kaempferol-3 D-glucoside; the O-3 linkage was suggested by u.v. spectrophotometric and paper-chromatographic data. Without supporting data, the authors proposed that the slow hydrolysis of the flavonoid glucoapioside with β-D-glucosidase suggests that the apiose is bound to O-2 of the glucose.[37] Kaempferol-3 β-glucoapioside is a new type of glycoside, as all apiose-containing flavonoid glycosides previously identified are linked through O-7 of the aglycon (see Table I, p. 141). Apiose, not L-rhamnose, appears to be the sugar constituent of frangulin B isolated from the bark of *Rhamnus frangula* (alder buckthorn).[41] The structure suggested for frangulin B is 6-(apio-furanosyloxy)-1,8-dihydroxy-3-methylanthraquinone.[41]

The role of the flavonoid glycosides of apiose in the plant from which they are isolated is unknown.[31] They may stimulate or retard growth, function as a plant hormone, or act as some nonfunctional end-product. Apiin has been reported to stimulate the growth of some young plants.[52] Apiin prevents the action of sodium taurocholate during early growth, although sodium taurocholate inhibits growth, even in the presence of apiin, 12 days after germination of the grain seeds.[52]

D-Apiose has been identified as being linked to D-glucose glycosidically bonded to a nonflavonoid aglycon in furcatin, a compound isolated from the leaves of *Viburnum furcatum*.[39] Furcatin was hydrolyzed under mildly acidic conditions, and the liberated sugar, apiose, was identified by paper chromatography and by the formation of its *p*-bromophenylosazone. The remaining glucoside was hydrolyzed by a β-D-glucosidase, indicating that the glucose is D-glucose and the linkage is β-D. Glucose was identified by paper chromatography and the formation of its phenylosazone. As furcatin consumes four moles of periodate per mole, and produces two moles of acid, the linkage[39] between D-apiose and D-glucose is (1 → 6). The D configuration of apiose in this disaccharide is based upon the work of Gorin and Perlin,[18] but the structure should be accepted with caution, as little is known about periodate oxidation of oligosaccharides containing nonreducing apiosyl end-groups, and because Hattori and Imaseki[39] only estimated the total acid produced. The authors[39] concluded that the aglycon was *p*-vinylphenol, even though they did not crystallize and rigorously identify it. Identification was based upon a variety of features that were similar to those reported for authentic *p*-vinylphenol.[39] Thus, furcatin appears to be *p*-vinylphenyl 6-O-D-apiofuranosyl-β-D-glucopyranoside. Distribution of furcatin appears to be very limited, and it was not detected by paper chromatography in eleven other species of *Viburnum* tested.[39] Later, there was isolated from *V. furcatum* a glucosidase that hydrolyzes furcatin into the aglycon (*p*-vinylphenol) and 6-O-apiofuranosyl-β-D-glucose.[53]

A free disaccharide containing apiose has been extracted from an Australian shrub, the native hop bush, *Daviesia latifolia*.[40] The disaccharide was isolated as a dibenzoate. Methylation followed by debenzoylation, as well as comparison with the disaccharide of D-apiose from apiin, suggested that they are similar in every aspect measured. Thus, the structure of the disaccharide without its two benzoyl groups is, tentatively, 2-O-D-apiofuranosyl-D-glucose.[40] The

(52) G. Di Maggio, *Sperimentale, Sez. Chim. Biol.*, **4**, 66–68 (1953).
(53) H. Imaseki and T. Yamamoto, *Arch. Biochem. Biophys.*, **92**, 467–474 (1961).

very mild conditions used in the extraction of this disaccharide into ether and water suggest that it is not a degradation product of a polysaccharide, as is the apiobiose disaccharide produced by the very mildly acidic conditions used in extraction of *L. minor*.[54]

Deoxypentoses occur commonly in Nature, and, at one time, it was supposed that cordycepose was a branched-chain deoxypentose.[17,55,56] Cordycepose (3-deoxy-D-*erythro*-pentose) is obtained by acid hydrolysis of a crystalline antibiotic, cordycepin, isolated from the culture fluids of *Cordyceps militaris*.[55] Subsequently, from the fermentation broth of *Aspergillis nidulans*, a deoxyadenosine was isolated whose infrared and nuclear magnetic resonance spectra were identical with the spectra of (*a*) the deoxyadenosine isolated from *Aspergillus nidulans*, (*b*) synthetic 3'-deoxyadenosine, and (*c*) two samples of cordycepin. Additional proof of the 3'-deoxyadenosine structure of cordycepin was supplied by the chemical and biosynthetic studies reported by Suhadolnik and coworkers.[57,58] Finally, mass-spectrometric evidence established the straight-chain, 3-deoxypentose structure of the sugar component of cordycepin.[59] The mass spectrum of 3'-deoxyadenosine is identical to that of cordycepin, showing that natural cordycepose is 3-deoxy-D-*erythro*-pentose (not 3-deoxyapiose)[59]

Raphael and Roxburgh[17] attempted to ascertain if cordycepose is a branched-chain sugar, and they synthesized 3-deoxyapiose. This branched-chain sugar synthesized[17] by them was incorrectly given the name cordycepose, and, although it has been suggested[60] that confusion in the chemical literature be avoided by abandoning the trivial name "cordycepose," the name "3-deoxyapiose" should be used for the branched-chain sugar synthesized by Raphael and Roxburgh;[17] this permits continued use of the trivial name "cordycepin" for the nucleoside antibiotic elaborated by *C. militaris* and *A. ni-*

(54) D. A. Hart and P. K. Kindel, *Biochemistry*, **9**, 2190–2196 (1970).

(55) H. R. Bentley, K. G. Cunningham, and F. S. Spring, *J. Chem. Soc.*, 2301–2305 (1951).

(56) E. A. Kaczka, N. R. Trenner, B. Arison, R. W. Walker, and K. A. Folkers, *Biochem. Biophys. Res. Commun.*, **14**, 456 (1964).

(57) R. J. Suhadolnik and J. G. Cory, *Biochim. Biophys. Acta*, **91**, 661–662 (1964).

(58) R. J. Suhadolnik, G. Weinbaum, and H. P. Meloche, *J. Amer. Chem. Soc.*, **86**, 948–949 (1964).

(59) S. Hanessian, D. C. DeJongh, and J. A. McCloskey, *Biochim. Biophys. Acta*, **47**, 480–482 (1966).

(60) J. J. Fox, K. A. Watanabe, and A. Bloch, *Progr. Nucleic Acid Res. Mol. Biol.*, **5**, 251–313 (1966).

dulans, and of "cordycepose" for 3-deoxy-D-*erythro*-pentose, the sugar moiety of cordycepin.

2. Polysaccharides Containing Apiose

The first report, by Bell and coworkers[61], identifying D-apiose as a component of a molecule other than a flavonoid glycoside was important for three reasons. (*1*) It was the first time that D-apiose had been described as a component of a "polysaccharide," (*2*) the useful and widely emulated preparation of the di-*O*-isopropylidene derivative was detailed, and (*3*) a theory equating the recalcitrance of *Posidonia australis* and *Zostera marina* to microbial decay with the presence of "some derivative of D-apiose" was proposed. Subsequently, the presence of apiose in the polysaccharides of a large number of plants has been reported.

In *Posidonia australis* (a seagrass),[5,61] *Tilia* sp., [62] *Zostera marina* (eel grass),[5,62–64] *Lemna gibba* (duckweed),[19] *L. minor* (duckweed),[5,6,65–69] *Z. nana* (a seagrass),[64] *Z. pacifica* (a seagrass),[64] *Phyllospadix* (a seagrass),[64] and *Taraxacum kok-saghyz* (a dandelion),[70] apiose appears to be primarily a component of polysaccharides. The proportions of total apiose vary both between and within species. For instance, *L. minor* grown in pond water contained 6 to 7 per cent of apiose, whereas that grown in nutrient solution contained[5] only about 4 per cent. The following species contain apiose: *Thalassioideae* (*Thalassia hemprichii*, a seagrass), *Potamogetoneae* [*Potamogeton pectinatus* (pond weed) and *Ruppia spiralis* (widgeon grass)], *Cymodoceae* (*Cymodocea nodosa*), *Posidonieae* (*Posidonia oceanica*, a seagrass), and *Lemnaceae* (*Spirodella polyrhiza*, great duckweed). The conditions of extraction from dried leaves, namely, hydrolysis in 0.2 *M* sulfuric acid for 1.5 h at 100°, allow no conclusions to be drawn regarding the presence of apiose in polysaccharides.[24] Other species of *Potomogetoneae*, as well as other

(61) D. J. Bell, F. A. Isherwood, and N. E. Hardwick, *J. Chem. Soc.*, 3702–3704 (1954).
(62) J. S. D. Bacon, *Biochem. J.*, **89**, 103P–104P (1963).
(63) D. T. Williams and J. K. N. Jones, *Can. J. Chem.*, **42**, 69–72 (1964).
(64) R. G. Ovodova, V. E. Vaskovsky, and Yu. S. Ovodov, *Carbohyd. Res.*, **6**, 328–332 (1968).
(65) E. Beck and O. Kandler, *Z. Naturforsch.*, **20B**, 62–67 (1965).
(66) J. Mendicino and J. M. Picken, *J. Biol. Chem.*, **240**, 2797–2805 (1965).
(67) E. Beck, *Ber. Deut. Bot. Ges.*, **77**, 396–397 (1966).
(65) E. Beck and O. Kandler, *Z. Naturforsch.*, **20B**, 62–67 (1965).
(69) D. A. Hart and P. K. Kindel, *Biochem. J.*, **116**, 569–579 (1970).
(70) J. Chrastil, *Chem. Listy*, **50**, 163–164 (1956).

plants tested, did not contain the sugar. Duff and Knight[5] assayed 175 plants (representing about 100 families of higher plants) for apiose. The authors, without direct evidence, stated[5] that it "seems likely that the sugar is a constituent unit of a polysaccharide in the high yielding plants." In a number (unspecified) of plants containing apiin, the cell-wall polysaccharide was found to contain no apiose.[24a]

Polysaccharide fractions containing apiose have been isolated from *Z. marina*,[62-64,71] *Tilia* sp., [62] *L. minor*, [6,66-69] *P. australis*,[63] *Z. pacifica*,[64] and *Phyllospadix*.[64] Zosterine, a pectic polysaccharide containing D-apiose, was isolated from three species of *Zosteraceae*: *Z. marina*, *Z. pacifica*, and *Phyllospadix*.[64] Air-dried, powdered plants were extracted with methanol, successively treated with 1% aqueous formaldehyde (12 to 16 h at 20°) and dilute hydrochloric acid (pH 4–5 for 3–4 h at 50°), and washed with cold water. Zosterine was then obtained by extracting the solid three times with 1% aqueous ammonium oxalate (3–5 h at 70°), and was precipitated from the extract with methanol. Precipitation, dialysis against water, and lyophilization yielded a remarkably homogeneous product having a molecular weight of 40,000–45,000. Hydrolysis of zosterine with sulfuric acid for 16 h at 75° yielded apiose, arabinose, galactose, galacturonic acid, xylose, and traces of an O-methylxylose.[64] Partial hydrolysis with acid afforded a galacturonan [poly(galacturonic acid)], suggesting the pectic nature of zosterine.[64] Digestion of zosterine with pectinase released an apiogalacturonan containing a large proportion of D-apiose.[71] Periodate oxidation and methylation studies indicated a linear, $(1 \rightarrow 4)$-linked D-galacturonan backbone, with apiose on O-2, O-3, or both. A 1–2% aqueous solution of zosterine (pH 2.0) at 20° gels in 24 h.[72] Apiogalacturonan, isolated from purified zosterine, gels[72] at pH 3.0 in 24 h.

Isolation of apiogalacturonans from the cell walls of *L. minor* was accomplished by using gentle extraction with 0.5% ammonium oxalate.[68,69] Beck[24a,68] isolated two apiogalacturonans from the cell walls of *L. minor;* one contained a third as much xylose as apiose, and the other was free from xylose. These procedures were further refined by Hart and Kindel.[69] On the dry-weight basis, the apiogalacturonans extracted from cell walls of *L. minor* constituted 14% of the total walls, and contained 20% of the D-apiose originally present in these

(71) Yu. S. Ovodov, R. G. Ovodova, O. D. Bondarenko, and I. N. Krasikova, *Carbohyd. Res.*, **18**, 311–318 (1971).

(72) Yu. S. Ovodov, V. D. Sorochan, B. K. Vasilev, G. M. Nedahskovskaya, and A. K. Dzizenko, *Khim. Prir. Soedin.*, 267–268 (1973).

walls.[69] In addition, 83% of the total D-apiose of *L. minor* was recovered in the cell walls.[69]

The authors[69] extended their elegant studies by first postulating, on the basis of low yields of formaldehyde from periodate oxidation,[69] a D-apiose disaccharide. They then succeeded in isolating apiobiose [O-β-D-apiofuranosyl-(1 → 3)-D-apiose].[54] Apiobiose is released from the isolated apiogalacturonan fractions by very mild hydrolysis at pH 4.5 for 3 h at 100°. Apiobiose appears to be a disaccharide of D-apiose having the glycosidic linkage between C-1 of the nonreducing moiety and O-3 or O-3^1 of the reducing D-apiose residue. Proton magnetic resonance and optical rotation data suggested, but did not confirm, that the glycoside linkage in apiobiose is of the β configuration.[54] Rapid hydrolysis of the glycoside under mildly acidic conditions explains why apiose has not been recovered from acid hydrolyzates of cell walls of *L. minor*.[65] Beck[68] isolated, from *L. minor*, an apiogalacturonan that contained a single apiose residue as a side chain. It is not known whether this apiogalacturonan is different from that isolated by Hart and Kindel,[68,69] or whether one apiose residue was lost from apiobiose during purification of the polysaccharide.[24a]

Beck and Kandler[65] studied the incorporation of $^{14}CO_2$ into apiose-containing polysaccharides of *L. minor* and *L. gibba*. The rate of incorporation of label into apiose paralleled that for xylose, not that for glucose. The proportion of the label in apiose increased during photosynthesis; this differs from the behavior for starch, in which some radioactivity is lost during photosynthesis. The authors concluded that apiose-containing polysaccharides have no energy-storage function, but, instead, are structural components of the plant cell-wall.

The large number of branched-chain sugars that have been isolated[3] and the widespread occurrence of apiose[73] led to speculation regarding the natural role of apiose-containing polysaccharides. A number of authors[61,68,69,73] have hypothesized that apiose-containing polysaccharides possess an increased resistance to microbial degradation, particularly to pectinases released by pathogenic fungi. Removal of apiose, by hydrolysis with dilute acid, from an apiogalacturonan fraction of *L. minor* cell-walls resulted in greatly increased hydrolysis by pectinase, whereas the intact apiogalacturonan is almost completely resistant to pectinase degradation.[69]

(73) J. S. D. Bacon and M. V. Cheshire, *Biochem. J.*, **124**, 555–563 (1971).

3. Metabolism of Apiose

Although apiose is widely distributed in the plant kingdom, little information is available on its metabolism.[74] There has been isolated a mutant of *Aerobacter aerogenes*, namely PRL-R₃, which uses D-apiose as the sole source of carbon.[74] Subsequently, from the surface of germinating parsley-seed was isolated a micrococcus of the genus *Neisseria* which metabolized D-apiose as the sole source of carbon, with the formation of carbon dioxide.[75] An inducible enzyme, D-apiose reductase[74] or D-apiitol dehydrogenase,[75] which catalyzes the NAD⁺-dependent reduction of D-apiose to D-apiitol, was isolated from both sources. The enzyme(s) are very substrate-specific for D-apiose and D-apiitol, although, as is customary in enzymology, their ability to use L-apiose as a substrate was not tested.[74,75] These products of the action of D-apiose reductase on D-apiitol and D-apiose, respectively, were identified by their rate of migration in paper chromatography in two different solvent systems.[74] The importance of either enzyme in the metabolism of apiose, either *in vivo* or *in vitro*, has not been demonstrated.[74,75] D-Apiitol dehydrogenase has a very low, apparent K_m for NAD⁺ and NADH, and a relatively high K_m for D-apiose and D-apiitol.[75] The dehydrogenase is highly specific for D-apiose and D-apiitol; among many cyclitols, sugars, and sugar alcohols tested, only *myo*-inositol and erythritol showed much ability to function as its substrate.[75]

L-Apiosyladenine (a synthetic compound; see p. 184) is bacteriostatic to *Escherichia coli* and *Staphylococcus aureus* at 50–100 μg/ml in liquid culture.[76] Growth inhibition might be due to greater inability to deaminate the compound (as compared to adenosine), whether by hydrogen bonding between the 3- and 3¹-hydroxyl groups of the apiose moiety, or because the 3¹-hydroxyl group prevents attachment to the active site of the enzyme.[76] An L-apiosyladenine nucleotide was found to be deaminated by adenosine deaminase.[76] From the point of view of therapeutic utilization, this observation is intriguing, as biological interest in nucleoside analogs centers on (*a*) the speed with which adenosine deaminase decomposes them, and (*b*) their susceptibility to phosphorylation for subsequent metabolism.[76]

A partially characterized apioside has been reported that has low toxicity when administered intravenously to the dog.[77] It did not af-

(74) D. L. Neal and P. K. Kindel, *J. Bacteriol.*, **101**, 910–915 (1970).
(75) R. Hanna, J. M. Picken, and J. Mendicino, *Biochim. Biophys. Acta*, **315**, 259–271 (1973).
(76) J. M. J. Tronchet and J. Tronchet., *Helv. Chim. Acta*, **54**, 1466–1479 (1971).
(77) R. A. Paris and J. Gueguen. *Ann. Pharm. Fr.*, **11**, 421–424 (1953).

fect blood pressure, although it increased the resistance, and decreased the permeability of the capillaries. Apiin, as well as 13 flavonoid glycosides not containing apiose, had anti-spasmodic activity greater than that of monosaccharides or disaccharides.[78] Metabolism of apiose or apiose-containing compounds has not been demonstrated in animals. The role, or importance, of apiose in mammalian diets is unknown, although it is a constituent of celery,[26,27] chick peas,[37] duckweed,[79] and parsley,[25,27] which may be present in the diet.

The metabolism of apiose in plants has not been investigated extensively. It has been suggested[80] that apiose could give rise to simple terpenes by condensation and reduction. Incubation of [?-14C]apiose in parsley plants resulted in negligible, but unspecified, amounts of carbon-14 incorporated into apiin.[66] The specific activity of the [?-14C]apiose administered was low (5×10^5 d.p.m./μmole); for this reason, the data suggested only that no highly active mechanism is available in parsley for the metabolism of this branched-chain sugar.

Incubation of D-[U-14C]apiose with sterile *Lemma minor* (duckweed) produced less than 0.01% incorporation into the cell-wall polysaccharides.[75] Most of the D-[U-14C]apiose appeared as $^{14}CO_2$; some remained in solution in the medium and in the duckweed plants, primarily as degradation products of D-[U-14C]apiose, but not as the branched-chain sugar.[75] There is an efficient synthesis of the [U-14C]apiose moiety of cell-wall polysaccharides from D-[U-14C]glucose under similar conditions.[81] Of the plant tissues tested, only *L. minor* contained an enzyme system able to metabolize free apiose. Carrot, lettuce, and spinach tissues are unable to metabolize the free, branched-chain sugar.[75]

IV. BIOSYNTHESIS OF APIOSE *in vivo*

The elucidation of the biosynthetic pathway of apiose was founded on extensive, isotope-incorporation experiments in *L. minor*,[65,66,82,83]

(78) B. Borkowski and K. Szpunar, *Pharm. Zentralh.*, **99**, 280–285 (1960); *Chem. Abstr.*, **55**, 899 (1961).

(79) R. M. Roberts, R. H. Shah, and F. Loewus, *Plant Physiol.*, **42**, 659–666 (1967).

(80) A. W. Stewart, "Recent Advances in Organic Chemistry," Longmans, Green and Co., London, 4th Edition, 1920, p. 270.

(81) J. M. Picken and J. Mendicino, *J. Biol. Chem.*, **242**, 1629–1634 (1967).

(82) H. Grisebach and U. Dobereiner, *Biochem. Biophys. Res. Commun.*, **17**, 737–742 (1964).

(83) E. Beck and O. Kandler, *Z. Pflanzenphysiol.*, **55**, 71–84 (1966).

L. gibba,[65,79,83] and *P. crispum.*[66,79,82,84,85] In general, the results of these *in vivo* investigations, as well as those of *in vitro* biosynthetic studies, support the hypothesis proposed for the synthesis of D-apiose polysaccharides and apiin (4) by Grisebach and Dobereiner[82] (see Fig. 1). They postulated that decarboxylation of a "nucleoside

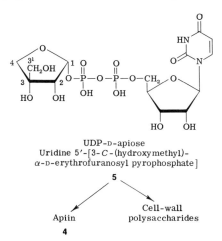

UDP-D-apiose
Uridine 5'-[3-*C*-(hydroxymethyl)-
α-D-erythrofuranosyl pyrophosphate]

5

Apiin   Cell-wall
polysaccharides
4

FIG. 1.–Transglycosylation of D-Apiose from UDP-D-apiose (**5**). (Studies *in vitro* showed that UDP-D-apiose {uridine 5'-[3-C-(hydroxymethyl)-α-D-erythrofuranosyl pyrophosphate]} functions in a transglycosylation reaction. The product is apiin, 4',5-dihydroxyflavon-7-yl 2-*O*-[3-C-(hydroxymethyl)-β-D-erythrofuranosyl]-β-D-glucopyranoside (**4**), formed through the aid of an acceptor molecule.[7] Additional products that may be formed by transglycosylation from an apiose donor-molecule include, principally, apiose-containing polysaccharides.)

diphosphate D-glucuronic acid" results in the formation of a "sugar nucleotide" containing D-apiose. It was further theorized that a rearrangement reaction must also occur after decarboxylation, in order to give the branched-chain D-apiose from the hypothetical pyranose intermediate. This suggestion was based on the hypothesis of Baddiley and coworkers[86] for the rearrangement of L-rhamnose to form L-streptose. The proposed end-product was a "nucleoside diphosphate D-apiose" suggested to function in the biosynthesis of apiose-containing polysaccharides and glycosides by participating as a substrate in transglycosylation reactions.[82]

(84) H. Grisebach and U. Dobereiner, Z. *Naturforsch.,* **21B,** 429–435 (1966).
(85) H. Grisebach and H. Sandermann, Jr., *Biochem. Z.,* **346,** 322–327 (1966).
(86) J. Baddiley, N. L. Blumson, A. Di Girolamo, and M. Di Girolamo, *Biochim. Biophys. Acta,* **50,** 391–393 (1961).

Other biosynthetic pathways presented for apiose have not yet been substantiated. An early hypothesis suggested that the metabolism of apiose and that of isoprene are related; this was based upon the presence of apiose in *Taraxacum kok-saghyz*.[70,87] A similar hypothesis arose from the observation that the bark from *Hevea brasiliensis* (rubber tree) yields apiose.[23] Inasmuch as apiose and isoprene have the same carbon skeleton, Patrick[23] conjectured that there is a metabolic link between apiose and the 5-carbon intermediates of isoprenoid synthesis. Others have postulated that the biosynthesis of apiose might result from condensation of 1,3-dihydroxy-2-propanone with glycolaldehyde.[88] This theory was supported by an observation of Utkin,[89] who isolated a homolog of apiose, namely, dendroketose [4-*C*-(hydroxymethyl)-DL-*glycero*-pentulose], from the self-condensation of two molecules of 1,3-dihydroxy-2-propanone. Finally, it was speculated that the biosynthesis of branched-chain carbohydrates may proceed by conversion of straight-chain carbohydrates into glycosuloses.[90] Further transformation thereof through a cyclic, acyloin state (R—CHOH—CO—R) can yield dicarbonyl compounds having a branched-chain structure, because cleavage of cyclic acyloin intermediates can produce either branched or unbranched carbohydrates. According to this scheme, apiose could be formed by the acyloin condensation of a pentos-4-ulose, with subsequent cleavage of the acyloin ring, and reduction.

The hypothesis of Grisebach and Dobereiner[82] was initially bolstered by the results of isotope-incorporation experiments in *Petroselinum crispum* (parsley) and *Lemma minor* (duckweed). The *in vivo* investigations of the biosynthesis of apiose were initiated with administration of [14C]formate, [1-14C]acetate, D-[U-14C]glucose, and D-[3,4-14C]glucose to young parsley-shoots.[82] Apiin was isolated, and its component parts were purified by gas–liquid chromatography. Although the purity of the products and the amount lost during purification were not measured, the data indicated that some carbon-14 from each labelled precursor appeared in D-apiose, D-glucose, and apigenin. When [1-4C]acetate and [14C]formate were supplied to the parsley shoots, twice as much carbon-14 was observed in the D-apiose as in the D-glucose; this suggested that [1-14C]formate might

(87) J. Chrastil, *Cesk. Biol.*, **3**, 359–362 (1954).
(88) L. Hough and J. K. N. Jones, *Nature*, **167**, 180–183 (1951).
(89) L. M. Utkin, *Dokl. Akad. Nauk SSSR*, **67**, 301–304 (1949); *Chem. Abstr.*, **44**, 3910a (1950); *Zh. Obshch. Khim.*, **25**, 530–536 (1955).
(90) M. M. Shemyakin, A. S. Kohkhlov, and M. N. Kolosov, *Dokl. Akad. Nauk SSSR*, **85**, 1301–1304 (1952); *Chem. Abstr.*, **47**, 4292a (1953).

be more intimately and directly involved in the biosynthesis of D-apiose than in that of D-glucose. During the long period of incubation of the radioactive precursors with the parsley shoots, metabolism of the precursor compounds occurred, as was evidenced by the extensive presence of carbon-14 in the apigenin.[82] Thus, the importance of these radioactive precursors in the direct biosynthesis of D-apiose was not demonstrated.

Equal amounts of carbon-14 from D-[U-^{14}C]glucose and D-[3,4-^{14}C]-glucose were recovered in D-glucose and D-apiose isolated from apiin, indicating that D-glucose is at least an indirect precursor of D-apiose. To determine whether carbon-14 from D-[3,4-^{14}C]glucose was specifically incorporated into apiin, apiin was oxidized with periodate in neutral solution.[82] About 40% of the D-[^{14}C]apiose of apiin from parsley was located in C-3', and 45% in C-3. Therefore, C-3' of D-apiose must originate either from C-3 or C-4 of D-glucose. The stereochemistry of apiose in apiin is known[1,15,16,20] to be D, and D-apiose does not occur in a free form after its synthesis and before incorporation into apiin. If it did, an equilibrium between C-3' and C-4 of D-apiose should occur by way of the aldehyde form. However, this does not occur,[2,82,84] thus providing additional support for the hypothesis of Grisebach and Dobereiner.[82] Subsequently, the biosynthesis of D-apiose *in vivo* was investigated[66] with [2-^{14}C]acetate, L-[3-^{14}C]serine, D-[1-^{14}C]glucose, and NaH^{14}CO$_3$. These labeled compounds were administered into holes drilled in parsley roots which were then incubated for 2 days. Apiin was isolated by precipitation, and purified by paper chromatography. Although the relative purity and the recovery of D-glucose, D-apiose, and apigenin was not noted, significant conversion of each radioactive precursor into D-glucose and D-apiose occurred.[66] Twice as much L-[^{14}CH$_3$]methionine was converted into D-apiose as into D-glucose, whereas more carbon-14 from the other precursors appeared in D-glucose and D-apiose. As a similar proportion of carbon-14 was detected in both D-glucose and D-apiose, some randomization had taken place on metabolism of carbon-14 in the labelled compound.[66] Thus, the data do not (*a*) show that any of these compounds are directly and preferentially incorporated into D-apiose, or (*b*) support the suggestion that the carbon atoms in either of the two hydroxymethyl groups of D-apiose are derived from L-[^{14}CH$_3$]methionine and L-[3-^{14}C]serine through the transfer of one carbon atom to a D-apiose precursor. Later, L-[^{14}CH$_3$]methionine was supplied to sterile *Lemma minor,* and less carbon-14 was incorporated into apiose than into glucose.[83] Also, there was no indication that apiose was more preferentially labelled than xylose by L-[^{14}CH$_3$]methionine. Clearly, the biosynthesis of apiose does not proceed by the transfer of a hydroxymethyl group.[75]

Mendicino and Picken[66] reported the key observation in isotope-incorporation experiments involving D-[6-^{14}C]glucose, namely, that the D-apiose moiety of apiin has a specific activity one-fifth that of the D-glucose moiety; this suggested that, in conversion of the hexose into D-apiose, C-6 of D-glucose is lost. It was then postulated that D-apiose is derived from D-glucose through the loss of the terminal carbon atom; if this is correct, the administration of D-[2-^{14}C]glucose to parsley should result in equal labelling of the D-glucose and D-apiose moieties of apiin. When the labelled compound was administered, the specific activities of the two monosaccharide units of apiin were approximately the same.[66,84] According to Grisebach and Dobereiner's hypothesis,[82] C-6 of D-glucose *should* be lost, and the hydrogen atom on C-4 *may* be lost when an aldos-4-ulose intermediate is formed during the biosynthesis of apiose.

To test this theory, D-[6-^{14}C, 4-^{3}H]glucose and D[U-^{14}C, 3-^{3}H]-glucose were separately furnished to parsley shoots.[82] When D-[6-^{14}C, 4-^{3}H]glucose was administered, less carbon-14 was incorporated into the D-apiose than into the D-glucose of apiin. Degradation of the D-apiose formed by D-[6-^{14}C, 3-^{3}H]glucose indicated that the carbon-14 in the D-apiose was derived from an equilibrium between C-6 and C-1 of D-glucose. This might be the result of (*a*) reversal of glycolysis, or (*b*) refixation of ^{14}CO$_2$ prior to transformation into D-apiose. In addition, 70% of the tritium in the D-[6-^{14}C, 4-^{3}H]-glucose was located on C-3 of the D-apiose. From the general hypothesis of Grisebach and Dobereiner,[82] this result may be accounted for by the assumption that the tritium was removed by a coenzyme–enzyme complex from UDP-D-[6-^{14}C, 4-^{3}H]glucuronic acid. The same tritium may be used by the coenzyme–enzyme complex for the reduction of the transient carbonium ion postulated to be formed during the rearrangement that results in the biosynthesis of UDP-apiose. Administration of D-[U-^{14}C, 3-^{3}H]glucose to parsley shoots resulted in the appearance of 60% of the tritium at C-3^1 of D-apiose, excluding the possibility that C-3^1 becomes a carboxyl group during the benzilic acid type of rearrangement.[2]

Mendicino and Picken[66] originally suggested that only negligible amounts of radioactivity from NaH^{14}CO$_3$, DL-[3-^{14}C]serine, and [2-^{14}C]acetate were incorporated into apiin in the leaves and stems of parsley. Tangential experiments showed that apiin was synthesized in the leaves and roots of parsley.[78,81,83,84] However, purified starch and sucrose from the same tissues contained large and significant proportions of carbon-14. Apiin isolated from the leaves of a whole parsley plant incubated in the presence of [2-^{14}C]acetate had incorporated only one-tenth of the ^{14}C/μmole of the apiin isolated from the roots. Therefore, it was suggested that apiin is not synthesized in excised leaves; instead, it is created in the root and transported throughout the whole plant.[66] However, other investigators demonstrated incorporation of carbon-14 from [^{14}C]formate, [1-^{14}C]acetate, D-[U-^{14}C]glucose, and D-[3,4-^{14}C]glucose into D-apiose in parsley shoots.[82,84] The same amount of carbon-14 from D-[U-^{14}C]glucuronic acid was incorporated into D-apiose in the whole parsley plant as in the excised stem and leaves.[85] Tritium from [2-^{3}H]*myo*-inositol was also recovered in the D-apiose of apiin in excised parsley-leaves.[79] Watson[31] showed that two enzymes exist for converting uridine 5'-(α-D-glucopyranosyluronic acid pyrophosphate) (UDP-D-glucuronic acid) and 7-D-glucopyranosylapigenin into apiin in both the leaves and roots of parsley. Several investigators[78,81,83,84] clearly demonstrated that D-apiose is synthesized in the leaves, stems, and roots of parsley.

Studies with *L. minor* (duckweed) confirmed much of the aforementioned work on parsley.[81] It had been reported that [^{3}H]isoleucine administered to *L. minor* is not efficiently incorporated into apiose, as the specific activity of the apiose decreased by two-thirds during the 7-day incubation period.[66] Three to four percent of the carbon-14 from D-[1-^{14}C]glucose, D-[2-^{14}C]glucose, and D-[3-^{14}C]glucose was incorporated into the apiose moieties purified from duckweed polysaccharides. Much less was incorporated when [2-^{14}C]acetate, L-[[14]CH_3]methionine, D-[6-^{14}C]glucose, and DL-[2-^{14}C]serine were administered (with incorporation of 0.1, 0.2, 1.1, and 1.2%, respectively).[81] The distribution of carbon-14 in the apiose of the cell-wall polysaccharides showed that carbon-14 from D-[1-^{14}C]glucose was incorporated[81] principally (63%) into C-1 of apiose. Carbon-14 from D-[2-^{14}C]glucose was found principally in C-2 of apiose (52%). When D-[3,4-^{14}C]glucose was furnished to *L. minor* (duckweed), carbon-14 was detected significantly in C-3 (44%) and C-4 (25%) of apiose. D-[6-^{14}C]Glucose was not efficiently incorporated into apiose, and its carbon-14 appeared principally in C-1 of apiose (49%). Much of this incorporation might be expected to occur during the long period of incubation with *L. minor*, with significant randomization of carbon-14 by the enzymes of the Embden–Meyerhof pathway, resulting in transfer of some of the carbon-14 to C-1 of D-glucose.[81] Beck and Kandler[83] also studied the incorporation of carbon-14 from D-[1-^{14}C]glucose, D-[2-^{14}C]glucose, and D-[6-

^{14}C]glucose into apiose, xylose, and glucose in the polysaccharides of another duckweed, *L. gibba.* To avoid refixation, $^{14}CO_2$ was trapped, and their data showed that cultures of duckweed grown under sterile conditions incorporate carbon-14 less efficiently into apiose and xylose from D-[6-^{14}C]glucose than from D-[1-^{14}C]glucose and D-[2-^{14}C]glucose. There was good agreement between the percentage of carbon-14 at C-3 and C-5 of glucose, and C-3^1 and C-4 of apiose. The same appeared to be true for C-4 of glucose and C-3 of apiose, suggesting that C-3 of glucose may become C-3^1 of apiose, and that C-4 yields C-3 of apiose. The foregoing results demonstrated that C-1 through C-4 of apiose are derived principally, and respectively, from C-1 through C-5 of glucose.[81-83]

Until 1966, no-one had demonstrated that D-glucuronic acid could be converted into D-apiose. D-[U-^{14}C]Glucuronic acid and D-[6-^{14}C]glucuronic acid were furnished[85] to the roots of parsley plants for 50 h. To lessen the refixation of $^{14}CO_2$ originating from the decarboxylation of D-glucuronic acid, the plants were kept in the dark for 2 h before the administration of the D-glucuronic acid. The parsley plants were irradiated during the experiment with far-infrared light (740 nm; 4000 erg. cm.$^{-2}$sec^{-1}), which stimulated the biosynthesis of flavonol in buckwheat[91] and was expected to stimulate the biosynthesis of apiin in parsley. Only 0.14% of the carbon-14 from D-[U-^{14}C]glucuronic acid was incorporated into D-apiose. Isolated C-3^1 contained 20% of the carbon-14 of D-apiose, the amount to be expected were C-1 through C-5 of D-[U-^{14}C]glucuronic acid converted into D-apiose. The low incorporation (0.005%) of carbon-14 from D-[U-^{14}C]glucuronic acid into the D-glucose residue in apiin showed marginal refixation of $^{14}CO_2$ from D-[U-^{14}C]glucuronic acid, and minimal conversion of D-[U-^{14}C]glucuronic acid into precursors of D-glucose;[85] this result implied that D-glucuronic acid is a relatively direct precursor of D-apiose. In contrast, the rate of incorporation from D-[6-^{14}C]glucuronic acid into D-apiose is lower by a factor of 0.005.

This result suggested that C-6 of D-glucuronic acid is lost during conversion of the acid into D-apiose. The relatively high incorporation of carbon-14 into D-glucose, and the low incorporation into D-apiose suggested that refixation of $^{14}CO_2$ occurred. When D-[U-^{14}C]glucuronic acid is furnished to parsley, it is unlikely that $^{14}CO_2$ would be refixed in such a way that carbon-14 is incorporated into a

(91) H. Harraschain and H. Mohr, Z. Bot., 51, 277–299 (1963).

direct precursor of D-apiose [resulting in high (95%) incorporation into D-apiose] and not into D-glucose and apigenin (6 and 3% incorporation, respectively),[85] especially because administration of D-[6-^{14}C]glucuronic acid to the plants resulted[85] in more carbon-14 in the D-glucose and apigenin (68%) than in the D-apiose (30%). A large amount of the carbon-14 was available for refixation as [14]CO$_2$, as it appeared in respiratory carbon dioxide (36 and 19%, respectively).

The foregoing work was confirmed by employing *myo*-inositol as the carbon-atom donor.[79] *myo*-Inositol is rapidly and specifically utilized in the biosynthesis of glycuronic acid and pentose residues, including D-glucuronic acid, 4-*O*-methyl-D-glucuronic acid, D-galacturonic acid, D-xylose, and L-arabinose in the higher plants.[92,93] This conversion results from a cleavage of *myo*-inositol between C-1 and C-6, with concomitant oxidation of C-1 to a carboxyl group, to form D-glucuronic acid.[92,93] Subsequent metabolism of D-glucuronic acid accounts for the appearance of specifically labelled glycuronic acid and pentose residues in cell-wall polysaccharides.[92] After incubation of [2-^{14}C]*myo*-inositol with sterile cultures of *L. gibba* (duckweed) for 4 to 18 weeks, both the apiose and xylose contained about equal proportions of carbon-14. Under these conditions, the glucose did not contain carbon-14, suggesting that the labeled carbon atom was not greatly metabolized before incorporation into apiose and xylose.[79] Periodate oxidation of the newly formed apiose resulted in liberation of carbon-14 primarily from C-3', C-4, or C-3' and C-4 (94%). In the xylose, 97% of the carbon-14 was found at C-5, suggesting that the C$_6$ precursor to xylose and, perhaps, apiose (presumably, glucuronic acid) contained carbon-14 at C-5 only. Previously, it had been shown that C-3 and C-4 of glucose become, principally, C-3 and C-3' of D-apiose. Thus, the carbon-14 of apiose formed from [2-^{14}C]*myo*-inositol was probably localized principally in C-4 not C-3'. These results independently provided direct evidence for the metabolism of D-glucuronic acid in the biosynthesis of apiose. The relatively direct and complete conversion of *myo*-inositol into pentoses and apiose was also shown in *P. crispum*.[79] [2-^{3}H]*myo*-Inositol was incorporated only into D-apiose, and not into apigenin and D-glucose.

When L-[1-^{14}C]arabinose was supplied to *L. gibba* (duckweed), both the arabinosyl and xylosyl residues of the cell-wall polysac-

(92) R. M. Roberts and F. Loewus, *Plant Physiol.*, **41**, 1489–1498 (1966).
(93) R. M. Roberts, R. Shah, and F. Loewus, *Arch. Biochem. Biophys.*, **119**, 590–593 (1967).

charides contained carbon-14, but carbon-14 was not found in the apiosyl groups.[79] Although some plants contain an epimerase that can convert L-arabinose into D-xylose (UDP-L-arabinose $\rightleftarrows$ UDP-D-xylose), there is no mechanism in L. gibba G_3 for transforming exogenously supplied L-arabinose, or endogenously formed D-xylose and L-rhamnose, into apiose. The biosynthesis of apiose from $^{14}CO_2$ was measured during photosynthesis in L. minor and L. gibba; apiose and xylose were mainly present in the cell-wall polysaccharides. As the amount of carbon-14 incorporated into apiose and D-xylose from $^{14}CO_2$ is very similar at 5, 10, 30, and 120 minutes of growth in $^{14}CO_2$, although it differs significantly thereafter, these two sugars might be branches of the same biosynthetic pathway.[65]

V. BIOSYNTHETIC REACTIONS *in vitro* INVOLVING APIOSE

Biosynthetic reactions forming apiose can be divided into decarboxylative reactions resulting in the biosynthesis of UDP-apiose, and transglycosylation reactions resulting in the transfer of apiose from one compound to another. Formation of apiose *in vitro* has been extensively investigated in only two species of plants, namely, parsley (*P. crispum*) and duckweed (*L. minor*). Enzymes from these sources have been used in demonstrating the formation of apiose *in vitro*.[94−103]

1. Biosynthesis of UDP-D-apiose by Decarboxylation

Early investigations of apiose already reviewed showed that its biosynthesis involves D-glucuronic acid as an intermediate. Apiose is formed from D-glucuronic acid by loss of C-6, with a concomitant, in-

(94) H. Sandermann, Jr., G. T. Tisue, and H. Grisebach, *Biochim. Biophys. Acta*, **165**, 550–552 (1968).

(95) D. L. Gustine and P. K. Kindel, *J. Biol. Chem.*, **244**, 1382–1385 (1969).

(96) H. Sandermann, Jr., and H. Grisebach, *Biochim. Biophys. Acta*, **208**, 173–180 (1970).

(97) P. K. Kindel, D. L. Gustine, and R. R. Watson, *Plant Physiol.*, **46**, *Suppl.*, 27 (1970).

(98) R. R. Watson and P. K. Kindel, *Plant Physiol.*, **46**, *Suppl.*, 28 (1970).

(99) K. Hahlbrock and E. Wellman, *Planta* (Berlin), **94**, 236–239 (1970).

(100) E. Wellmann, D. Baron, and H. Grisebach, *Biochim. Biophys. Acta*, **244**, 1–6 (1971).

(101) E. Wellmann and H. Grisebach, *Biochim. Biophys. Acta*, **235**, 389–397 (1971).

(102) P. K. Kindel, D. L. Gustine, and R. R. Watson, *Fed. Proc.*, **30**, 1117 Abstr. (1971).

(103) H. Grisebach, H. Sandermann, Jr., and G. T. Tisue, *Angew. Chem. Intern. Ed. Engl.*, **6**, 971 (1967).

ternal rearrangement. It was suggested that these reactions might occur with nucleotide-bound sugars.[82] In 1968, Sandermann and Grisebach[104] first presented evidence for the existence of an apiose-containing "sugar nucleotide." After hydrolysis of a mixture of sugar nucleotides isolated from parsley, 0.13% of the sugars liberated migrated like apiose. It was tentatively suggested that the mixture of sugar nucleotides contained an apiose "sugar nucleotide(s)." This conclusion was tentative, as the liberated apiose might have come from some of the large proportion of apiin present in parsley. Furthermore, one of the solvents used in paper chromatography to purify the "sugar nucleotides," namely 7:5:3 (v/v) ethanol–M ammonium acetate–water (pH 7.5) was later shown[31] to degrade UDP-apiose (5) rapidly to uridine 5'-phosphate and 3-C-(hydroxymethyl)-α-D-erythrofuranosyl (cyclic)1,2-phosphate (6); thus, much of the UDP-

3-C-(Hydroxymethyl)-
α-D-erythrofuranosyl
1, 2-phosphate

6

apiose present would be expected to be degraded upon purification. The presence of a compound that migrated like apiose was observed after treatment of part of the "sugar nucleotide" fraction with phosphoric diesterase plus alkaline phosphatase, but none of the untreated material had been chromatographed in order to verify the absence of apiose before the enzymic hydrolysis.[104] Based upon these considerations, neither the isolation nor the identification of a "sugar nucleotide" of apiose had been demonstrated.

Very preliminary data indicated[94] that UDP-apiose (5) is synthesized by cell-free extracts of *L. minor* and *P. crispum*, but the validity of the work could not be appraised, because of (*a*) the marginal conversion of UDP-D-glucuronic acid into a "sugar nucleotide" of apiose (0.025%), and (*b*) the small percentage (3%) and amount (giving 260 c.p.m.) of the UDP-C_5 sugars formed, that, after hydrolysis, gave com-

(104) H. Sandermann, Jr., and H. Grisebach, *Biochim. Biophys. Acta*, **156**, 435–436 (1968).

pounds that migrated like apiose. Later, Gustine and Kindel[95] prepared, from *L. minor*, a cell-free system that carried out the synthesis of apiose, and 22% of the UDP-D-glucuronic acid was converted into an apiose compound. No other naturally occurring sugar is known that forms formaldehyde, glycolic acid, and formic acid in the yields that were observed upon periodate oxidation.[95] Subsequently, it was shown[7] that the D-apiose compound is formed from UDP-apiose *during the paper chromatography* at 22° with 7:3 (v/v) 95% aqueous ethanol–*M* ammonium acetate (pH 7.5).

Incubation of a cell-free extract of *L. minor* in the presence of NAD$^+$ and UDP-D-[U-^{14}C]glucuronic acid generated two compounds, each containing a different radioactive C_5 sugar.[96] The presence of apiose after hydrolysis was diagnosed by (*a*) paper chromatography, (*b*) specific complex-formation with benzeneboronic acid, (*c*) electrophoretic mobility, and (*d*) formation of its isopropylidene acetal. Treatment with ammonia at 0° showed that the compound was formed by a nonenzymic cyclization.[96] Treatment with phosphoric diesterase plus phosphatase further supported an earlier report of the existence of a D-apiosyl cyclic phosphate.[95]

Kindel and Watson[7] definitively showed the formation of a D-apiose cyclic phosphate, the α-D-apiofuranosyl 1,2-phosphate (**6**), from UDP-apiose, and rigorously characterized it and the conditions necessary for its formation. The alkaline degradation product of UDP-apiose was conclusively identified by (*a*) an apiose to phosphate ratio of 1:1.02, (*b*) the presence of only one ionizable hydrogen atom in the phosphate group, (*c*) the production of D-apiose 2-phosphate by acid hydrolysis of the alkaline degradation product, and (*d*) the reaction of α-D-[U-^{14}C]apiofuranosyl 1,2-phosphate with sodium metaperiodate, with the release of [^{14}C]formaldehyde.[7] This work confirmed earlier evidence that the apiose cyclic phosphate is a breakdown product of UDP-apiose.[105] Mendicino and Hanna[105] had shown that an apiose cyclic 1,2-phosphate is formed by attachment to O-2 of the phosphate group of 3-*C*-(hydroxymethyl)-α-D-erythrofuranosyl phosphate (**7**). It was apparently identical to the cyclic compound formed during paper chromatography, after enzymic conversion of UDP-D-glucuronic acid into an apiose derivative. The authors also described[105] the synthesis of 3-*C*-(hydroxymethyl)-α-D-erythrofuranosyl phosphate (**7**) and 3-*C*-(hydroxymethyl)-β-L-threofuranosyl phosphate (**8**) from β-D-apiose tetraacetates. The nu-

(105) J. Mendicino and R. Hanna, *J. Biol. Chem.*, **245**, 6113–6124 (1970).

3- C-(Hydroxymethyl)-
α-D-erythrofuranosyl
phosphate

7

3-C -(Hydroxymethyl)-
β-L-threofuranosyl
phosphate

8

clear magnetic resonance spectra of α-D-xylopyranosyl phosphate and α-D-glucopyranosyl phosphate were determined, to aid in assignment of the configuration of the carbon atom bearing the phosphate group; little, or no, β anomer was formed, due, perhaps, to the greater lability of the β anomer. The configuration at C-3 of the D-apiose moiety in UDP-apiose (**5**) and 3-C-(hydroxymethyl)-α-D-erythrofuranosyl 1,2-phosphate (**6**) was assumed to be that found[7,16] in the D-apiose in apiin (**4**). As D-apiosyl 1,2-phosphate is formed from UDP-apiose, the configuration of C-1 in UDP-apiose (**5**) and 3-C-(hydroxymethyl)-α-D-erythrofuranosyl 1,2-phosphate (**6**) must be based upon steric considerations.[7] The formation of 3-C-(hydroxymethyl)-α-D-[U-^{14}C]erythrofuranosyl 1,2-phosphate is in accord with the fact that "sugar nucleotides" having the proper stereochemistry at C-1 and C-2 of the glycosyl group, such as that in the apiosyl group of UDP-apiose, are degraded under alkaline conditions to yield the corresponding nucleoside 5′-monophosphate and a glycosyl cyclic phosphate.[54] D-[U-^{14}C]Apiose 2-phosphate was identified as the principal product of the acid hydrolysis of 3-C (hydroxymethyl)-α-D-[U-^{14}C]erythrofuranosyl 1,2-phosphate by (*a*) the presence of two ionizable hydrogen atoms, (*b*) the susceptibility to the alkaline phosphatase from *E. coli*, (*c*) the acid stability, which eliminated the possibility that it was α-D-[U-^{14}C]apiosyl phosphate, and (*d*) the production of 2 moles of [^{14}C]formaldehyde for each mole of sugar on oxidation with sodium metaperiodate.[7]

UDP-apiose (**5**) is very unstable under a variety of conditions of pH and temperature; this instability had previously prevented [7] the isolation of sufficient quantities of UDP-apiose for definitive identification. UDP-[U-^{14}C]apiose is degraded at pH 8.0 to uridine 5′-phosphate and 3-C-(hydroxymethyl)-α-D-[U-^{14}C]erythrofuranosyl 1,2-phosphate at 80, 25, and 4°. The half-lives of UDP-[U-^{14}C]apiose under these conditions are 31.6 seconds, 97.2 minutes, and 16.5 hours, respectively.[7] The half-life of UDP-[U-^{14}C]apiose at pH 3.0 and 40° is 4.67 minutes. It is degraded[7] to uridine 5′-pyrophosphate and D-[U-^{14}C]apiose. At pH 6.2–6.6 and 4°, degradation (of both the acid and the base type) of UDP-[U-^{14}C]apiose occurs. After 20 days under these conditions, 23% of the starting UDP-[U-^{14}C]apiose was hydrolyzed to uridine 5'-pyrophosphate and D-[U-^{14}C]apiose, and 17% was degraded to uridine 5'-phosphate and 3-C-(hydroxymethyl)-α-D-[U-^{14}C]erythrofuranosyl 1,2 phosphate.[7] UDP-[U-^{14}C]apiose was found to be stable at pH 6.4 and $-20°$ in 35% ethanol; after 120 days, 4% of the starting UDP-[U-^{14}C]apiose was hydrolyzed, and 2% was degraded by cyclization.[7]

Based upon these observations, doubly labelled UDP-apiose was synthesized, and identified, through the use of nmolar quantities of a new substrate, [^{3}H]UDP-D-[U-^{14}C]glucuronic acid.[7] The ^{3}H:^{14}C ratio in [^{3}H]UDP-[U-^{14}C]apiose was that to be expected if [U-^{14}C]apiose remains attached to the [^{3}H]uridine, and one carbon atom of D-[U-^{14}C]glucuronic acid is lost as ^{14}CO$_2$. The reaction products containing [^{3}H]UDP-[U-^{14}C]apiose migrated in paper chromatography and electrophoresis with the mobility of UDP-D-xylose.[7] The mixture of [^{3}H]UDP-[U-^{14}C]apiose and [^{3}H]UDP-D-[U-^{14}C]xylose was hydrolyzed at 4° at pH 2.0 with the formation of [^{3}H]UDP, [U-^{14}C]apiose, and D-[U-^{14}C]xylose. [U-^{14}C]Apiose and D-[U-^{14}C]xylose were not detected prior to hydrolysis, establishing that the acid hydrolysis products of [^{3}H]UDP-[U-^{14}C]-apiose are [^{3}H]uridine 5'-pyrophosphate and D-[U-^{14}C]apiose, identified by paper chromatography and by radioactivity measurements.[7]

[^{3}H]UDP-[U-^{14}C]apiose was further characterized by base-catalyzed, intramolecular cyclization. Extensive chromatography on paper showed the products to be [^{3}H]uridine 5'-phosphate, 3-C-(hydroxymethyl)-α-D-[U-^{14}C]erythrofuranosyl 1,2-phosphate, and undegraded [^{3}H]UDP-D-[U-^{14}C]xylose (originally present with the [^{3}H]UDP-[U-^{14}C]apiose). [^{3}H]Uridine 5'-phosphate and 3-C-(hydroxymethyl)-α-D-[U-^{14}C]erythrofuranosyl 1,2-phosphate were not detected in the uncyclized, [^{3}H]UDP-[U-^{14}C]apiose control solution treated identically, except at 4°. Finally, [^{3}H]UDP-[U-^{14}C]apiose was shown to act as a D-[U-^{14}C]apiosyl donor. Incubation of the substrate with apiin synthetase and apigenin-7 β-D-gluco-pyranoside resulted in the formation of [^{3}H]uridine 5'-pyrophos-phate, [^{14}C]-apiin (^{14}C in the D-apiosyl group only), and some [^{3}H]uridine 5'-phosphate and 3-C-(hydroxymethyl)-α-D-[U-^{14}C]ery-throfuranosyl 1,2-phosphate from intramolecular cyclization of [^{3}H]-UDP-[U-^{14}C]apiose. [^{3}H]UDP-[U-^{14}C]xylose had not reacted, and remained unchanged.[7] These experiments conclusively showed that UDP-apiose (and, presumably, carbon dioxide) is formed from

UDP-D-glucuronic acid by the uridine 5'-(α-D-glucopyranosyluronic acid pyrophosphate) cyclase of *L. minor.*

The foregoing observations, and many of the results of the tracer studies *in vivo*, have been confirmed with conversion of UDP-D-[3-^{14}C]glucuronic acid into UDP-[3^1-^{14}C]apiose; this result directly demonstrated that C-3 of D-glucuronic acid becomes C-3^1, the carbon atom in the hydroxymethyl group of apiose. Transfer of D-[3^1-^{14}C]apiose by apiin synthetase to form apiin allowed[106] C-3^1 to be "isolated." Periodate oxidation revealed that all of the carbon-14 was located in C-3^1. These results further emphasized the similarity between the rearrangement of the sugar chain leading to apiose formation and that leading to formation of L-streptose.[107] The fate of the hydrogen atoms at C-3 and C-4 of UDP-D-glucuronic acid was studied in the same way.[107] In both instances, all of the tritium in the labelled UDP-apiose formed from UDP-D-[U-^{14}C, 3-^{3}H]glucuronic acid and UDP-D-[U-^{14}C, 4-^{3}H]glucuronic acid was located in C-3^1 of D-apiose (except the 50% lost in the formation of the former "sugar nucleotide").[108] As C-3 of D-glucuronic acid becomes C-3^1 of D-apiose, these results suggested a hydride shift from C-4 of D-glucuronic acid to C-3^1 of D-apiose.[106-108] An alternative explanation for these results is the theory of biosynthesis of D-apiose proposed by Mendicino and Picken.[81] It involves an aldol cleavage between C-2 and C-3 of the aldos-4-ulose intermediate formed from UDP-D-glucuronic acid, followed by isomerization and intramolecular, aldol reaction; this produces, in theory, an intermediate that, upon reduction by NADH, yields UDP-apiose.[106]

In these experiments, several doubly and triply labelled compounds were reported.[108] So-called "UDP-D-[U-^{14}C, 3,4-^{3}H]glucuronic acid" was formed by mixing UDP-D-[U-^{14}C]glucuronic acid, UDP-D-[3-^{3}H]glucuronic acid, and UDP-D-[4-^{3}H]glucuronic acid; it was not, of course, a triply labelled compound, but a mixture of 3 singly labelled ones.[108] This factor prevented any conclusions about intramolecular transfer with doubly or triply labelled compounds to be reached.

UDP[U-^{14}C]apiose is extremely labile in the presence of alkali at 4°. UDP-D-glucose, on the other hand, was unchanged[109] after 18 h at

(106) W. J. Kelleher, D. Baron, R. Ortmann, and H. Grisebach, *FEBS Lett.*, 203–204 (1972).

(107) D. J. Candy and J. Baddiley, *Biochem. J.*, **96**, 526–529 (1965).

(108) W. J. Kelleher and H. Grisebach, *Eur. J. Biochem.*, **23**, 136–142 (1971).

(109) A. C. Paladini and L. F. Leloir, *Biochem. J.*, **51**, 426–430 (1952).

pH 8.0 and 18°; under more vigorous conditions, it cyclized, affording[109] α-D-glucopyranosyl (cyclic) 1,2-phosphate and uridine 5'-phosphate. In contrast, UDP-[U-^{14}C]apiose containing D-apio-furanose is degraded to 3-C-(hydroxymethyl)-α-D-[U-^{14}C]erythro-furanosyl 1,2-phosphate and uridine 5'-phosphate after 2 h at pH 8.0 and 25°. The UDP-D-[U-^{14}C]xylose present was undegraded.[7] The following explanation was suggested[7,31] for the greater lability, in alkali, of a "sugar nucleotide" containing a glycofuranosyl group. The five-membered, phosphorus-containing ring that is formed on breakdown of "sugar nucleotides" in alkali has a conformation that is planar or nearly so. In order to accommodate this new ring, the C–O bonds of C-1 and C-2 of the furanose and pyranose rings of sugars must be skewed, or completely eclipsed. In furanoses, complete eclipsing of these two bonds occurs when the furanose is in the planar conformation. Generally, as the various conformations of a furanose are readily interconvertible, very little energy is needed to afford the conformation that is necessary to accommodate the new five-membered ring. On the other hand, the pyranoid forms must change, partially or completely, from a stable, chair conformation into an "unstable," half-chair conformation. A substantial expenditure of energy is needed to overcome nonbonded interactions between the substituents on C-1 and C-2 of the pyranose ring. Therefore, cyclization of UDP-D-glucose and other "sugar nucleotides" containing pyranosyl groups does not occur readily. A relatively large expenditure of energy is needed for this conversion between the chair and boat forms of pyranoses (to overcome the energy barrier). The large difference in the amount of energy needed to form a favored conformation of the sugar portion of each type of "sugar nucleotide," so that degradation can proceed, could account for the differences in alkali-lability of the two types of "sugar nucleotide." Certain "sugar nucleotides" cannot yield a sugar cyclic phosphate when degraded by alkali, because of stereochemical considerations. This is the situation when the sugar portion is in the pyranoid form and the hydroxyl groups on C-1 and C-2 are in the axial–axial orientation, or when the sugar portion is in the furanoid form and the hydroxyl groups on C-1 and C-2 are *trans*-disposed.[7]

The ready formation of the 1,2-phosphate **6** (see p. 163) at pH 8 may also be explained, in part, by its stability, which may be attributable to the presence of two 5-membered rings.[7] This structure is in contrast to that of the 1,2-phosphates of L-xylopyranose and D-glucopyranose, which contain a 5-membered ring condensed to a 6-membered ring. The facile degradation of 5-O-phosphono-D-ribofuranosyl pyrophosphate was cited as an analogous reaction resulting in the formation of two condensed, 5-membered rings. Both reactions are essentially irreversible and, hence, the relative stability of the product seems to have little influence on the tendency of the reaction to occur.[7] Thus, the foregoing suggestion does not seem to explain the different tendency of UDP-apiose to cyclize, as compared to that of UDP-D-glucose and UDP-D-xylose.

Because, in general, furanosides are hydrolyzed by acid more readily than pyranosides, it was expected[7] that UDP-[U-^{14}C]apiose would be more labile in acid than "sugar nucleotides" containing a glycopyranosyl group. The mechanism of acid hydrolysis of furanosides has not yet been established;[110] however, assuming that the mechanism involves the formation of a carbonium–oxonium ion on the ring, and that its formation is the rate-limiting step, the same general argument can be used to explain the greater instability, in acid, of "sugar nucleotides" containing a glycofuranosyl group as was used to explain their greater instability in alkali.[7] Formation of a carbonium–oxonium ion requires that the furanose ring be close to the planar conformation, which requires less energy than the pyranose form to bring it into a half-chair conformation for hydrolysis.[110] This requirement could account for the difference in stability,[7] in acid, of the two types of "sugar nucleotide." The isolation from *Penicillium charlesii* of another "sugar nucleotide" containing a glycofuranosyl group, namely, UDP-D-galactofuranose, has been described.[111,112] Data on the stability of this "sugar nucleotide" in acid and alkali have not yet been reported; however, it was stated that the "sugar nucleotide" is very labile in acid. UDP-apiose and UDP-D-galactofuranose are the only sugar nucleotides described to date that contain glycofuranosyl groups.

2. UDP-D-glucuronic Acid Cyclase

While UDP-apiose was being identified, several workers began the purification and characterization of the enzyme(s) in cell-free extracts of *L. minor* and *P. crispum* that produce UDP-apiose.[31,94,96,97,100–102] It had earlier been inferred that only one enzyme might be necessary to form UPD-apiose and UDP-D-xylose

(110) B. Capon and D. Thacker, *J. Chem. Soc.*, 185–189 (1967).
(111) A. G. Trejo, G. J. F. Chittenden, J. G. Buchanan, and J. Baddiley, *Biochem. J.*, **117**, 637–639 (1970).
(112) A. G. Trejo, J. W. Haddock, G. J. F. Chittenden, and J. Baddiley, *Biochem. J.*, **122**, 49–57 (1971).

from UDP-D-glucuronic acid, because of the similarity of the intermediates postulated in their formation.[82] A 55-fold purification of the
new enzyme from a cell-free extract of *L. minor* was reported.[102] The
enzyme catalyzed the formation of UDP-apiose from UDP-D-
glucuronic acid, and was given the name "uridine diphosphate glucuronic acid" (UDP-GlcA) cyclase. The systematic name suggested[31]
for this enzyme is "UDP-glucuronic carboxy-lase (cleaving, cy-
clizing)." It catalyzes the conversion of UDP-D-[U-^{14}C]glucuronic
acid into UDP-[U-^{14}C]apiose in the presence of catalytic quantities
of exogenous NAD$^+$. In the absence of exogenous NAD$^+$, UDP-GlcA
cyclase is inactive. Enzymic activity catalyzing formation of UDP-D-
xylose from UDP-D-glucuronic acid [which may be UDP-D-
glucuronate carboxylase (EC 4.1.1.35)] was also present.[31,102] In
contrast, other workers found that both activities were almost absent
in the absence of NAD$^+$, with 3 and 17%, respectively, of the UDP-
GlcA cyclase and UDP-GlcA decarboxylase remaining.[96] Both enzymes were inhibited identically by NADH, suggesting that they
might be a single protein, or, at least, very similar.[96]

During purification, the UDP-GlcA decarboxylase activity decreased with each purification step, relative to the UDP-GlcA cyclase
activity[102] from *L. minor*. Therefore, part of the UDP-GlcA decarboxylase may be an entity separate and distinct from the UDP-GlcA
cyclase. In contrast, other workers have partially purified UDP-GlcA
cyclase and UDP-GlcA decarboxylase from *L. minor*. They were
unable to separate the two enzymes during column-chromatographic
purification, analytical disc-electrophoresis, or isoelectric focusing.[101]
The effect of various ions, nucleotides, and sulfhydryl reagents was
the same on both enzymic activities. However, as NH$_4^+$ inhibited
UDP-GlcA cyclase, and stimulated UDP-GlcA decarboxylase, two
enzymes may have been present,[101] rather than one protein synthesizing both "sugar nucleotides." Subsequently, this question of one
enzyme or two was studied with cell-suspension cultures of
parsley.[100,113,114] An enzyme(s) having the characteristics of the UDP-
GlcA cyclase and UDP-GlcA decarboxylase of *L. minor* was isolated
and partially purified. It required NAD$^+$, and was not inhibited by
NADH. It had a pH optimum of 5.5, it was inhibited by NH$_4^+$ ion, its
biosynthesis was not affected by light, and it was clearly distinct

(113) D. Baron, E. Wellmann, and H. Grisebach, *Biochim. Biophys. Acta,* **258,**
310–318 (1972).
(114) D. Baron, U. Streitberger, and H. Grisebach, *Biochim. Biophys. Acta,* **293,**
526–533 (1973).

from the enzyme(s) forming UDP-apiose. With improved purification procedures, UDP-GlcA cyclase and UDP-GlcA decarboxylase were purified about 1077- and 967-fold, respectively.[114] The ratio of UDP-apiose to UDP-D-xylose formed did not change significantly during the purification, going only from 1.3 to 1.4 in five purification steps. The activities of the two, partially purified enzymes were not separated by disc electrophoresis or density-gradient centrifugation. The molecular weight determined[114] by density-gradient centrifugation is 101,000. Inability to separate the two activities does not, of course, exclude the possible presence of two proteins, one forming UDP-D-xylose and one forming UDP-apiose. The effect of inhibitors on the two activities suggested the presence of two separate entities. It appears that two enzymes are involved, at least in part, in the biosynthesis of UDP-D-xylose and UDP-apiose, either as a multi-enzyme complex or as two enzymes having very similar properties.

The only significant differences observed between the UDP-GlcA cyclases of parsley and duckweed were the energy of activation and the temperature at which heat denaturation of the respective enzyme began to occur, that is, the temperature for optimal activity.[31] The temperature of optimal activity of each enzyme is quite distinctive. It is 27° for duckweed UDP-GlcA cyclase, and 42° for parsley UDP-GlcA cyclase. These data also suggested that UDP-GlcA cyclase should be assayed at 25° (where heat denaturation should not affect either enzyme) and *not* at 30°, the standard temperature suggested by the Enzyme Commission for most enzyme-assays.

It may be of physiological significance that uridine 5'-pyrophosphate, a product of such transglycosylation reactions as those predicted for formation of cell-wall polysaccharide,[115] inhibits duckweed UDP-GlcA cyclases by 41%, and duckweed UDP-GlcA decarboxylase by 57% at low concentrations (1 μM).[31] This result suggested that, for transglycosylation reactions to proceed, uridine 5'-pyrophosphate must be removed. Parsley apiin synthase is also significantly inhibited at low concentrations[31] of uridine 5'-pyrophosphate. The intracellular function of UDP-GlcA cyclase and UDP-GlcA decarboxylase is to catalyze the biosynthesis of UDP-apiose and UDP-D-xylose, respectively. The low K_m value for UDP-D-glucuronic acid suggested that it is the substrate for parsley UDP-GlcA cyclase.[31] It would be of interest to determine the activity of the parsley and duckweed UDP-GlcA cyclases towards UDP-D-

(115) (a) J. H. Pazur, in "The Carbohydrates: Chemistry and Biochemistry," W. Pigman and D. Horton, eds., Academic Press, Inc., New York, 1970, Vol. IIA, p. 99; (b) W. Z. Hassid, *ibid.*, pp. 302–336.

galacturonic acid as a possible substrate. Both UDP-D-glucuronic acid and UDP-D-galacturonic acid, a possible substrate, are probably present in parsley and duckweed.[116]

Light-induced, increased activity of UDP-GlcA cyclase was demonstrated in parsley cell-suspension cultures.[99] The activity increased rapidly in white light, and reached maximal activity in one day, decreasing thereafter. The activity of another enzyme involved in the biosynthesis of flavonoid glycosides increased more rapidly, reached maximal activity earlier, and decreased more rapidly than that of UDP-GlcA cyclase.[99] Little or no activity of either enzyme was found in cells grown in the dark. Subsequently, the activity of eight enzymes involved in the biosynthesis of flavonoid glycosides was investigated, before and after exposure of the cell suspension to white light.[117] All eight enzymes were detectable shortly after application of light. L-Phenylalanine ammonia-lyase, *trans*-cinnamic acid 4-hydroxylase, and *p*-coumarate:CoA ligase were distinguishable from the others on the basis of their response to light: chalcone–flavanone isomerase, UDP-D-glucose:apigenin-7 glucosyltransferase, apiin synthetase, UDP-GlcA cyclase, and S-adenosyl-L-methionine:luteolin 3'-O-methyltransferase. Both UDP-GlcA cyclase and apiin synthetase are present in the greatest proportions in very young parsley leaves, decreasing similarly to low levels with increasing age.[118]

3. Transglycosylation Reactions of D-Apiose

A wide variety of phenolic compounds, including the flavonoids and related compounds, exists in the higher plants.[115] The enzymic glycosylation of phenolic compounds and flavonoids has been described only a very few times for plants,[115] and the glycosylation of a flavanone isomerase UDP-D-glucose:apigenin-7 glucosyltransferase, apiin synthetase, UDP-GlcA cyclase, and S-adenosyl-L-bean).[119] Further incubation of the enzyme preparation in the presence of thymidine 5'-(L-rhamnopyranosyl pyrophosphate) (TDP-L-rhamnose) catalyzed the transfer of L-rhamnose, to form 3',4',5,7-tetrahydroxyflavon-3-yl 6-O-β-L-rhamnopyranosyl-β-D-glucopyrano-

(116) H. Sandermann, Jr., and H. Grisebach, *Eur. J. Biochem.*, **6**, 404–410 (1968).
(117) K. Hahlbrock, J. Ebel, R. Ortmann, A. Sutter, E. Wellmann, and H. Grisebach, *Biochim. Biophys. Acta*, **244**, 7–15 (1971).
(118) K. Hahlbrock, A. Sutter, E. Wellmann, R. Ortmann, and H. Grisebach, *Phytochemistry*, **10**, 109–116 (1971).
(119) G. A. Barber, *Biochemistry*, **1**, 463–468 (1962).

side. In 1970, Watson and Kindel[98] and Grisebach and coworkers[120] first presented evidence that apiin is synthesized from a cell-free extract of parsley, 4,5-dihydroxyflavon-7-yl β-D-glucopyranoside. and an apiose-containing compound later shown[7] to be UDP-apiose The latter authors[120] incubated the cell-free extract with [U-14C]-apiose, and D-apiose 1,2-phosphate, or a mixture of UDP-D-[U-14C]-glucuronic acid, UDP-D-[U-14C]xylose, and UDP-[U-14C]apiose. Incubation of the latter mixture resulted in the formation of apiin.[7,98] The formation of a compound having the chromatographic characteristics of petroselinin from 3',4',5-trihydroxyflavon-7-yl β-D-glucopyranoside was also reported.[121] These reactions were later studied with a 100-fold purified, apiin synthetase from parsley.[7,31] By incubation of [3H]UDP-D-[U-14C]glucuronic acid with NAD+ and purified duckweed UDP-GlcA cyclase, 65% of the substrate was converted into the desired product, [3H]UDP-[U-14C]apiose. Some was converted into [3H]UDP-D-[U-14C]xylose (30%) plus a little (<5%) α-D-[U-14C]apiosyl 1,2-phosphate, the degradation product of [3H]-UDP-[U-14C]apiose. Purified, parsley, apiin synthetase was incubated with this mixture plus recrystallized 4',5-dihydroxyflavon-7-yl β-D-glucopyranoside. The products of the reaction were [3H]uridine 5'-pyrophosphate and apiin, containing carbon-14 in the D-apiose portion, only, of the flavonoid glycoside.[7]

4. Apiin Synthetase

The enzyme catalyzing the transfer of D-apiose from UDP-apiose to 4',5-dihydroxyflavon-7-yl β-D-glucopyranoside is commonly called apiin synthetase.[7] Activity is measured by the formation of [14C]apiin from UDP-[U-14C]apiose. Apiin synthetase can be measured by the rapid separation and isolation, by poly(ethylenimine)-paper chromatography, of a product of the reaction, namely, [14C]apiin, from UDP-D-[U-14C]xylose and degradation products of UDP-[U-14C]apiose.[31] There are reports of the isolation and purification of apiin synthetase from parsley leaves,[31] from cell-suspension cultures of parsley,[121] and from foxglove (Digitalis purpurea).[31] Apiin synthetase isolated from parsley does not require metal ions, NAD+, or other cofactors, and is soluble. It is inhibited by several heavy metals, but not by tetra-N-

(120) R. Ortmann, H. Sandermann, Jr., and H. Grisebach, FEBS Lett., 7, 164–166 (1970).

(121) R. Ortmann, A. Sutter, and H. Grisebach, Biochim. Biophys. Acta, 289, 293–302 (1972).

acetylethylenediamine.[31] The pH optimum for the apiin synthetase isolated from cell-suspension cultures of parsley was reported[121] to be pH 7.0. When isolated from parsley leaves,[31] it has a broad, optimal range of activity of pH 7.6–8.4. The enzyme is specific for UDP-apiose as the glycosyl donor.[31,120] UDP-D-xylose, UDP-D-glucuronic acid, UDP-D-galactose, UDP-arabinose (chirality unspecified), TDP-D-glucose, and UDP-D-glucose do not function as glycosyl donors.[31,121] 7-D-Glucosides of flavones, flavanones, and isoflavones, as well as apigenin-7 D-glucosiduronic acid and glucosides of p-substituted phenols, can function as acceptors. 4',5-Dihydroxyflavon-7-yl β-D-glucopyranoside and biochanin A [5-hydroxy-4'-methoxy-isoflavon-7-yl β-D-glucopyranoside] are the best acceptors. No transfer of D-apiose from UDP-apiose was observed[121] with the following compounds included as glycosyl-acceptor molecules: flavonol-3 D-glucoside, flavonol-7 D-glucoside, 8-C-D-glucosylapigenin, aglycons of flavonoids, or free D-glucose. In chromatography on both DEAE-Sephadex and Sephadex G-100, apiin synthetase migrated distinctly from UDP-GlcA cyclase and UDP-GlcA decarboxylase, and could readily be separated from them.[31] Its molecular weight is 50,000, and it has[121] an isoelectric point of 4.8.

Apiin synthetase is inhibited by agents reacting with sulfhydryl groups.[121] Glycerol, low temperatures, and compounds containing free thiol groups help stabilize the enzyme.[31] Apiin synthetase has been isolated from the leaves, stems, and roots of parsley. The activity found in the stems and roots was about a third and a quarter, respectively, of the activity found in the same weight of leaves.[31] The K_m for UDP-apiose was [31] 6 μM. The K_m value for 4',5-dihydroxyflavon-7-yl β-D-glucopyranoside was reported as[31] 70 μM and[120] 66 μM. Uridine, its 5'-phosphate, 5'-pyrophosphate, and 5'-triphosphate, and D-galactose, UDP-D-glucuronic acid, and UDP-D-xylose inhibit apiin synthetase to some extent, but much lower concentrations of uridine 5'-phosphate and 5'-pyrophosphate inhibit it.[31,121] UDP-D-glucuronic acid, UDP-D-xylose, and uridine 5'-triphosphate at 100 μM concentration decrease the apiin synthetase activity by only about 20%; uridine 5'-pyrophosphate was just as inhibitory when present at 1/100th of this concentration.[31] UDP-D-glucuronic acid and UDP-D-xylose at 100 μM concentration decrease the apiin synthase activity only ~10% from that obtained with 10 μM and 1 μM concentrations of these compounds. Based upon these results, it was concluded that UDP-D-xylose, formed with the UDP-apiose in the standard assay, or added with exogenous UDP-apiose in the modified, standard assay, probably does not significantly affect the activity of apiin transferase.[31]

A requirement for adenosine 5'-triphosphate (ATP) in the glycosylation reactions of some flavonoids has been noted.[119] It was suggested that the reaction requires ATP, in order to prevent the degradation of the "sugar nucleotides" involved in the transglycosylation reactions by enzymes present in the crude, mung-bean extract employed.[119] Formation of apiin by apiin synthetase was not stimulated[31,121] by the addition of uridine or adenosine 5'-triphosphates. It may be of physiological importance that uridine 5'-pyrophosphate significantly inhibits apiin synthetase activity, even at very low concentrations.[31] Apiin synthetase activity was inhibited 13% by 1 μM uridine 5'-pyrophosphate. This result suggested that the uridine 5'-pyrophosphate formed by apiin synthetase must be removed in order that transglycosylation reactions may proceed at high rates.[115] Control of the biosynthesis of polysaccharides may also occur through the regulation of the production of precursor "sugar nucleotides," or through modulation of the activity of transferase enzymes.[114] It should be noted that "sugar nucleotides" of D-apiose that contain bases other than uridine have not yet been prepared and tested as substrates.[7,31,121]

The isolation of a "glucoapiosyl-apigenin," perhaps apiin, from *Digitalis purpurea* (foxglove)[66] has been accomplished. This result is supported by the observation that a compound that migrated identically to apiin, and required 4',5-dihydroxyflavon-7-yl β-D-glucopyranoside and UDP-apiose as substrates, was formed by a partially purified extract isolated from *D. purpurea*.[31] A second type of transglycosylation reaction involving D-apiose was discovered with a glycosidase from *Viburnum furcatum*.[53] Furcatin, an apiose-containing glycoside, is hydrolyzed by a glycosidase, found in the leaves of *Viburnum furcatum*, into a disaccharide (a 6-O-apiofuranosyl-β-glycopyranose) and *p*-vinylphenol. The K_m value for furcatin is high (3.6 mM). Like some other glycosidases, furcatinase is sensitive to sulfhydryl reagents; this result was verified by reversal, or prevention, of inhibition by cysteine.[53] Furcatinase acts like other glycosidases when incubated with furcatin and an acceptor molecule that has a phenolic or alcoholic hydroxyl group; the sugar moiety is transferred to the hydroxyl group of the acceptor. The rate of transglycosylation varies with the chain length of the acceptor alcohol, the rate of the transglycosylation being in the order propyl alcohol > ethanol > methanol. The primary hydroxyl group of butyl alcohol is a more effective acceptor than a secondary alcoholic group, which is more effective than a tertiary alcoholic group.[53] Acid hydrolysis followed by paper chromatography confirmed that each new transglycosylation product contained both glycose and apiose. Furcatinase is specific for both the $(1 \rightarrow 6)$ bond of the 6-O-apiofuranosyl-β-D-glucopyranoside moiety and for the aglycon (as the enzyme will not hydrolyze rutin and apiin). It only slowly hydrolyzes alkyl 6-O-apiofuranosyl-β-D-glucopyranosides.[53]

VI. CHEMICAL SYNTHESIS OF D-APIOSE AND ITS ANALOGS

1. Chemical Synthesis of D-Apiose

Naturally occurring apiose appears to be D-apiose.[18,20,122] DL-Apiose was first chemically synthesized[17] in 1955.

Chemical synthesis of D-apiose has been accomplished by a number of methods.[18,20,63] A one-step conversion of the branched-chain pentose 4-C-(hydroxymethyl)-D-*threo*-pentose (9) yielded[18]

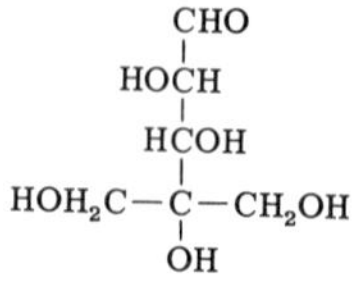

$$
\begin{array}{c}
\text{CHO} \\
|\\
\text{HOCH} \\
|\\
\text{HCOH} \\
|\\
\text{HOH}_2\text{C}-\overset{|}{\underset{|}{\text{C}}}-\text{CH}_2\text{OH} \\
\text{OH}
\end{array}
$$

9

only trace amounts of D-apiose (1; see p. 138). However, Gorin and Perlin[18] succeeded in synthesizing D-apiose by another pathway, starting with the benzylation of 1,2:4,5-di-O-isopropylidene-D-fructose (10) to give 3-O-benzyl-1,2:4,5-di-O-isopropylidene-D-fructose (11). Acid hydrolysis of 11 (first with 50% acetic acid and then 0.05 M sulfuric acid) yielded syrupy 3-O-benzyl-D-fructose (12). This product was treated with sodium cyanide, the resulting nitriles (13) were subjected to alkaline hydrolysis, the mixture of products was acidified, and the acids (14) were lactonized, to give a mixture of the 3-O-benzyl lactones (15). Reduction of the mixture (15) with sodium borohydride gave 3-O-benzyl-2-C-(hydroxymethyl)-D-*arabino*-hexitol (16), which was oxidized (with lead tetraacetate in acetic acid). The resulting aldose (17) was then treated with methanolic hydrogen

(122) O. T. Schmidt, *Ann.*, 483, 115–123 (1930).

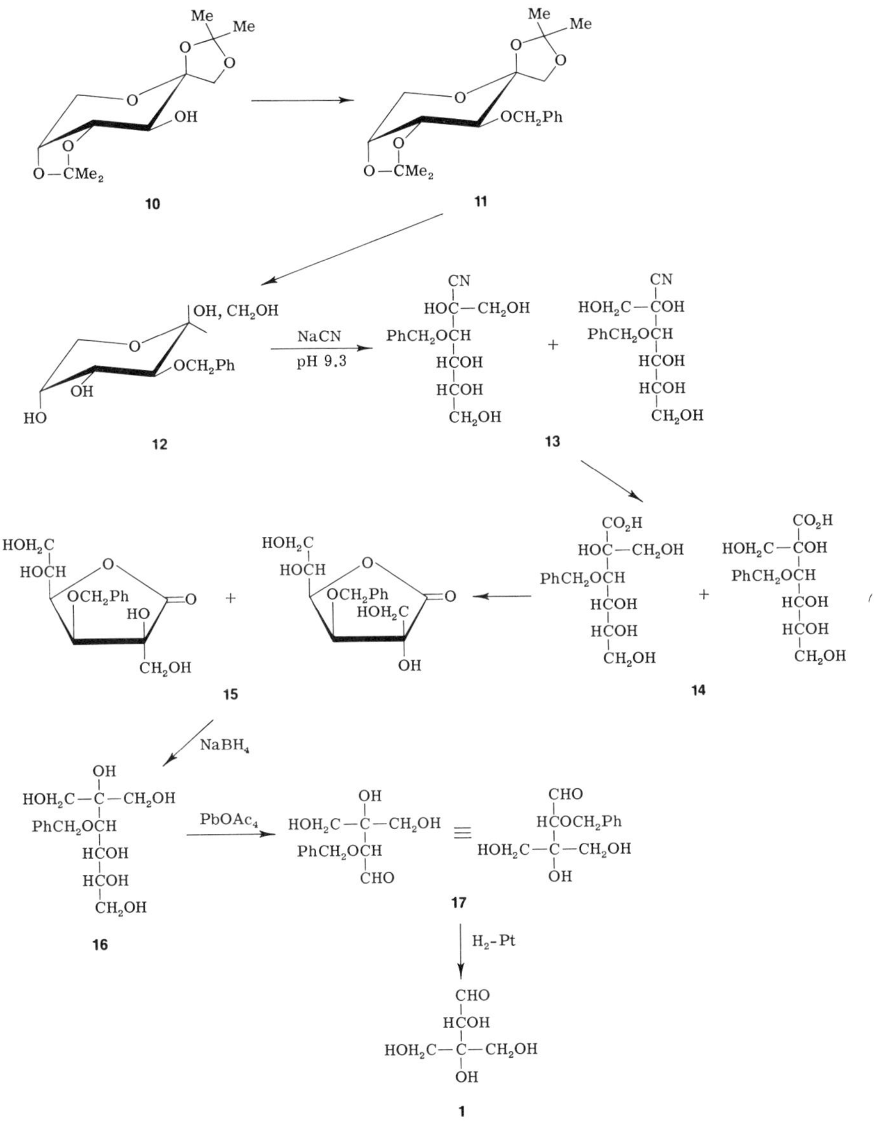

chloride, the resulting glycoside debenzylated by hydrogenolysis in the presence of palladium, and the glycoside hydrolyzed with 0.15 M sulfuric acid. The resulting product contained three components as detected by paper chromatography; the component corresponding to D-apiose (1) was separated from the other two products by column chromatography. D-Apiose was characterized[18] by optical rotatory data, by its infrared absorption spectrum, and by treatment with (2,5-dichlorophenyl)hydrazine to give the characteristic, substituted osazone.

Khalique[20] synthesized D-apiose by treating keto-D-fructose pentaacetate (18) with diazomethane, and deacetylating the product (19) with barium methoxide to give epoxide (20), which, by oxidation with 2 equivalents of sodium metaperiodate, was converted into a syrupy epoxy-aldehyde (21), characterized as its crystalline phenylosazone. Hydrolysis of compound 21 with aqueous acid produced

$$
\begin{array}{ccc}
\text{CH}_2\text{OAc} & \text{CH}_2\text{OAc} & \text{CH}_2\text{OH} \\
| & |\!\diagdown\!\text{O} & |\!\diagdown\!\text{O} \\
\text{C=O} & \text{C}\!-\!\text{CH}_2 & \text{C}\!-\!\text{CH}_2 \\
| & | & | \\
\text{AcOCH} & \text{AcOCH} & \text{HOCH} \\
| & | & | \\
\text{HCOAc} & \text{HCOAc} & \text{HCOH} \\
| & | & | \\
\text{HCOAc} & \text{HCOAc} & \text{HCOH} \\
| & | & | \\
\text{CH}_2\text{OAc} & \text{CH}_2\text{OAc} & \text{CH}_2\text{OH} \\
\textbf{18} & \textbf{19} & \textbf{20}
\end{array}
$$

CH$_2$N$_2$; Ba(OMe)$_2$

$$
\begin{array}{cc}
\text{CH}_2\text{OH} & \text{CHO} \\
|\!\diagdown\!\text{O} & | \\
\text{C}\!-\!\text{CH}_2 \quad\equiv\quad \text{H}_2\text{C}\!\diagdown\!\overset{\text{O}}{-}\text{C} & \text{HCOH} \\
| & | \\
\text{HOCH} & \text{CH}_2\text{OH} \\
| & \\
\text{CHO} & \\
\textbf{21} &
\end{array}
\qquad
\begin{array}{c}
\text{CHO} \\
| \\
\text{HCOH} \\
| \\
\text{HOH}_2\text{C}\!-\!\text{C}\!-\!\text{CH}_2\text{OH} \\
| \\
\text{OH} \\
\textbf{1}
\end{array}
$$

D-apiose (1), characterized by optical rotatory data, paper-chromatographic mobility, and the melting point of the (p-bromophenyl)-osazone and the diisopropylidene acetal.[20]

A third method[63] of synthesizing D-apiose consists in treating 2,3:4,5-di-O-isopropylidene-aldehydo-L-arabinose (22) with formaldehyde and sodium hydroxide to yield 4-C-(hydroxymethyl)-1,2:3,4-di-O-isopropylidene-L-lyxitol (23). Partial hydrolysis of 23 with acid to give 24, with subsequent acetylation of compound 24, gave a tetra-O-acetyl-O-isopropylidene derivative (25). The latter, upon oxidation with periodate, produced 26; deacetylation then gave 3-C-(hydroxymethyl)-2,3-O-isopropylidene-aldehydo-L-threose (27). Hydrolysis of compound 27 with aqueous acid then gave D-apiose (1), which was characterized as its 1-(2-benzyl-2-phenyl)hydrazone.

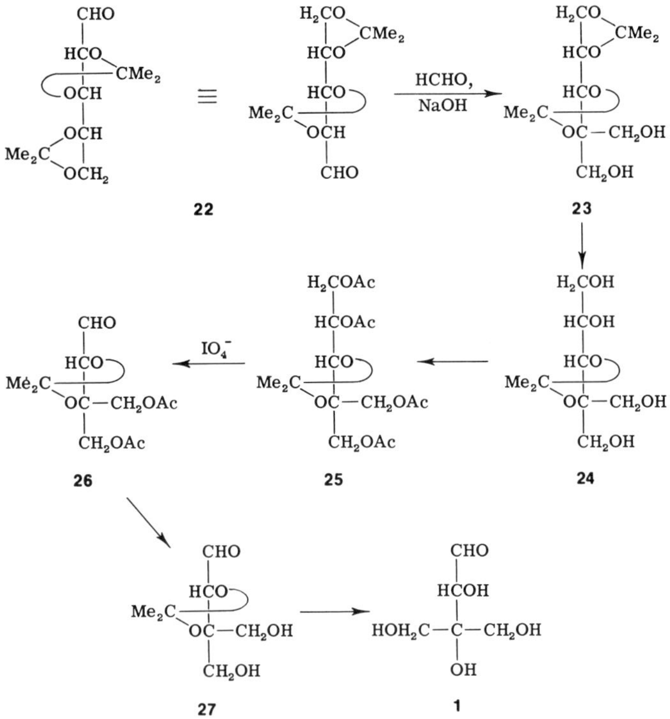

D-Apiose has been called a "Type A" branched-chain sugar; the branching is by substitution of a hydrogen atom, not replacement of a hydroxyl group ("Type B").[123] A general route for the synthesis of branched-chain sugars of Type A has been proposed.[8,124,125] Treatment of a 5-deoxy-1,2-O-isopropylidene-glycofuranos-3-ulose with (cyanomethylene)triphenylphosphorane yields the two isomers of a branched-chain, unsaturated sugar. cis-Dihydroxylation of these compounds, and subsequent reduction with borohydride, yields the corresponding hydroxymethyl, branched-chain sugars.[124]

Synthesis of D-apiose from 1,2-O-isopropylidene-β-L-threofuranose (28) has been accomplished.[126] Oxidation with ruthenium tetraoxide

(123) W. G. Overend, Chem. Ind. (London), 342–354 (1963).
(124) J. M. J. Tronchet, R. Graf, and R. Gurny, Helv. Chim. Acta, 55, 613–623 (1972).
(125) J. M. J. Tronchet, J. M. Bourgeois, J. M. Chalet, R. Graf, R. Gurny, and J. Tronchet, Helv. Chim. Acta, 54, 687–691 (1971).
(126) A. D. Ezekiel, W. G. Overend, and N. R. Williams, Tetrahedron Lett., 1635–1638 (1969).

yielded the crystalline tetros-3-ulose derivative **29** which, on subsequent treatment with diazomethane in 1:1 methanol–ether, gave a
mixture of epoxides. The major one was 3,3^1-anhydro-3-C-(hydroxymethyl)-1,2-O-isopropylidene-α-D-erythrofuranose (**30**), which, when
treated with aqueous methanolic sodium hydroxide for 2 days at room
temperature, gave[126] 3-C-(hydroxymethyl)-1,2-O-isopropylidene-α-D-
erythrofuranose (**31**). Hydrolysis of **31** with acid then yielded[126,127] D-
apiose (**1**).

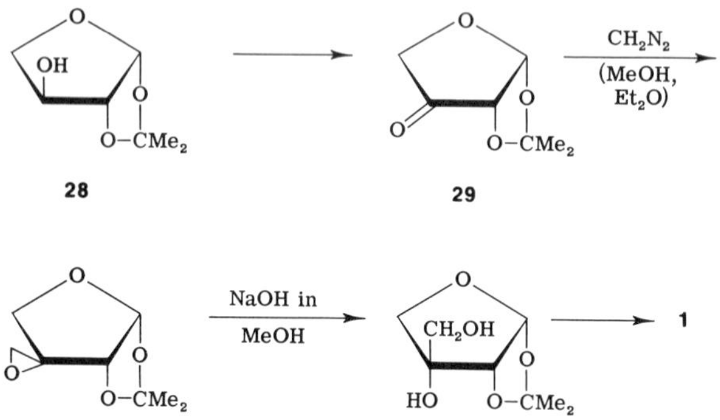

2. Chemical Synthesis of L-Apiose

The chemical synthesis of L-apiose has been accomplished by
starting with (*a*) L-threaric [(+)-tartaric] acid,[19] (*b*) 2,3-O-
isopropylidene-β-L-*erythro*-pentopyranosid-4-ulose,[128] or (*c*) an isopropylidene acetal of a pentodialdo-1,4-furanose.[129]

Synthesis of L-apiose from L-threaric acid [(+)-tartaric acid] (**32**)
was achieved[19] by first lengthening the carbon chain by forming,
through a series of steps, methyl 2,3-O-acetyl-5-amino-5-deoxy-L-
threo-4-pentulosonate (**33**). Compound **33** was, in turn, converted
into methyl 2,3,5-tri-O-acetyl-4,4^1-anhydro-4-C-(hydroxymethyl)-L-
threo-tetronate (**34**). De-esterification and opening of the epoxide ring
of compound **34** gave 4-C-(hydroxymethyl)-L-*threo*-pentonic acid
(**35**), and subsequent degradation of **35** yielded L-apiose (**36**).

(127) J. M. J. Tronchet and J. Tronchet, *Compt. Rend., Ser. C*, **267**, 626–629 (1968).
(128) W. G. Overend, A. C. White, and N. R. Williams, *Carbohyd. Res.*, **15**, 185–195
       (1970).
(129) R. Schaffer, *J. Amer. Chem. Soc.*, **81**, 5452–5454 (1959).

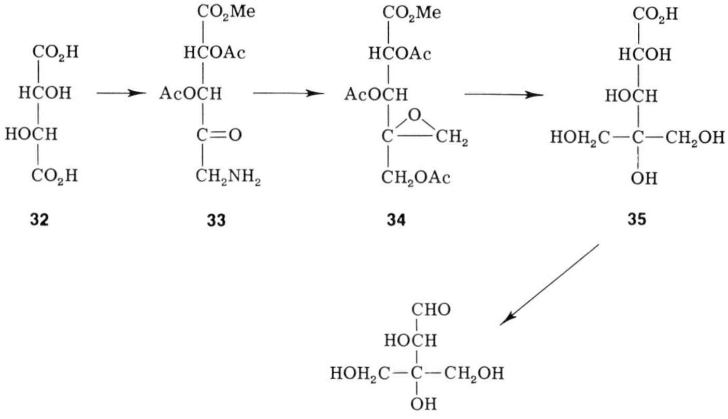

$$
\begin{array}{cccc}
 & CO_2Me & CO_2Me & CO_2H \\
CO_2H & HCOAc & HCOAc & HCOH \\
HCOH & AcOCH & AcOCH & HOCH \\
HOCH & C{=}O & C\!\!\begin{array}{c}O\\ \diagdown CH_2\end{array} & HOH_2C-\overset{\displaystyle OH}{\underset{\displaystyle |}{C}}-CH_2OH \\
CO_2H & CH_2NH_2 & CH_2OAc & \\
\mathbf{32} & \mathbf{33} & \mathbf{34} & \mathbf{35}
\end{array}
$$

$$
\begin{array}{c}
CHO \\
HOCH \\
HOH_2C-\overset{\displaystyle}{\underset{\displaystyle OH}{C}}-CH_2OH \\
\text{L-Apiose} \\
\mathbf{36}
\end{array}
$$

A mixed aldol reaction[129] of formaldehyde with 1,2-*O*-isopropylidene-α-D-*xylo*-pentodialdo-1,4-furanose (**37**) in sodium hydroxide solution gave 4-*C*-(hydroxymethyl)-1,2-*O*-isopropylidene-L-*threo*-pentofuranose (**38**). Hydrolysis of compound **38** with aqueous acid, to give **39**, followed by oxidation with bromine in the presence of calcium carbonate, resulted in the formation of calcium 4-*C*-(hydroxymethyl)-L-*threo*-pentonate pentahydrate (**40**) which, in turn, was converted into L-apiose (**36**). These almost simultaneous experiments of Weygand and Schmiechen[19] and Schaffer[129] were the first published syntheses of L-apiose.

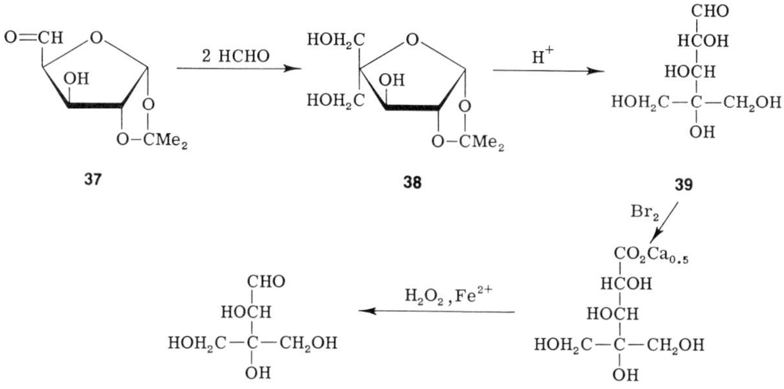

37 → (2 HCHO) → **38** → (H⁺) → **39**

39 → (Br₂) → **40** (CO₂Ca₀.₅) → (H₂O₂, Fe²⁺) → **36**

A different scheme for the synthesis of L-apiose, yielding a large array of branched-chain sugar derivatives, has been reported.[128] The oxidation of methyl 2,3-O-isopropylidene-α-D-lyxopyranoside (41) with ruthenium tetraoxide in carbon tetrachloride resulted in the formation of colorless, syrupy methyl 2,3-O-isopropylidene-β-L-erythro-pentopyranosid-4-ulose (42). Treatment of compound 42 with phenylethynylmagnesium bromide in ether, followed by deacetonation, gave methyl 4-C-(phenylethynyl)-β-L-ribopyranoside (43). Partial hydrogenation of the alkyne group in the presence of Lindlar catalyst, followed by ozonolysis, yielded methyl 4-C-formyl-β-L-ribopyranoside (44). Hydrogenation of compound 44 in the presence of Adams' catalyst gave methyl 4-C-(hydroxymethyl)-β-L-ribopyranoside (45). An alternative route to compound 45 consisted in treatment of the pyranosidulose 42 with ethereal diazomethane to give a mixture of two spiro-epoxides which, on alkaline hydrolysis followed by hydrolysis with dilute acid, afforded a mixture of 4-C-(hydroxymethyl)pentosides, separable by column chromatography on cellulose to yield compound 45. Hydrolysis of compound 45 with acid gave 4-C-(hydroxymethyl)-L-erythro-pentose (46) which, upon oxidation with bromine in the presence of barium benzoate, yielded 4-C-(hydroxymethyl)-L-erythro-pentono-1,4-lactone (47), Ruff degradation of which gave L-apiose (36) in low yield.

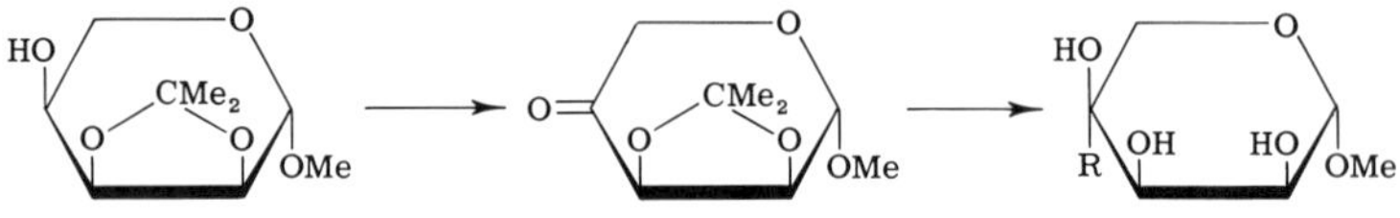

41   42   43 R = C≡C—Ph
                                                        44 R = CHO
                                                        45 R = CH₂OH

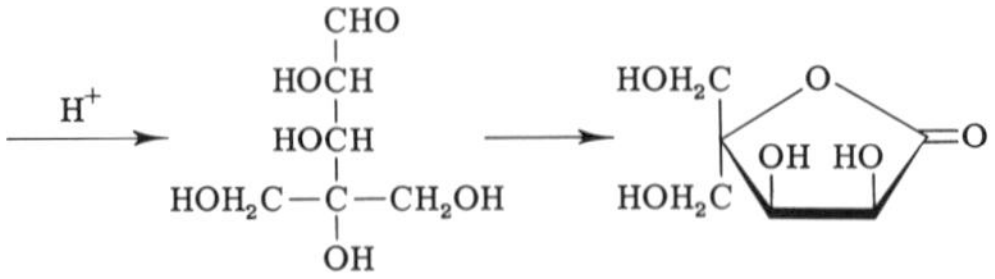

46   47

3. Chemical Synthesis of Derivatives of Apiose

The di-O-isopropylidene derivatives of D-apiose are prepared by treating the sugar with acetone containing 5% (v/v) of concentrated sulfuric acid.[61,63,130] The structures of these derivatives were first

(130) D. J. Bell, Methods Carbohyd. Chem., 1, 260–263 (1962).

postulated[61,63] and subsequently confirmed,[131,132] to be 1,2:3,3'-di-*O*-isopropylidene-[3-*C*-(hydroxymethyl)-β-L-threofuranose][131] (**48**) and 1,2:3,3'-di-*O*-isopropylidene-[3-*C*-(hydroxymethyl)-α-D-erythrofuranose][132] (**49**). If the reaction mixture is allowed to reach equilibrium, only compound **48** can be isolated. However, if the reaction is stopped (by neutralization) before equilibrium is reached, compounds **48** and **49** can both be isolated.[132]

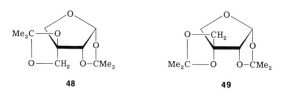

48   **49**

Methylation of D-apiose[9,63,124] and of derivatives of apiose[5,9,128,133,134] has been reported. Methyl glycosidation of D-apiose is accomplished by heating the sugar under reflux with methanolic hydrogen chloride (17%) for 16 hours;[63] methylation yields four isomers of methyl tri-*O*-methyl-D-apioside.[124] 3-*C*-(Hydroxymethyl)-1,2-*O*-isopropylidene-α-D-erythrofuranose,[134] 4'-*O*-methylapiin,[133] 3-*C*-(hydroxymethyl)-2,3-*O*-isopropylidene-β-D-erythrofuranose,[9] and various other D-apiose derivatives have been prepared.

Osazones have been synthesized from D-apiose,[61,63,130,132] and are useful for identifying the sugar.

The value of some nucleosides for use as flavor enhancers,[134a] and the possible antiviral or antitumor activity of certain nucleosides,[135] have led to the synthesis of purine nucleosides containing analogs of D-apiose in which the ring-oxygen atom of the furanose forms has been replaced.[136] Monomolar *p*-toluenesulfonylation of 3-*C*-(hydroxymethyl)-1,2-*O*-isopropylidene-β-L-threofuranose (**50**) yielded 3-*C*-(hydroxymethyl)-1,2-*O*-isopropylidene-3'-*O*-*p*-tolylsulfonyl-β-L-threofuranose (**51**) which, through a series of steps, was converted into either methyl 2,3-*O*-isopropylidene-[3-*C*-(hydroxymethyl)-4-thio-β-D-

(131) F. A. Carey, D. H. Ball, and L. Long, Jr., *Carbohyd. Res.*, 3, 205–213 (1966).
(132) D. H. Ball, F. A. Carey, I. L. Klundt, and L. Long, Jr., *Carbohyd. Res.*, 10, 121–128 (1969).
(133) W. G. Overend, A. D. Ezekiel, and N. R. Williams., *J. Chem. Soc.* (*C*), 2907–2911 (1971).
(134) A. D. Ezekiel, W. G. Overend, and N. R. Williams, *Carbohyd. Res.*, 20, 251–257 (1971).
(134a) D. H. Ball, Personal communication.
(135) F. Perini, F. A. Carey, and L. Long, Jr., *Carbohyd. Res.*, 11, 159–161 (1969).
(136) M. H. Halford, D. H. Ball, and L. Long, Jr., *Carbohyd. Res.*, 8, 363–365 (1968).

erythrofuranoside][136] (**52**) or methyl 2,3-O-isopropylidene-[N-acetyl-4-amino-4-deoxy-3-C-(hydroxymethyl)-β-D-erythrofuranoside][137] (**53**).

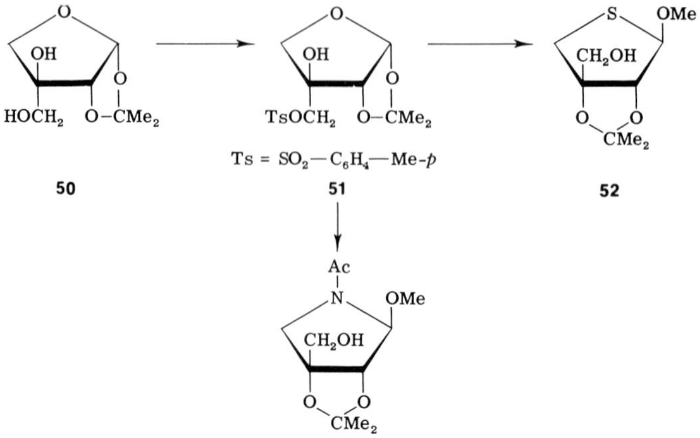

Ts = SO_2—C_6H_4—Me-p

50   **51**   **52**

53

The first synthetic apiosyl nucleoside having a defined structure[135] was 2-chloro-9-[3-C-(hydroxymethyl)-α-L-threofuranosyl]adenine (**54**). The synthesis of 9-[3-C-(hydroxymethyl)-β-L-threofuranosyl]-adenine (**55**) has since been accomplished,[138] and the bacteriostatic effects of the compound assayed. Inhibition of growth of both *Escherichia coli* and *Staphylococcus aureus* occurred in liquid cultures.[76]

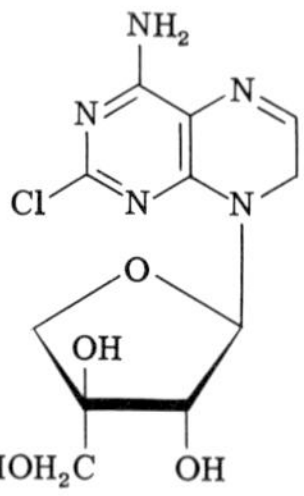

54 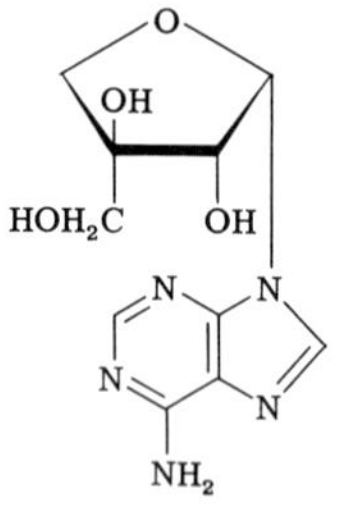   **55**

(137) M. H. Halford, D. H. Ball, and L. Long, Jr., *Chem. Commun.*, 255–256 (1969).
(138) J. M. J. Tronchet and J. Tronchet, *Helv. Chim. Acta*, **53**, 853–856 (1970).

SPECIFIC DEGRADATION OF POLYSACCHARIDES

By Bengt Lindberg, Jörgen Lönngren, and Sigfrid Svensson*

Department of Organic Chemistry, Arrhenius Laboratory, University of Stockholm,
S-104 05 Stockholm, Sweden

I. Introduction

The elucidation of the structure of polysaccharides and of the carbohydrate portion of glycoconjugates involves sugar analysis, linkage analysis, and the determination of anomeric configurations and the sequences of the sugar residues. When non-sugar substituents are present, these should also be identified and located.

There are several good methods for sugar analysis. Linkage analy-

* Present address: Department of Clinical Chemistry, University Hospital, S-221 85 Lund, Sweden.

sis, by methylation analysis,[1] has now been developed into a rapid and accurate method. The determination of anomeric configurations and sequences is more complicated, and there is no generally applicable procedure. Anomeric configurations may be determined by chemical, spectroscopic, or biochemical methods, often in conjunction with the characterization of fragments obtained by degradation.

The method most commonly used to determine sequences is graded hydrolysis with acid, followed by fractionation and identification of the resulting oligosaccharides. The information obtained by this method is qualitative only, and it may be difficult to assess the structural significance of oligosaccharides formed in low yield. There is, therefore, a need for more selective or, preferably, specific methods for the degradation of polysaccharides.

For polysaccharides composed of oligosaccharide repeating units, the structural studies may ultimately lead to a complete structure. Other polysaccharides have less-ordered structures, and the aim of the structural studies should be to define the factors governing these structures. Polysaccharide methodology has been summarized[2] in two review articles.

Only a few methods for the specific degradation of polysaccharides were known when an earlier article on structural polysaccharide chemistry appeared in this Series.[3] The Smith degradation,[4] which has become the most frequently used, had only just been introduced. Since then, a number of specific degradation techniques have been developed. In addition, there have also been some modifications and improvements to existing methods. In this article, chemical methods for specific or selective degradation of polysaccharides, and their applications in structural analysis, will be discussed. Enzymic methods, which may be of considerable value, have already been treated in this Series.[5]

(1) B. Lindberg, Methods Enzymol., 28B, 178–195 (1972).

(2) G. O. Aspinall and A. M. Stephen, in "M T P International Review of Science," Org. Chem. Ser. 1, G. O. Aspinall, ed., Butterworths, London, 1973, Vol. 7, pp. 285–317; and G. O. Aspinall, in "Elucidation of Organic Structures by Physical and Chemical Methods," K. W. Bentley and G. W. Kirby, eds., Wiley, New York, 1973, Vol. IV, part II, pp. 379–450.

(3) H. O. Bouveng and B. Lindberg, Advan. Carbohyd. Chem., 15, 53–89 (1960).

(4) I. J. Goldstein, G. W. Hay, B. A. Lewis, and F. Smith, Abstr. Papers Amer. Chem. Soc. Meeting, 135, 3D (1959).

(5) J. J. Marshall, Advan. Carbohyd. Chem. Biochem., 30, 257–370 (1974).

II. Degradations by the Action of Acids

1. General Remarks

The degradation of polysaccharides by the action of acids, followed by the fractionation and identification of the oligosaccharides formed, is one of the methods used frequently in structural polysaccharide chemistry, and it gives information both on sequences and anomeric configurations. Often, the rates of hydrolysis of the various glycosidic linkages do not differ significantly, and complex mixtures of oligosaccharides are obtained. There are, however, several polysaccharides in which some of the glycosidic linkages differ considerably in lability; alternatively, it is sometimes possible to create such linkages by chemical modification of the polysaccharide. Under such conditions, fragmentation can be performed with a high degree of selectivity, and it is mainly these examples that will be discussed.

2. Hydrolysis

Weak glycosidic linkages in polysaccharides are most frequently associated with furanose and deoxy sugar residues.[6] Ribose, fructose, and apiose occur only as furanosides in Nature, and arabinose and galactose may be either furanosidic or pyranosidic. 6-Deoxyhexoses are common, but their glycosidic linkages are hydrolyzed only about 5 times faster than those of the corresponding hexoses. 3,6-Dideoxyhexopyranosidic linkages, which occur in certain bacterial polysaccharides,[7] are hydrolyzed more readily. Glycosides having a deoxy function in the position vicinal to the glycosidic carbon atom are very sensitive to acid hydrolysis. Representatives of this class, occurring in polysaccharides and glycoconjugates, are the 3-deoxyglyculosonic acids, the most common being 3-deoxy-D-*manno*-octulosonic acid and neuraminic acid (5-amino-3,5-dideoxy-D-*glycero*-D-*galacto*-nonulosonic acid). Finally, 3,6-anhydrogalactopyranosidic linkages, occurring in galactans from red algae, are also readily hydrolyzed. Results from studies on the hydrolysis of glycosides of low molecular weight cannot always be extrapolated to polysaccharides, as emphasized in an article by BeMiller.[6]

(6) J. N. BeMiller, *Advan. Carbohyd. Chem.*, **22**, 25–108 (1967).
(7) O. Lüderitz, K. Jann, and R. Wheat, in "Comprehensive Biochemistry," M. Florkin, ed., Elsevier, Amsterdam, 1968, Vol. 26A, pp. 105–228.

The controlled, acid hydrolysis of polysaccharides containing a limited number of acid-labile glycosidic linkages will produce mono-saccharides or oligosaccharides, or both, as a result of cleavage of the labile linkages. Investigation of these products, and of any residual polysaccharide backbone, will therefore provide information of structural significance.

Early examples involving furanosides are the isolation of 3-O-β-L-arabinopyranosyl-L-arabinose[8] and 3-O-β-L-arabinopyranosyl- and 3-O-α-D-xylopyranosyl-L-arabinose[9] by mild, acid hydrolysis of larch arabinogalactan and golden-apple gum, respectively. There are several examples from structure elucidations of gums and hemicelluloses in which graded, acid hydrolysis has been used to split off side chains linked to the main chains by way of a furanosidic sugar residue.[2,10–12] Partial hydrolysis with acid was applied in studies of the pentasaccharides obtained by dephosphorylation from the *Pneumococcus* types 29 (Ref. 13) and 34 (Ref. 14) capsular polysaccharides. The latter yielded D-galactose, 2-O-α-D-glucopyranosyl-D-galactose, and 2-O-α-D-galactopyranosyl-D-ribitol, thereby accounting for all the sugar residues in the pentasaccharide repeating unit (**1**).

$$\beta\text{-D-Gal}f\text{-}(1{\rightarrow}3)\text{-}\alpha\text{-D-Glc}p\text{-}(1{\rightarrow}2)\text{-}\beta\text{-D-Gal}f\text{-}(1{\rightarrow}3)\text{-}\alpha\text{-D-Gal}p\text{-}(1{\rightarrow}2)\text{-D-Ribitol}$$

1

An apiofuranosyl-apiose, apiobiose, is released on mild, acid hydrolysis of the apiogalacturonan from *Lemna minor*.[15]

Terminal 3,6-dideoxyhexoses that occur in lipopolysaccharides from *Salmonella* and *Yersinia* (*Pasteurella*) species[7] could be hydrolyzed off with a high degree of selectivity. They may, therefore, be located by methylation analysis of the original lipopolysaccharide and of a partially hydrolyzed sample. Thus, for the *Salmonella typhimurium* 395 MS lipopolysaccharide, composed[16] of oligosaccharide

(8) J. K. N. Jones, *J. Chem. Soc.*, 1672–1675 (1953).

(9) P. Andrews and J. K. N. Jones, *J. Chem. Soc.*, 4134–4138 (1954).

(10) G. O. Aspinall, *Advan. Carbohyd. Chem.*, **14**, 429–468 (1959).

(11) T. E. Timell, *Advan. Carbohyd. Chem.*, **19**, 249–302 (1964); **20**, 409–483 (1965).

(12) G. O. Aspinall, *Advan. Carbohyd. Chem. Biochem.*, **24**, 333–379 (1969).

(13) E. V. Rao, M. J. Watson, J. G. Buchanan, and J. Baddiley, *Biochem. J.*, **111**, 547–556 (1969).

(14) W. K. Roberts, J. G. Buchanan, and J. Baddiley, *Biochem. J.*, **88**, 1–7 (1963).

(15) D. A. Hart and P. K. Kindel, *Biochemistry*, 9, 2190–2196 (1970).

(16) C. G. Hellerqvist, B. Lindberg, S. Svensson, T. Holme, and A. A. Lindberg, *Carbohyd. Res.*, **8**, 43–55 (1968).

repeating units (2), all of the 4,6-di-O-methyl-D-mannose obtained

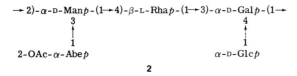

$$\rightarrow 2)\text{-}\alpha\text{-}\text{D-Man}p\text{-}(1\rightarrow4)\text{-}\beta\text{-}\text{L-Rha}p\text{-}(1\rightarrow3)\text{-}\alpha\text{-}\text{D-Gal}p\text{-}(1\rightarrow$$

2

from the original material appeared as 3,4,6-tri-O-methyl-D-mannose in the hydrolyzed sample, demonstrating that the 3,6-dideoxy-D-*xylo*-hexose (abequose) (3) was linked to O-3 of the D-mannosyl residue (4).

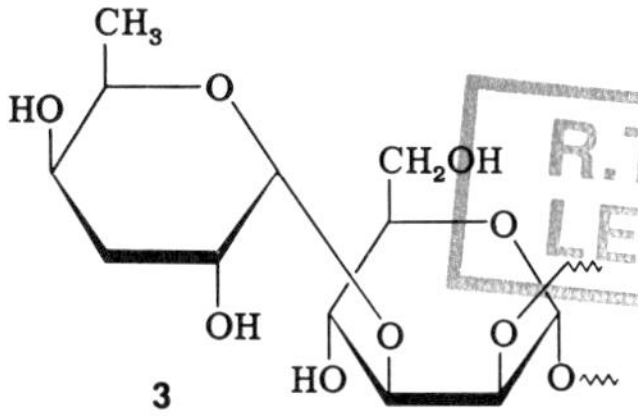

Similarly, methylation analysis of a partially hydrolyzed sample of the *Klebsiella* type 38 capsular polysaccharide, composed of pentasaccharide repeating units (5), revealed that the terminal 3-deoxy-L-*glycero*-pentulosylonic acid group (6) was located at O-3 of a D-galactosyl residue (7), as the 6-O-methyl-D-galactose in the methylation

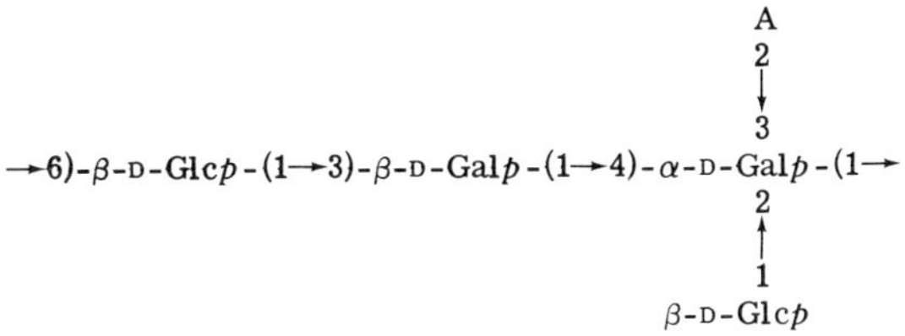

$$\rightarrow 6)\text{-}\beta\text{-}\text{D-Glc}p\text{-}(1\rightarrow3)\text{-}\beta\text{-}\text{D-Gal}p\text{-}(1\rightarrow4)\text{-}\alpha\text{-}\text{D-Gal}p\text{-}(1\rightarrow$$

5 (A = 6)

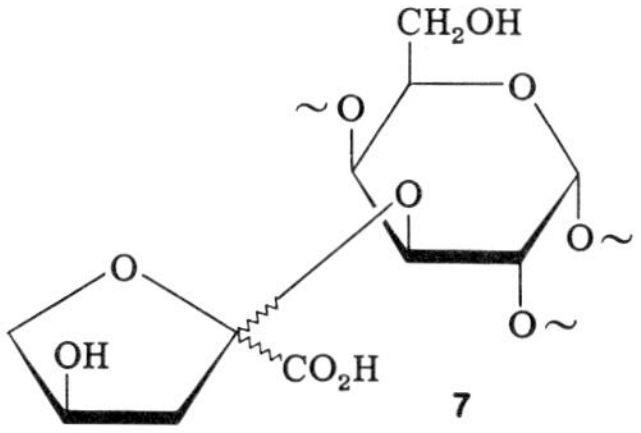

7

6

analysis of the original material was replaced by 3,6-di-O-methyl-D-galactose.[17] From the analysis, it was also evident that no other linkages had been hydrolyzed.

The polysaccharide component of a lipopolysaccharide can be separated from the lipid component by selective hydrolysis of the glycosidic linkages of the 3-deoxy-D-*manno*-octulosonic acid residues connecting these two components. The conditions for the hydrolysis are mild, namely, 0.1 M acetic acid for 1.5 h at 100° (Ref. 18). Similar conditions, namely, M formic acid for 1 h at 100° or 0.05 M hydrogen chloride in methanol for 1 h at 85°, were used to split off the sialic acid residues from gangliosides.[19,20]

The 3,6-anhydrogalactopyranose end-groups released on mild, acid hydrolysis of some galactans are acid-labile, and different procedures have been devised in order to protect them. Thus, Araki and Hirase and others have used mercaptolysis or methanolysis. With these procedures, good yields of agarobiose, 3,6-anhydro-4-O-β-D-galactopyranosyl-L-galactose, isolated as the diethyl dithioacetal[21] (**8a**) or dimethyl acetal[22] (**8b**), were obtained from agar. Mercaptolysis of an

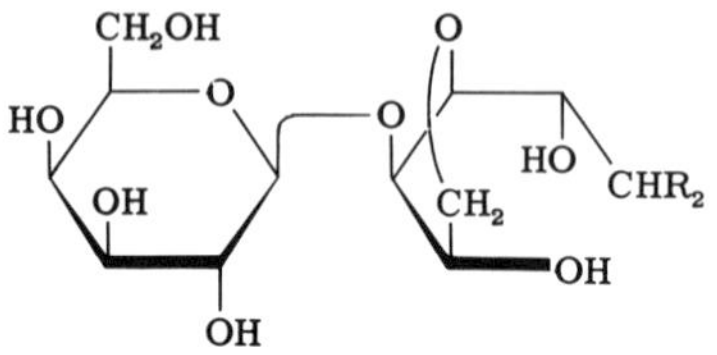

8a R = SEt
8b R = OMe

acid-labile sugar was also used by Wolfrom and coworkers in their studies on streptomycin.[23]

Wolfrom and Rice hydrolyzed heparin in the presence of bromine in order to convert the uronic acids released into aldaric acids.[24] The same method, oxidative hydrolysis, was used by Rees and coworkers in order to protect liberated 3,6-anhydrohexose residues. The

(17) B. Lindberg, K. Samuelsson, and W. Nimmich, *Carbohyd. Res.*, **30**, 63–70 (1973).
(18) A. M. Staub, *Methods Carbohyd. Chem.*, **5**, 93–95 (1965).
(19) M. Holm, J.-E. Månsson, and L. Svennerholm, *FEBS Lett.*, **38**, 271–273 (1964).
(20) K. Puro, *Biochim. Biophys. Acta*, **187**, 401–413 (1969).
(21) S. Hirase and C. Araki, *Bull. Chem. Soc. Jap.*, **27**, 105–109 (1954).
(22) C. Araki and S. Hirase, *Bull. Chem. Soc. Jap.*, **27**, 109–112 (1954).
(23) I. R. Hooper, L. H. Klemm, W. J. Polglase, and M. L. Wolfrom, *J. Amer. Chem. Soc.*, **68**, 2120–2121 (1946).
(24) M. L. Wolfrom and F. A. H. Rice, *J. Amer. Chem. Soc.*, **68**, 532 (1946).

method was first used for the complete hydrolysis of methylated carrageenans,[25] without decomposition of labile sugars. It was later used for partial hydrolysis of carrageenans, and good yields of oligosaccharide sulfates [for example, carrabonic acid 4-sulfate (9)] were obtained.[26]

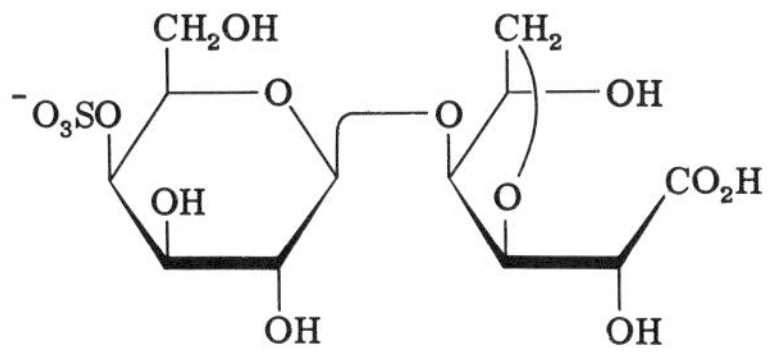

9

Structural studies on the oligosaccharide derivatives obtained by partial, acid hydrolysis of fully methylated polysaccharides often furnish valuable information on the positions at which the oligosaccharides were linked in the original polysaccharide. When the fully methylated *Klebsiella* O group 9 lipopolysaccharide,[27] which contains D-galactopyranose and D-galactofuranose residues, was subjected to mild, acid hydrolysis, and the product reduced with lithium aluminum deuteride and remethylated with trideuteriomethyl iodide, a good yield of the disaccharide derivative **10** was obtained.

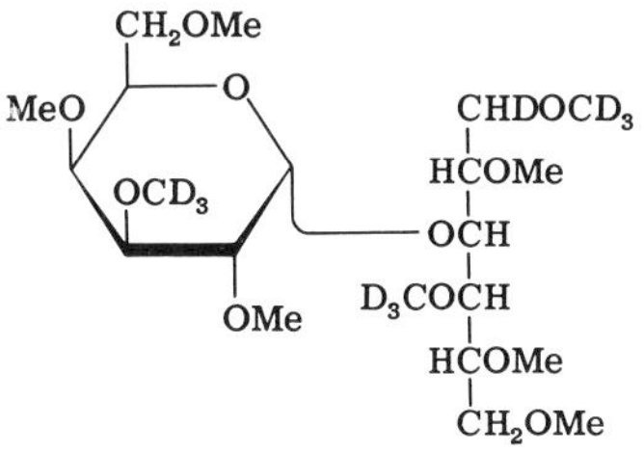

10

The trideuteriomethyl group at O-3 in the D-galactopyranose residue was located by mass spectrometry (m.s.), and its position demonstrated that it derived from a chain residue, substituted at this position; this was an important piece of evidence in determining the

(25) N. S. Anderson, T. C. S. Dolan, and D. A. Rees, *J. Chem. Soc., C*, 596–601 (1968).
(26) A. Penman and D. A. Rees, *J. Chem. Soc. Perkin I*, 2191–2196 (1973).
(27) B. Lindberg, J. Lönngren, and W. Nimmich, *Carbohyd. Res.*, **23**, 47–55 (1972).

sequence of sugar residues in the pentasaccharide repeating-unit (11). The fragment giving the disaccharide derivative is set off in the formula by dashes.

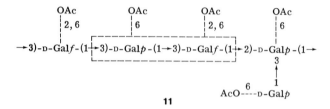

11

Partial hydrolysis of a fully methylated polysaccharide and investigation of the products as just described may provide valuable information, even when the partial hydrolysis is not very selective. The method has been used in studies of the *Klebsiella* type 38 (Ref. 17) and type 52 (Ref. 28) capsular polysaccharides. In their studies of the *Klebsiella* type 56 capsular polysaccharide, Choy and Dutton used a modification of this method.[29]

It is possible to introduce acid-labile linkages in polysaccharides by chemical modification. The most important example, the Smith degradation, will be treated separately (see Section III,3, p. 203). In an investigation by Gorin and Spencer, the α-D-glucopyranose (**12**) and 2-O-α-D-glucopyranosyl-α-D-mannopyranose side-chains were selectively removed from the (1 → 6)-linked chain (**13**) of a fungal glucomannan by converting them into 3,6-anhydro derivatives (**14**)

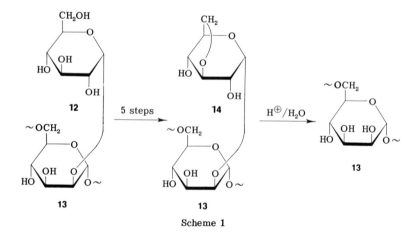

Scheme 1

(28) H. Björndal, B. Lindberg, J. Lönngren, M. Mészáros, J. L. Thompson, and W. Nimmich, *Carbohyd. Res.*, **31**, 93–100 (1973).

(29) Y. M. Choy and G. G. S. Dutton, *Can. J. Chem.*, **51**, 3021–3026 (1973).

by successive tritylation, acetylation, detritylation, p-toluenesul-
fonylation, and treatment with sodium methoxide, followed by mild,
acid hydrolysis (see Scheme 1).[30] The conditions needed for the
cleavage of the modified side-chains (0.33 M sulfuric acid at 100° for
5 h) were, however, more vigorous than expected.

Theander demonstrated that methyl hexodialdo-1,5-pyranosides,
which form a hemiacetal with the 3-hydroxyl group (as in **15**) are

15

hydrolyzed 10–100 times faster than their parent glycosides.[31,32] The
presence of such residues (formed during the isolation) may offer an
explanation of the so-called "weak linkages" in cellulose. Introduction of aldehyde groups at C-5 of hexose residues (for example, by
oxidation of terminal, α-D-galactopyranose residues with D-galactose
oxidase, or by controlled, periodate oxidation of heptopyranose residues unsubstituted at O-6 and O-7) could also give rise to acid-labile
linkages. This possibility does not, however, seem to have been
exploited.

The two most common types of acid-resistant glycosidic linkages
in polysaccharides are those of uronic acid residues and 2-amino-2-
deoxyhexose residues. Thus, in polysaccharides containing isolated
uronic acid residues, it is nearly always possible to obtain fair to
good yields of the corresponding aldobiouronic acids by acid hydrolysis. Numerous examples have been given in articles in this
Series.[10-12] The increased stability of the aldobiouronic acids to acid
hydrolysis has been discussed by BeMiller,[6] but is still not well understood. The differences in the rates of hydrolysis of alkyl hexo-
pyranosiduronic acids and the corresponding alkyl hexopyranosides
are not very large.[33] Capon and Ghosh[34] showed that the relative
rates of hydrolysis of 2-naphthyl β-D-glucopyranosiduronic acid and
2-naphthyl β-D-glucopyranoside are 1 : 45 in M hydrochloric acid, but
35 : 1 at pH 4.79; they attributed this effect to the different inductive
effects of carboxyl and carboxylate groups. The practical value of per-

(30) P. A. J. Gorin and J. F. T. Spencer, *Carbohyd. Res.*, **13**, 339–349 (1970).
(31) O. Theander, *Acta Chem. Scand.*, **18**, 1297–1300 (1964).
(32) B. Petterson and O. Theander, *Acta Chem. Scand.*, in press.
(33) M. D. Saunders and T. E. Timell, *Carbohyd. Res.*, **5**, 453–460 (1967).
(34) B. Capon and B. C. Ghosh, *Chem. Commun.*, 586–587 (1965).

forming graded hydrolysis, at pH 4–5, of polysaccharides containing uronic acid residues seems, however, to be limited, as it is accompanied by extensive epimerization and degradation.[35,36]

Uronic acid residues may be introduced into certain polysaccharides by oxidation of the primary hydroxyl groups with oxygen in the presence of a platinum catalyst, as described by Aspinall and coworkers.[37,38] The oxidation, which takes a considerable time and is incomplete, is accompanied by some depolymerization. The method has, nevertheless, furnished valuable structural information. Hydrolysis of an oxidized, larch arabinogalactan yielded 6-O-(L-arabinofuranosyluronic acid)-D-galactose, thus showing the location of the L-arabinofuranose residues in the polysaccharide.[37] Similar treatment of an arabinoxylan yielded 3-O-(L-arabinofuranosyluronic acid)-D-xylose, revealing that the terminal L-arabinofuranose residues are attached directly to the xylan chain.[38] Three groups of workers have used the method in studies of bacterial dextrans, and demonstrated that at least some of the side chains in these polysaccharides consist of single α-D-glucopyranose residues.[39–41]

Oxidation of terminal α-D-galactopyranose residues to α-D-*galacto*-hexodialdo-1,5-pyranose residues, using galactose oxidase, followed by treatment with hypoiodite, introduces D-galacturonic acid residues. This sequence has been applied to guaran; about 45% of the end groups were oxidized, and 6-O-(α-D-galactopyranosyluronic acid)-D-mannose was isolated after acid hydrolysis of the product.[42]

The stability of the glycosidic linkage in 2-amino-2-deoxyaldosides is due to the inductive effect of the ammonium ion formed in acid solution. In some polysaccharides, the amino groups are sulfated, and are readily desulfated during mild, acid hydrolysis (compare Ref. 43). In most polysaccharides containing 2-amino-2-deoxyaldose residues, the amino group is N-acetylated. In order to take advantage of the acid resistance of the 2-amino-2-deoxyaldosidic linkages, the polysaccharides have to be N-deacetylated. Different methods for N-deacetylation have been described by Hanessian.[44] Only N-

(35) T. Popoff and O. Theander, *Carbohyd. Res.*, **22**, 135–149 (1972).
(36) O. Theander, personal communication.
(37) G. O. Aspinall and A. Nicolson, *J. Chem. Soc.*, 2503–2507 (1960).
(38) G. O. Aspinall and I. M. Cairncross, *J. Chem. Soc.*, 3998–4000 (1960).
(39) D. Abbott, E. J. Bourne, and H. Weigel, *J. Chem. Soc.*, C, 827–831 (1966).
(40) B. Lindberg and S. Svensson, *Acta Chem. Scand.*, **22**, 1907–1912 (1968).
(41) H. Miyaji, A. Misaki, and M. Torii, *Carbohyd. Res.*, **31**, 277–287 (1973).
(42) J. K. Rogers and N. S. Thompson, *Carbohyd. Res.*, **7**, 66–75 (1968).
(43) F. Danishefsky, *Methods Carbohyd. Chem.*, **5**, 407–409 (1965).
(44) S. Hanessian, *Methods Carbohyd. Chem.*, **6**, 208–214 (1972); see also, D. Horton and D. R. Lineback, *ibid.*, **5**, 403–406 (1965).

deacetylation under basic conditions (for example, with aqueous sodium or barium hydroxide or with hydrazine) can be used for polysaccharides. Substituents at O-3 may decrease the extent of N-deacetylation.

Degradation of polysaccharides on treatment with hydrazine has been observed.[45-47] Kochetkov and coworkers[48,48a] compared various basic reagents used for this purpose, and obtained quantitative N-deacetylation with anhydrous hydrazine plus a catalytic amount of hydrazine sulfate, a reagent introduced by Yosizawa and coworkers.[49] They[48,48a] investigated two model substances, benzyl 2-acetamido-2-deoxy-3-O-β-D-galactopyranosyl-α-D-glucopyranoside (16) and benzyl 2-acetamido-2-deoxy-6-O-α-D-mannopyranosyl-α-D-glucopyranoside (18). In agreement with previous experience,[50] the derivative 16, containing the 3-O-substituted 2-acetamido-2-deoxy-D-glucose residue, was found more resistant to N-deacetylation than the other (18). After N-deacetylation, hydrolysis of 16 with M hydrochloric acid at 100° for 2 h resulted in a quantitative yield of D-galactose and benzyl 2-amino-2-deoxy-α-D-glucopyranoside (17). On similar treatment, compound 18 afforded, also in quantitative yield, a mixture of D-mannose and 17.

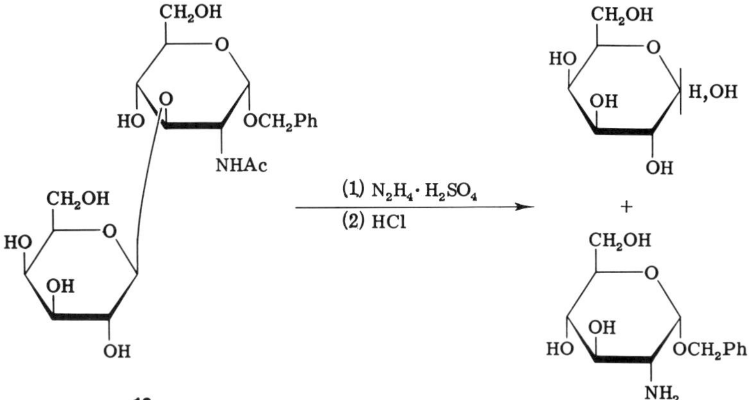

(45) M. L. Wolfrom and B. O. Juliano, *J. Amer. Chem. Soc.*, 82, 2588–2592 (1960).

(46) Z. Yosizawa and T. Sato, *Biochim. Biophys. Acta*, 52, 591–593 (1961).

(47) Z. Yosizawa and T. Sato, *J. Biochem.* (Tokyo), 51, 233–241 (1962).

(48) B. A. Dmitriev, Yu. A. Knirel, and N. K. Kochetkov, *Carbohyd. Res.*, 29, 451–457 (1973).

(48a) B. A. Dmitriev, Yu. A. Knirel, and N. K. Kochetkov, *Carbohyd. Res.*, 30, 45–50 (1973).

(49) Z. Yosizawa, T. Sato, and K. Schmid, *Biochim. Biophys. Acta*, 121, 417–420 (1966).

(50) M. Fujinaga and Y. Matsushima, *Bull. Chem. Soc. Jap.*, 39, 185–190 (1966).

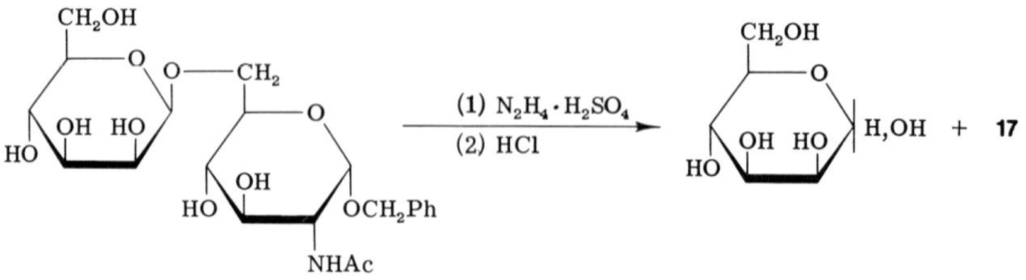

18

In the literature, there are several examples of partial hydrolysis of polysaccharides or glycoconjugates containing 2-amino-2-deoxyhexose residues. One example is the hydrolysis of carboxyl-reduced, N-deacetylated chondroitin, with the formation of 4-O-(2-amino-2-deoxy-β-D-galactopyranosyl)-D-glucose.[51] Another is the isolation of 6-O-(2-amino-2-deoxy-β-D-galactopyranosyl)-D-galactose after N-deacetylation and partial hydrolysis of the pentasaccharide (**19**) obtained

$$\beta\text{-}D\text{-}GalNAc}p\text{-}(1\rightarrow 6)\text{-}\beta\text{-}D\text{-}Gal}f\text{-}(1\rightarrow 3)\text{-}\beta\text{-}D\text{-}Gal}p\text{-}(1\rightarrow 6)\text{-}\beta\text{-}D\text{-}Gal}f\text{-}(1\rightarrow 1)\text{-}D\text{-}Ribitol$$

19

on dephosphorylation of *Pneumococcus* type 29 capsular polysaccharide.[13] With the improvements in the technique already discussed, the method will probably become more widely used.

The rate of hydrolysis of a glycopyranoside is considerably affected[52] by the substituent at C-5. Introduction of an electron-attracting group at C-6 in hexopyranose residues may, therefore, favor increased selectivity during partial, acid hydrolysis. Modification at C-2 may give similar results. The method has been used with a highly branched dextran,[53] which was subjected to successive tritylation, methylation, detritylation, p-toluenesulfonylation, and methanolysis. A good yield of p-toluenesulfonylated isomaltose derivatives was obtained, indicating that the major part of the branches in this dextran contains at least two D-glucose residues. It was difficult to separate and characterize the O-p-tolylsulfonyl derivatives, and, unless improvements are made, for example, by the use of gas–liquid chromatography–mass spectrometry (g.l.c.–m.s.), the technique will probably not become very important.

(51) K. Onodera, T. Komano, and S. Hirano, *Biochim. Biophys. Acta*, **83**, 20–26 (1964).
(52) T. E. Timell, W. Enterman, F. Spencer, and E. J. Soltes, *Can. J. Chem.*, **43**, 2296–2305 (1965).
(53) D. A. Rees, N. G. Richardson, N. J. Wight, and Hirst, E. L., *Carbohyd. Res.*, **9**, 451–462 (1969).

3. Acetolysis

Acetolysis, which has been treated in a previous Volume of this Series,[54] is generally performed in a mixture of acetic anhydride, acetic acid, and sulfuric acid. It is complementary to acid hydrolysis, as the relative rates of cleavage of the glycosides in the two reactions are sometimes reversed. Thus, whereas $(1 \rightarrow 6)$-linkages in oligo- and poly-saccharides are relatively stable to acid hydrolysis, they are preferentially cleaved during acetolysis—an observation that has been applied advantageously in structural studies.[2] Several such reactions have been performed with dextrans and, in a notable example, a high yield of nigerose was isolated.[55] Yeast mannans are composed of $(1 \rightarrow 6)$-linked chains of α-D-mannopyranose residues to which different side-chains are attached. Useful structural information on these mannans has been adduced by Ballou, Gorin, and coworkers by preferential cleavage of the $(1 \rightarrow 6)$-linkages by acetolysis, and isolation of the oligosaccharides representing the side chains. Several examples were given in an article in this Series.[56]

Systematic studies of the acetolysis of disaccharides have been performed.[57–60a] Lindberg[61,62] showed that anomerization and acetolysis of acetylated glycosides is retarded by the presence of electronegative groups in the aglycon. A mechanism involving an acyclic intermediate (**20**) was proposed for these reactions (see Scheme 2).

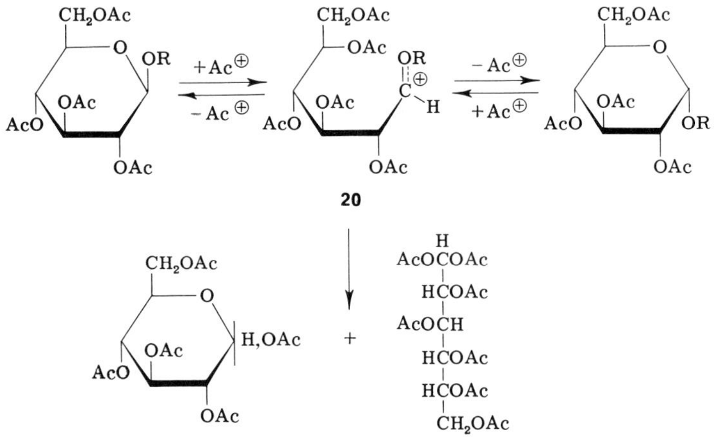

Scheme 2

(54) R. D. Guthrie and J. F. McCarthy, *Advan. Carbohyd. Chem.*, **22**, 11–23 (1967).
(55) I. J. Goldstein and W. J. Whelan, *J. Chem. Soc.*, 170–175 (1962).
(56) P. A. J. Gorin and J. F. T. Spencer, *Advan. Carbohyd. Chem.*, **23**, 367–417 (1968).
(57) K. Matsuda, H. Watanabe, K. Fujimoto, and K. Aso, *Nature*, **191**, 278 (1961).
(58) V. I. Govorchenko and Yu. S. Ovodov, *Khim. Prir. Soedin.*, 256–258 (1972).

Rosenfeld and Ballou demonstrated that this principle can be used to rationalize the relative rates of acetolysis of disaccharides.[60a] The reaction should be retarded by O-acetyl groups in the aglycon which are in α-position to the glycosidic linkage, and, to a lesser extent, by the less electronegative, ring-oxygen atom when it is in the same position. Were this the only factor influencing the reactivity, disaccharides that differ only in the position of attachment of the sugar residues should show the following order of reactivities: $(1 \rightarrow 6) \gg (1 \rightarrow 4) > (1 \rightarrow 3) > (1 \rightarrow 2)$. However, the anomeric nature of the glycosidic linkage and the configurations of the sugar residues also influence the rate of acetolysis. The rate of acetolysis of different disaccharides is given in Table I. It is to be expected that disac-

TABLE I

Acetolysis of Disaccharides[60a]

Disaccharide	Relative rate[a]
α-D-Glc$p$-(1 → 2)-D-Glc	0.18
α-D-Glc$p$-(1 → 3)-D-Glc	0.27
α-D-Glc$p$-(1 → 4)-D-Glc	**1.0**
α-D-Glc$p$-(1 → 6)-D-Glc	7.9
β-D-Glc$p$-(1 → 2)-D-Glc	1.4
β-D-Glc$p$-(1 → 3)-D-Glc	1.9
β-D-Glc$p$-(1 → 4)-D-Glc	1.2
β-D-Glc$p$-(1 → 6)-D-Glc	58[b]
α-D-Man$p$-(1 → 2)-D-Man	0.21
β-D-Man$p$-(1 → 3)-D-Man	2.9[c]
α-D-Man$p$-(1 → 4)-D-Man	1.4
α-D-Man$p$-(1 → 6)-D-Man	60[d]
β-D-Gal$p$-(1 → 4)-D-Gal	4.1
α-D-Gal$p$-(1 → 6)-D-Gal	41[d]
α-D-Glc$p$-(1 → 3)-D-Fru	71[d]

[a] Rate of acetolysis, relative to that of maltose, $k = 0.12$ h^{-1}, in 5:5:2 acetic anhydride–acetic acid–sulfuric acid at 40°.
[b] Simultaneous anomerization.
[c] Determined by using the tetrasaccharide α-D-Man$p$-(1 → 3)-α-D-Man$p$-(1 → 2)-α-D-Man$p$-(1 → 2)-D-Man.
[d] Rate relative to maltose in 10:10:1 acetic anhydride–acetic acid–sulfuric acid at 40°

(59) V. I. Govorchenko, V. I. Gorbatch, and Yu. S. Ovodov, *Khim. Prir. Soedin.*, 258–260 (1972).
(60) V. I. Govorchenko, V. I. Gorbatch, and Yu. S. Ovodov, *Carbohyd. Res.*, **29**, 421–425 (1973).
(60a) L. Rosenfeld and C. E. Ballou, *Carbohyd. Res.*, **32**, 287–298 (1974).
(61) B. Lindberg, *Acta Chem. Scand.*, **3**, 1153–1169 (1949).
(62) B. Lindberg, *Acta Chem. Scand.*, **3**, 1350–1354 (1949).

charides containing pentoses should react much faster than those containing hexoses, by analogy with results for simple glycosides.[63] The high reactivity observed for turanose [α-D-Glc$p$-(1 → 3)-D-Fru] is difficult to explain, unless a different mechanism operates.

Feather and Harris[64] demonstrated that the relative rates of acid hydrolysis of the glycosidic linkages in cellotriose (21) are $k_a:k_b =$

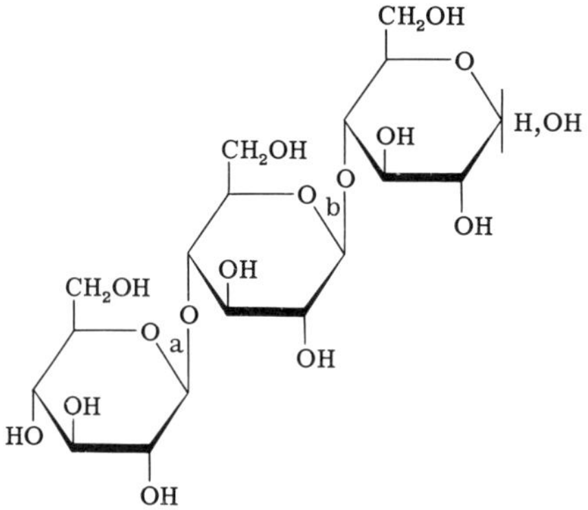

21

3:2, but that the corresponding rates for acetolysis are 1:3. Increased stability of a nonreducing, terminal, glycosyl residue was also observed during the acetolysis of carboxyl-reduced gum arabic, when a disaccharide containing L-rhamnose was obtained.[65] Unexpectedly high stability of sialic acid linkages during acetolysis has been observed in gangliosides[66] and in a hexasaccharide from human milk.[67] Anomerization may occur during acetolysis,[61] but is retarded by electronegative substituents. It should, therefore, be most important in the case of oligosaccharides containing (1 → 6)-linkages. That anomer having an axially attached aglycon in the most stable conformation (generally, the α anomer) constitutes approximately 90% of the equilibrium mixture. Complications should, therefore, arise only when the less-stable anomers are present in the starting material. Anomerization has been observed during the acetolysis of a gum containing (1 → 6)-linked β-D-galactopyranose residues.[68]

(63) B. Lindberg, *Acta Chem. Scand.*, **6**, 949–952 (1952).

(64) M. S. Feather and J. F. Harris, *J. Amer. Chem. Soc.*, **89**, 5661–5664 (1967).

(65) G. O. Aspinall, A. J. Charlson, E. L. Hirst, and R. Young, *J. Chem. Soc.*, 1696–1702 (1963).

(66) R. Kuhn and H. Wiegandt, *Chem. Ber.*, **96**, 866–880 (1963).

(67) L. Grimmonprez, S. Bouquelet, B. Bayard, G. Spik, M. Monsigny, and J. Montreuil, *Eur. J. Biochem.*, **13**, 484–492 (1970).

(68) G. O. Aspinall and J. P. McKenna, *Carbohyd. Res.*, **7**, 244–254 (1968).

Jerkeman observed that D-mannofuranose residues are partially epimerized into D-glucose residues during acetolysis.[69] He predicted that other furanosides having the *cis* arrangement at C-2 and C-3 should also be partially epimerized; this prediction has been confirmed by Sowa[70] and by Chittenden and coworkers.[71,72] The only furanosidic sugar residue having this configuration found in polysaccharides is D-ribose, which is, however, found only infrequently. Complications caused by epimerization of sugar residues should, therefore, not be very important in acetolysis studies of natural polysaccharides.

III. DEGRADATIONS BASED UPON PERIODATE OXIDATION

1. Periodate Oxidation of Polysaccharides

The periodate oxidation of carbohydrates, an important analytical technique, has been reviewed earlier in this Series.[73] The "dialdehydes" obtained on periodate oxidation of polysaccharides have also been discussed.[74] The requisite for the degradations to be discussed here is that part of the sugar residues in a polysaccharide are not oxidized by periodate and can be obtained separated from the oxidized residues as mono-, oligo-, or poly-saccharide derivatives after some chemical treatment. Characterization of these products may then give significant structural information.

It is essential that the oxidation of the glycol groupings in the polysaccharide be complete, and that no "overoxidation" occurs. The latter is excluded by performing the reaction with the exclusion of light, at low temperature, and at a pH of 3–3.5, as described by Bobbitt.[73] The use of oxygen-free water and the addition of propyl alcohol, as a radical scavenger, also lessen radical-induced depolymerization during oxidation.[75]

Painter and coworkers demonstrated that inter-residue hemiacetals may be formed during the periodate oxidation of polysaccharides,

(69) P. Jerkeman, *Acta Chem. Scand.*, **17**, 2769–2771 (1963).
(70) W. Sowa, *Can. J. Chem.*, **49**, 3292–3298 (1971).
(71) G. J. F. Chittenden, *Carbohyd. Res.*, **22**, 491–493 (1972).
(72) P. J. Boon, A. W. Schwartz, and G. J. F. Chittenden, *Carbohyd. Res.*, **30**, 179–182 (1973).
(73) J. M. Bobbitt, *Advan. Carbohyd. Chem.*, **11**, 1–41 (1956).
(74) R. D. Guthrie, *Advan. Carbohyd. Chem.*, **16**, 105–158 (1961).
(75) T. Painter and B. Larsen, *Acta Chem. Scand.*, **24**, 813–833 (1970).

thus retarding the oxidation. As expected, hemiacetals involving 6-membered rings, for example, from a partially oxidized xylan (**22**),

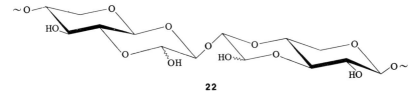

22

are most stable. Thus, in a $(1 \rightarrow 4)$-β-D-linked xylan,[76] the initial second-order rate-constant for the oxidation was lessened to 2% of its original value after the consumption of ~0.60 mol. of periodate per pentose residue. During the oxidation of amylose,[77] the rate constant was lowered to 4% of its original value after 0.64 mol. of periodate for every nonterminal residue has been consumed. The smaller effect for amylose is due to competition between intra- and inter-residue hemiacetal formation, the latter by reaction with the 6-hydroxyl group. For alginic acid[75] and guaran,[78] the inter-residue hemiacetals are so stable that achievement of the theoretical consumption of periodate is virtually impossible under normal conditions. The protected residues may, however, be exposed to oxidation by first subjecting them to borohydride reduction; as incomplete oxidation may result in serious misinterpretation of results, it was recommended[75] that the product be reduced with borohydride and re-oxidized, when eventual underoxidation should be detected. In order to decrease depolymerization of the oxidized polysaccharide due to β-elimination, a concentrated solution of borohydride should be used.[79] The oxidation can be monitored by methylation analysis of the polyalcohol obtained after reduction with borohydride.[80]

Periodate oxidation is generally performed in aqueous solution, but Yu and Bishop studied oxidation with periodic acid in dimethyl sulfoxide.[81] They observed limited oxidation, probably attributable to rapid formation of hemiacetals. Concentrated solutions of periodic acid in dimethyl sulfoxide may explode;[82] therefore, only dilute solu-

(76) T. Painter and B. Larsen, *Acta Chem. Scand.*, **24**, 2366–2378 (1970).

(77) T. Painter and B. Larsen, *Acta Chem. Scand.*, **24**, 2724–2736 (1970).

(78) M. F. Isnak and T. Painter, *Acta Chem. Scand.*, **27**, 1268–1276 (1973).

(79) T. Painter and B. Larsen, *Acta Chem. Scand.*, **27**, 1957–1962 (1973).

(80) B. Lindberg, J. Lönngren, and W. Nimmich, *Acta Chem. Scand.*, **26**, 2231–2236 (1972).

(81) R. J. Yu and C. T. Bishop, *Can. J. Chem.*, **45**, 2195–2203 (1967).

(82) J. J. M. Rowe, K. B. Gibney, M. T. Yang, and G. G. S. Dutton, *J. Amer. Chem. Soc.*, **90**, 1924 (1968).

tions should be used, and these should be prepared by adding the reagent to the dimethyl sulfoxide. Another glycol-cleaving agent, lead tetraacetate, which decomposes in water, may also be used in dimethyl sulfoxide.[83,84] Oxidations conducted in dimethyl sulfoxide may prove of practical value, as this is a good solvent for several polysaccharides.

2. The Barry Degradation

The Barry degradation[85] involves treatment of the periodate-oxidized polysaccharide with phenylhydrazine in dilute acetic acid. Oxidized residues are split off as bis(phenylhydrazones) of, for example, glyoxal, (hydroxymethyl)glyoxal, and a tetrosulose. Depending upon whether they are isolated or adjacent, the non-oxidized residues give phenylosazones of mono-, oligo-, or poly-saccharides. The experimental conditions, and examples of application of the method, are given in an article by O'Colla.[86]

In Scheme 3, the degradation is exemplified by the specific removal of a terminal D-galactose residue (23) linked to O-6 of a 3,6-disubstituted D-galactosyl residue (24). Several applications of the

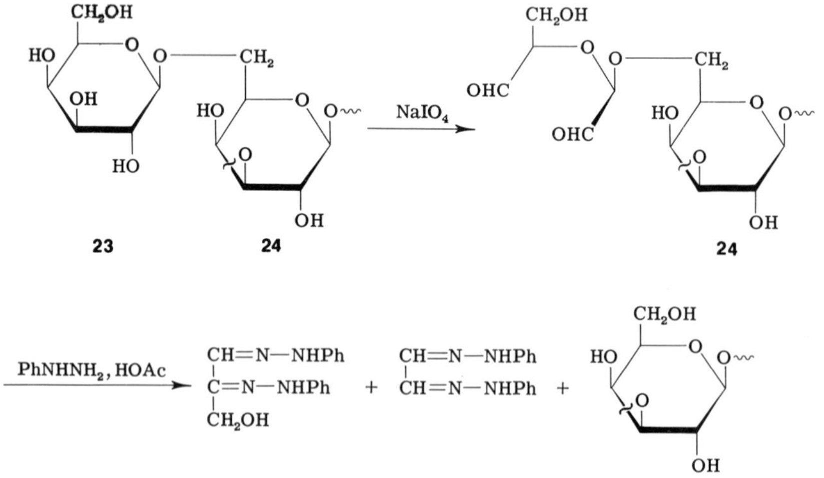

Scheme 3

(83) V. Zitko and C. T. Bishop, *Can. J. Chem.*, 44, 1749–1756 (1966).
(84) C. T. Bishop, *Methods Carbohyd. Chem.*, 6, 350–352 (1972).
(85) V. C. Barry, *Nature*, 152, 537–538 (1943).
(86) P. S. O'Colla, *Methods Carbohyd. Chem.*, 5, 382–392 (1965).

Barry degradation are reported in the literature; for example, to arabic acid,[87,88] snail (*Helix pomatia*) galactan,[89] larch arabinogalactans,[11] and nigeran.[90] In several of these examples, the polysaccharide backbone remaining after the first degradation was subjected to consecutive degradations. The Barry degradation has lost some of its importance, as similar (sometimes, more detailed) results are obtained by using the Smith degradation, which, furthermore, gives a cleaner product.

In a modification of the Barry degradation,[91,92] N,N-dimethylhydrazine is used instead of phenylhydrazine. The advantages of the modified procedure were not actually stated, but it is conceivable that the milder conditions employed and less discoloration produced result in an improved yield of the degraded product. In one application, the glycerol glycoside **25** of a tetrasaccharide obtained on limited per-

α-L-Ara$f$-(1$\rightarrow$3)-α-D-Glc$p$-(1$\rightarrow$2)-α-L-Ara$f$-(1$\rightarrow$3)-α-D-Gal$p$-(1$\rightarrow$2)-Glycerol

25

iodate oxidation–borohydride reduction of the pentasaccharide **1** (from the *Pneumococcus* type 34 specific polysaccharide) was subjected to three consecutive degradations.[92] Each degradation removed only the terminal sugar residue, and, by measuring the optical rotations of the products, the anomeric configurations of all of the sugar residues in **25** (and, thereby, also in **1**) were determined.

3. The Smith Degradation

The Smith degradation[4] involves reduction of the periodate-oxidized polysaccharide with borohydride, followed by mild hydrolysis with acid. (The periodate oxidation has already been treated in Section III,1, p. 200.) The modified sugar residues contain acyclic acetal groupings, which are hydrolyzed much faster than the glycosidic linkages. The product therefore contains small fragments (such

(87) T. Dillon, D. F. O'Ceallachain, and P. S. O'Colla, *Proc. Roy. Irish Acad. (B)*, **55**, 331–345 (1952).
(88) T. Dillon, D. F. O'Ceallachain, and P. S. O'Colla, *Proc. Roy. Irish Acad. (B)*, **57**, 31–38 (1954).
(89) P. S. O'Colla, *Proc. Roy. Irish Acad. (B)*, **55**, 165–170 (1952).
(90) S. A. Barker, E. J. Bourne, and M. Stacey, *J. Chem. Soc.*, 3084–3090 (1953).
(91) J. LeCocq and C. E. Ballou, *Biochemistry*, 3, 976–980 (1964).
(92) J. R. Dixon, J. G. Buchanan, and J. Baddiley, *Biochem. J.*, **100**, 507–577 (1966).

as C_2–C_4 polyols, glycolaldehyde, and glyceraldehyde) deriving from oxidized residues. The non-oxidized fragments occur as glycosides of mono-, oligo-, or poly-saccharides, the aglycons of which are the aforementioned small fragments. Characterization of the degradation products often affords considerable structural information, and the Smith degradation is the specific degradation technique most frequently used in structural polysaccharide chemistry. Recommended conditions for the degradation have been published by Goldstein and coworkers.[93] (Complete hydrolysis of the polyalcohol and characterization of the product may also give results of structural significance. It should be noted that some authors have, however, confused this nonspecific degradation with the Smith degradation.)

In the following, the different steps in the Smith degradation, after periodate oxidation, will be discussed. The reduction of the periodate-oxidized material to the polyalcohol can be performed catalytically with hydrogen over platinum, but the use of sodium borohydride is more convenient. The advantage of using a concentrated solution of borohydride, in order to lessen competing β-elimination reactions, has been emphasized.[79] The aldehyde groups in the oxidized polysaccharide are not present as such, but in the form of hemiacetals with suitably situated groups. The rate of hydrolysis of these cyclic structures is of the same order of magnitude as that of the intact glycosidic linkages. On reduction to polyalcohols, however, they become parts of non-cyclic structures, and are hydrolyzed at a much higher rate. Kuznetsova and Ivanov[6,94] determined the rate of hydrolysis of **26**, prepared from methyl α-D-glucopyranoside by periodate oxidation followed by reduction. The rate in M hydrochloric acid at 5°, namely, $K = 52 \cdot 10^{-5}$ sec^{-1}, should be compared with the value $K = 0.29 \cdot 10^{-5}$ sec^{-1} for the parent glycoside at 60°.

In a similar study,[95] the rates of acid hydrolysis of **26, 27,** and **28** in 125 mM sulfuric acid were determined (see Table II). These structures are typical for D-glucose residues that are unsubstituted (**26**),

(93) I. J. Goldstein, G. W. Hay, B. A. Lewis, and F. Smith, *Methods Carbohyd. Chem.,* **5,** 361–370 (1965); see also G. G. S. Dutton, *Advan. Carbohyd. Chem. Biochem.,* **28,** 12–160 (1973), especially pp. 98–100.

(94) Z. I. Kuznetsova and V. I. Ivanov, *Izv. Akad. Nauk SSSR, Otdel. Khim. Nauk,* 2044–2045 (1960).

(95) B. Erbing, O. Larm, B. Lindberg, and S. Svensson, *Acta Chem. Scand.,* **27,** 1094–1096 (1973).

TABLE II

Rate Constants at 40° and Activation Energies for the Hydrolysis,[95] in 125 mM Sulfuric Acid, of the Mixed Acetals 26, 27, and 28

Compound	$k \cdot 10^{-4}$ sec	kJ. mol^{-1}
26	11.3	98
27	0.94	82
28	12.9	98
Methyl α-D-glucopyranoside	0.00005[a]	147

[a] Extrapolated value.

substituted at O-2 (27), or substituted at O-4 (28), in a Smith degradation. The difference in the rate of hydrolysis of 26 and 28 is negli-

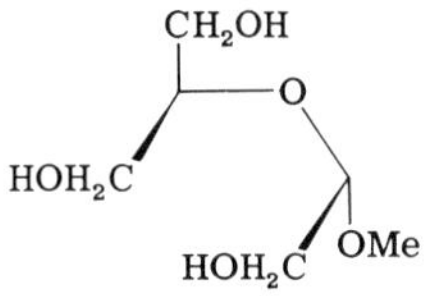

26 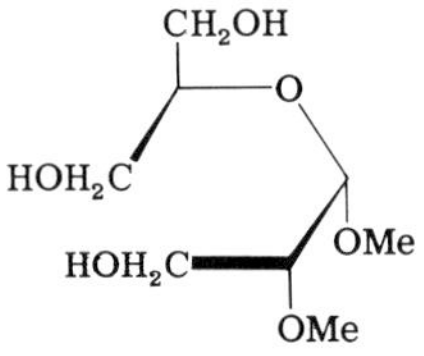   27 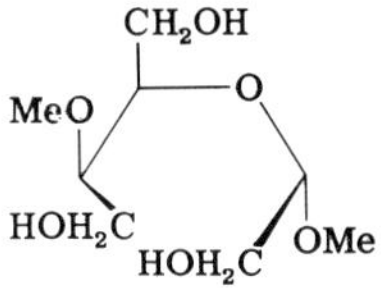   28

gible, but 27 is hydrolyzed approximately 10 times more slowly. The most probable reason for this behavior is the inductive effect of the hydroxymethyl group containing the original C-3, present in 27 but not in 26 and 28. All three are, however, much more readily hydrolyzed than the parent methyl α-D-glucopyranoside. From these results, it should, therefore, be possible to hydrolyze all of the non-cyclic acetal linkages in a modified polysaccharide while leaving intact the glycosidic linkages, except those that are very sensitive to acid hydrolysis. Different conditions have been used for the mild, acid hydrolysis. The activation energy is higher for the glycosidic than for the non-cyclic acetals. The differences in rates of hydrolysis are, therefore, more pronounced at lower temperatures. Consequently, the Smith hydrolysis should preferably be performed with a comparatively strong acid at low temperature, as also recommended by Smith and coworkers.[93]

In order to select the optimal conditions for the Smith hydrolysis, Dutton and Gibney[96] monitored the hydrolysis by g.l.c. of the tri-

(96) G. G. S. Dutton and K. B. Gibney, *Carbohyd. Res.*, 25, 99–105 (1972).

methylsilyl derivatives of the products. The hydrolysis can, alternatively, be followed by methylation analysis.[80]

One advantage of the Smith degradation is that the aglycon of a
low-molecular weight glycoside formed provides structural information concerning the sugar residue from which it is derived. Thus, a 2-
O-substituted tetritol would generally be derived from a 4-substi-
tuted hexopyranose residue (29), a 2-O-substituted glycerol from the
corresponding pentopyranose residue (30), and a 1-O-substituted

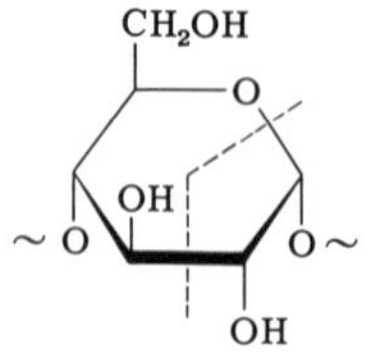

29

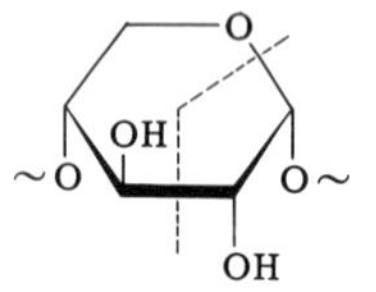

30

glycerol from a 6-substituted hexopyranose (31) or a 5-substituted
pentofuranose residue. A 2-O-substituted glyceraldehyde residue,
likewise, would be derived from a pyranose residue substituted at
C-2 (32).

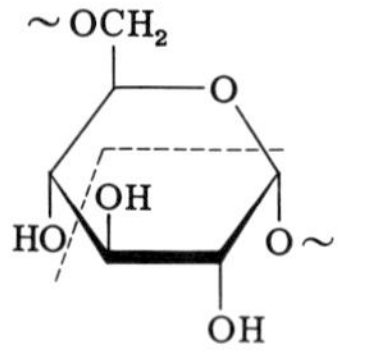

31

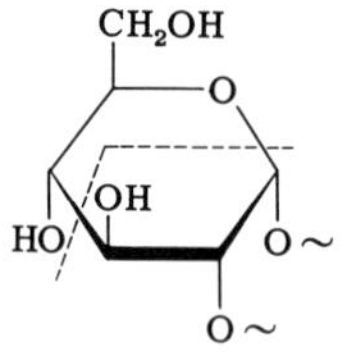

32

The formation of degradation products of low molecular weight
may be complicated by acetal migration, with the formation of more
stable, cyclic acetals. Thus, Smith and coworkers,[93] on degradation of
the β-D-glucan from oat flour, obtained (in addition to the expected 2-
O-β-D-glucopyranosyl-D-erythritol) the 1,3-O-(2-hydroxyethylidene)
derivative (33) of this compound. Gorin and Spencer[97] demonstrated

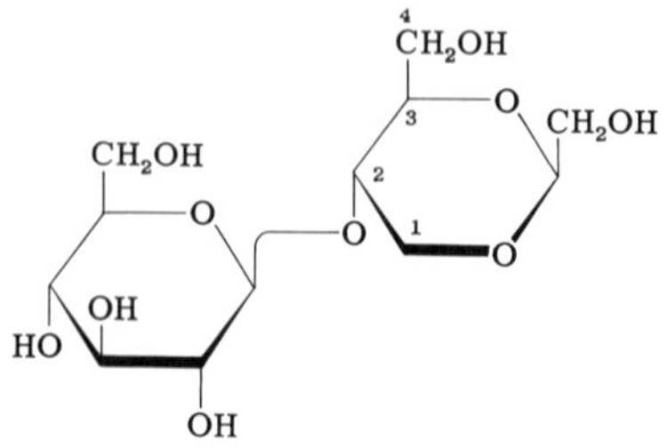

33

(97) P. A. J. Gorin and J. F. T. Spencer, Can. J. Chem., 43, 2978–2984 (1965).

that the corresponding, 5-membered 3,4-*O*-(2-hydroxyethylidene) acetal is also formed in small proportion from polysaccharides containing β-(1 → 4)-linked D-glucose or D-mannose residues, and that such acetals (for example, **34**) become the main byproducts when the cor-

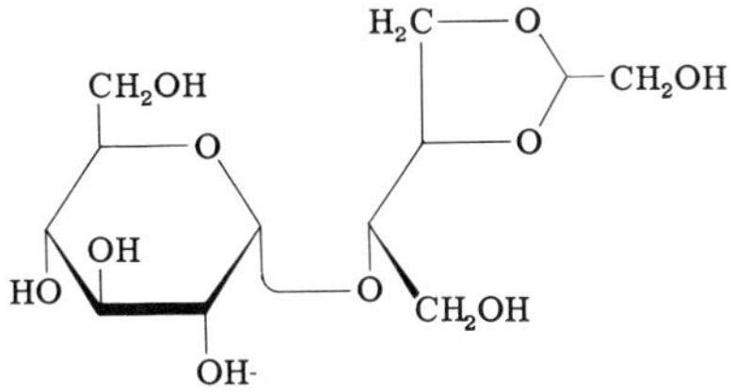

34

responding polysaccharide containing α-D-glycosidic linkages is degraded. They considered that the acetal migration is, most probably, kinetically controlled, and that the result is due to differences in steric compression in the transition states leading to 5- and 6-membered acetals, respectively.

The formation of cyclic acetals involving a sugar residue, which should theoretically be possible, seems to be less important.[98] Thus, only methyl 3-*O*-methyl-α-D-glucopyranoside (**36**) [and none of the 4,6-*O*-(2-hydroxyethylidene) acetal of this compound] was formed when the gentiobiose derivative **35** was subjected to a Smith degra-

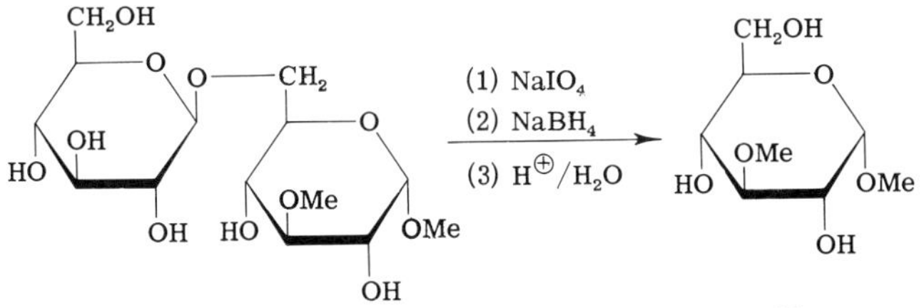

35   **36**

dation. Similarly, on Smith degradation, the disaccharide derivative **37** yielded only 2-*O*-α-D-galactopyranosyl-glycerol (**38**) and no 3,4-*O*-

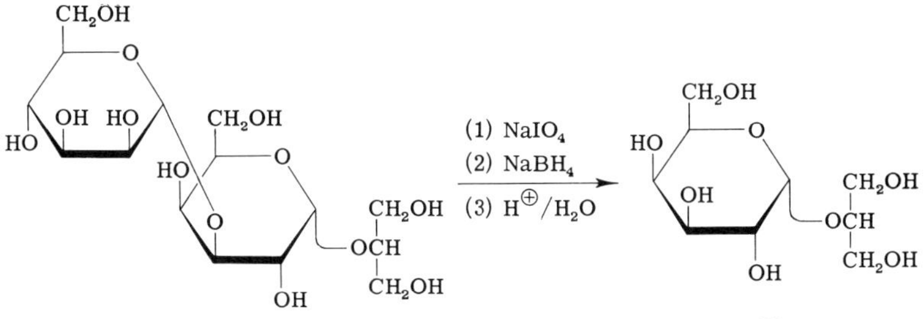

37   **38**

(98) B. Erbing, B. Lindberg, and S. Svensson, *Acta Chem. Scand.*, **B28**, 1180–1184 (1974).

(2-hydroxyethylidene) acetal. The rates of acid hydrolysis of the periodate-oxidized and reduced oligosaccharides, and the corresponding activation-energies, were found similar to the values for the simpler model compound **26** (see p. 205).

A plausible explanation for the occurrence of acetal migration with the alditol but not with the sugar moiety of the degradation product is that the primary reaction of the acetal (for example, **39**) is the fis-

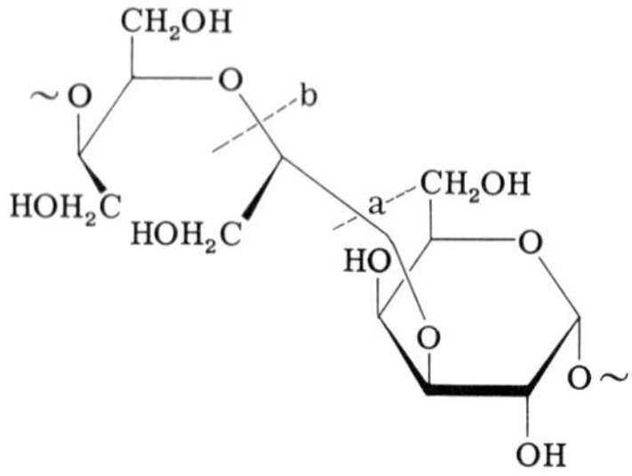

39

sion of the linkage (*a*) to the sugar moiety rather than that (*b*) to the alditol moiety.

The low-molecular weight glycosides formed in a Smith degradation are generally isolated by chromatography and identified by conventional methods. It is, however, also possible to convert the whole product into volatile derivatives (for example, by trimethylsilylation or methylation) and analyze this by g.l.c.–m.s. Only small amounts of polysaccharide are needed for this procedure, but information concerning the anomeric nature of the sugar residues is difficult to assess, unless the retention times can be compared with those of authentic derivatives. This method was used in studies of the *Klebsiella* O group 5 (Ref. 80) and O group 3 (Ref. 99) lipopolysaccharides. The O-antigens of these are composed of linear, D-mannopentaose repeating units containing three (1 → 2)- and two (1 → 3)-linkages. The Smith degradation product was reduced with sodium borodeuteride and the product trimethylsilylated. G.l.c.–m.s. revealed the presence of the 2-deoxyglycerol-2-yl α-D-mannoside (**40**) for O group 5, and the 2-deoxyglycerol-2-yl D-mannobioside (**41**)

(99) M. Curvall, B. Lindberg, J. Lönngren, and W. Nimmich, *Acta Chem. Scand.*, **27**, 2645–2649 (1973).

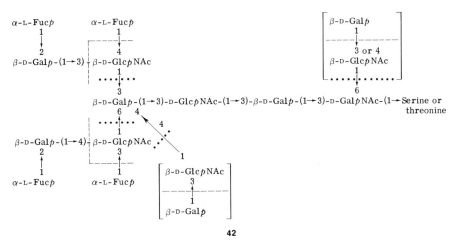

for O group 3, demonstrating that the 3-substituted D-mannopyranose residues are separated in the former but adjacent in the latter lipopolysaccharide.

Often, a polysaccharide gives a polymeric residue after a Smith degradation. Methylation analyses of this material and of the original polysaccharide may nonetheless furnish valuable structural information. It is also possible to subject the material to successive Smith degradations, a notable example being the stepwise degradation of the ovarian-cyst, blood-group H substance (**42**) by Lloyd and Kabat.[100]

[The dashed (---) and dotted (⋯) lines indicate the sugar residues removed by the first and second degradation, respectively.]

Duarte and Jones studied the galactan from the snail *Strophocheilus oblongus*, and obtained a polymeric residue in a yield of 8% after

(100) K. O. Lloyd and E. A. Kabat, *Proc. Nat. Acad. Sci. U. S.*, **61**, 1470–1477 (1968).

three successive degradations, indicating a high degree of multiple branching.[101]

An alternative procedure consists in methylating the polyalcohol before the mild, acid hydrolysis, and remethylating with trideuteriomethyl iodide, or ethylating, the product obtained after the hydrolysis. The hydroxyl groups liberated during the mild hydrolysis (which are subsequently labelled with alkyl groups) may be located when the glycosides formed, or the sugars in the hydrolyzate of the polymeric residue, or both, are investigated by g.l.c.–m.s. Thereby, structural information may be obtained which might be lost in a conventional Smith degradation. Another advantage is that acetal migration is precluded. This technique has been used in studies of the *Klebsiella* O group 7 lipopolysaccharide.[102] Methylation analysis of the original lipopolysaccharide indicated that its O-specific sidechains are composed of linear, tetrasaccharide repeating-units containing one 2-linked D-ribofuranose residue, two 3-linked L-rhamnopyranose residues, and one 2-linked L-rhamnopyranose residue, respectively. After the treatment just discussed, 2-O-ethyl-3,5-di-O-methyl-D-ribose (**43**) was obtained, thus establishing the sequence in the tetrasaccharide repeating-unit (**44**), in which the L-rhamnose resi-

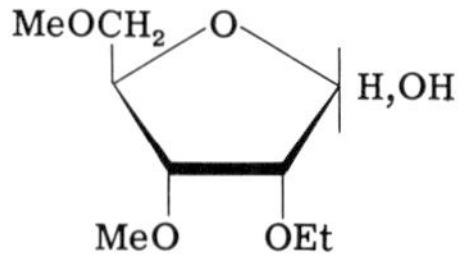

43

$$\rightarrow 2)\text{-}\alpha\text{-L-Rha}p\text{-}(1\rightarrow2)\text{-}\beta\text{-D-Rib}f\text{-}(1\rightarrow3)\text{-}\alpha\text{-L-Rha}p\text{-}(1\rightarrow3)\text{-}\alpha\text{-L-Rha}p\text{-}(1\rightarrow$$

44

due that is oxidized by periodate is linked to O-2 of the D-ribofuranose residue.

4. Other Methods

The polyaldehydes obtained on periodate oxidation of polysaccharides are alkali-labile, as discussed in previous articles in this Series.[74,103] The products are, however, generally intractable mixtures, and this mode of degrading polysaccharides (with alkali) has

(101) J. H. Duarte and J. K. N. Jones, *Carbohyd. Res.*, **16**, 327–335 (1971).
(102) B. Lindberg, J. Lönngren, U. Rudén, and W. Nimmich, *Acta Chem. Scand.*, **27**, 3787–3790 (1973).
(103) R. L. Whistler and J. N. BeMiller, *Advan. Carbohyd. Chem.*, **13**, 289–329 (1958).

not found application in structural studies, with the exception of polysaccharides containing hexofuranose or heptopyranose residues. It is possible to transform these residues into *aldehydo*-aldose residues by partial oxidation with periodate. On mild treatment with base, the substituent in the β-position to the aldehyde group will be eliminated specifically. This approach was used by Lüderitz, West-phal, and their coworkers in studies of the *Salmonella* T$_1$ lipopolysaccharide[104] and the heptose region of *Salmonella* lipopolysaccharides.[105]

An alternative approach is to oxidize the aldehyde groups to car-boxylic acid groups, and then to apply one of the different methods for carboxyl degradation to the product. This method was used by O'Colla and coworkers,[106] who used the Weermann reaction (a spe-cial case of the Hoffman degradation, with a hydroxyl or an alkoxyl group in the α-position to the carboxyl group) on a snail galactan. Their results were largely similar to those obtained by using the Barry[89] or Smith degradation.[106] The reaction is illustrated in Scheme 4 for a terminal D-galactopyranose residue (**45**).

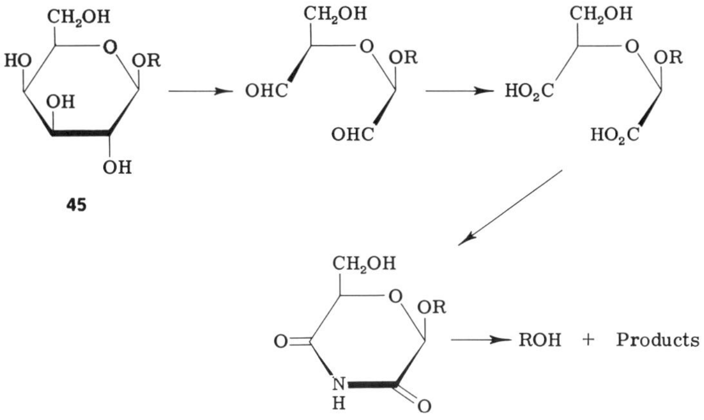

Scheme 4

The related Lossen reaction was used in studies on the core of a *Salmonella* lipopolysaccharide.[105] Aldehyde groups formed on perio-date oxidation were oxidized to carboxylic acid groups, these were esterified, the esters were treated with hydroxylamine, and the prod-ucts finally subjected to the Lossen reaction by treatment with a water-soluble carbodiimide.

(104) M. Berst, O. Lüderitz, and O. Westphal, *Eur. J. Biochem.*, **18**, 361–368 (1971).
(105) V. Lehmann, O. Lüderitz, and O. Westphal, *Eur. J. Biochem.*, **21**, 339–347 (1971).
(106) P. S. O'Colla, J. J. O'Donnell, and J. A. Mulloy, *Proc. Chem. Soc.*, 300 (1961).

IV. Degradations Based upon β-Elimination

1. General Remarks

Several kinds of groups (for example, alkoxyl and hydroxyl groups) in the β-position to an electron-withdrawing group, such as carbonyl, carboxylic ester, amide, or sulfone, are eliminated on treatment with base. Presence of a hydrogen atom in the α-position to these groups is essential, and alkoxyl groups are eliminated more readily than hydroxyl groups. There are numerous examples of β-elimination in carbohydrate chemistry, and the reaction constitutes the initial reaction in the alkaline degradation of reducing sugars, discussed earlier in this Series, for oligo- and poly-saccharides, by Whistler and Be-Miller[103] (see also, Ref. 116).

Eliminations at the reducing sugar residue may be useful in structural studies of oligosaccharides. A 3-substituent on the reducing sugar residue is readily recognized because of its alkali-lability. When O-2 is protected (for example, in a branched oligosaccharide), a 3-deoxy-2-enopyranose is formed. Linkages at O-2 and O-4 of this latter residue may be specifically cleaved by mild treatment with acid, as demonstrated by Aspinall and coworkers;[107] the sophorose derivative **46** was treated with base, and the product reduced; the reduction product yielded the unsaturated ketose **47** on subsequent treatment with acid (see Scheme 5).

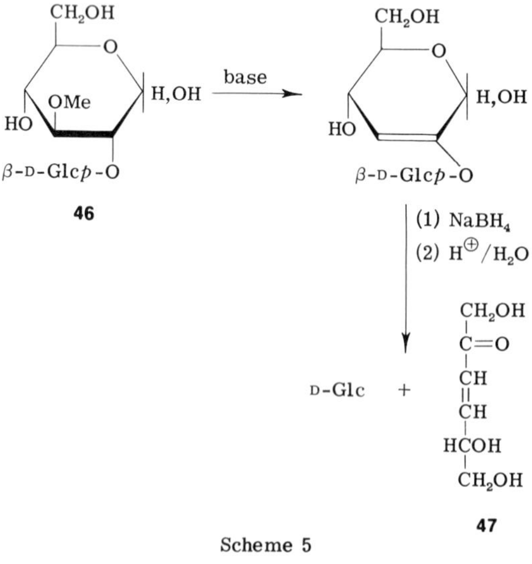

Scheme 5

(107) G. O. Aspinall, R. Khan, and Z. Pawlak, Can. J. Chem., **49**, 3000–3003 (1971).

On consecutive treatment with base and acid, the disaccharide derivative **48** yielded the β-D-xylopyranoside of 5-(hydroxymethyl)-2-furaldehyde (**49**) (see Scheme 6).[108] S. A. Barker and coworkers

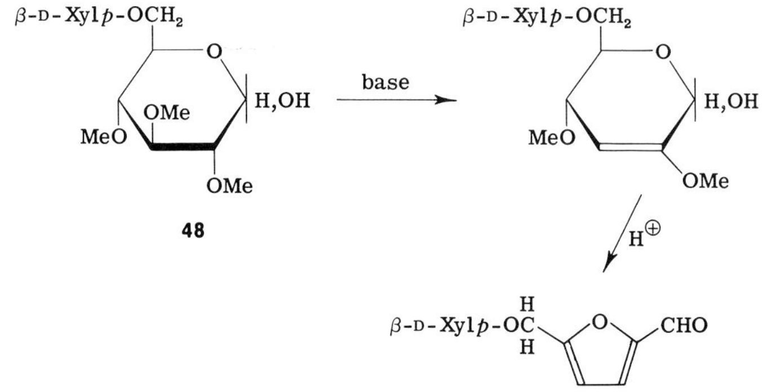

Scheme 6

developed methods for the analysis of the different types of saccharinic acid released successively on treatment of oligosaccharides with base.[109,109a]

Even though β-elimination from the reducing terminal may be a useful tool in structural studies of oligosaccharides it is, in the authors' opinion, of only limited value when applied directly to polysaccharides. Side reactions may completely obscure the results after the first three or four residues have been eliminated from a polysaccharide, and, in some instances, the degradation may stop after only a few residues have been eliminated. For polysaccharides of irregular structure, the structural composition in the vicinity of the reducing terminal may not be the same in the different molecules. As a result of the biosynthesis, or the isolation procedure, there may not even be any reducing terminals in the polysaccharide molecules or, at most, only in some of them.

In glycoproteins containing a carbohydrate–L-serine linkage, such as **50**, or a carbohydrate–L-threonine linkage, the carbohydrate is released from the protein on treatment with base. The carbohydrate released is then subjected to further degradation, until an alkali-stable structure is formed. This reaction was used by Kabat, Lloyd,

(108) G. O. Aspinall, R. Khan, R. R. King, and Z. Pawlak, *Can. J. Chem.*, **51**, 1359–1362 (1973).

(109) S. A. Barker, A. R. Law, P. J. Somers, and M. Stacey, *Carbohyd. Res.*, **3**, 435–444 (1967).

(109a) R. F. Burns and P. J. Somers, *Carbohyd. Res.*, **31**, 289–300; 301–309 (1973).

and their coworkers in studies on human blood-group substances.[110–114] The reaction, which was performed in the presence of
a low concentration of sodium borohydride, results in the formation
of oligosaccharides, many of which are terminated with 3-hexene-
1,2,5,6-tetrol residues. In order to prevent further elimination, the
reaction may be performed in the presence of a high concentration of
sodium borohydride, such that the carbohydrate is reduced before it
reacts further (see Scheme 7).[115]

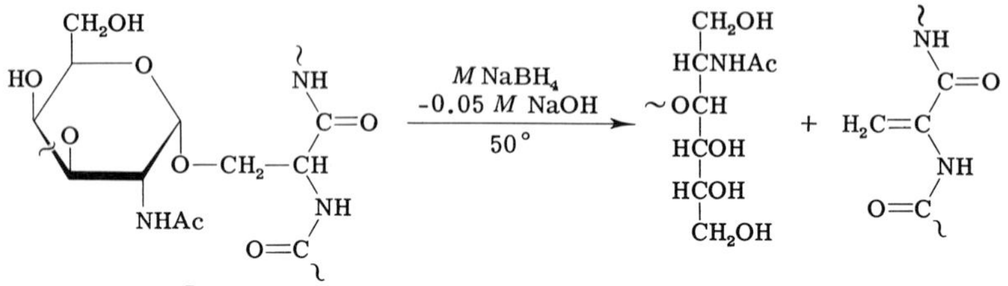

Scheme 7

In the β-elimination reactions discussed next, the electron-
withdrawing groups are first introduced chemically into the polysaccharide; this may be done by esterification of a carboxylic acid group,
by oxidation of an alcohol group to a carbonyl group, or by nucleo-
philic-displacement reactions.

2. Degradation of Polysaccharides Containing Uronic Acid Residues

On esterification, the carboxyl group in uronic acid residues becomes a carboxylate that is electron-withdrawing, and, on subsequent treatment with base, β-elimination will occur. The β-
elimination reaction for uronic acids and their derivatives has been
reviewed in this Series,[116] and only its applications to structural
polysaccharide chemistry will be discussed here.

(110) K. O. Lloyd and E. A. Kabat, *Biochem. Biophys. Res. Commun.*, **16**, 385–390
    (1964).
(111) G. Schiffman, E. A. Kabat, and W. Thompson, *Biochemistry*, **3**, 113–120 (1964).
(112) G. Schiffman, E. A. Kabat, and W. Thompson, *Biochemistry*, **3**, 587–593 (1964).
(113) K. O. Lloyd, E. A. Kabat, E. J. Layug, and F. Gruezo, *Biochemistry*, **5**,
    1489–1501 (1966).
(114) K. O. Lloyd, E. A. Kabat, and E. Licero, *Biochemistry*, **7**, 2976–2990 (1968).
(115) R. N. Iyer and D. M. Carlsson, *Arch. Biochem. Biophys.*, **142**, 101–105 (1971).
(116) J. Kiss, *Advan. Carbohyd. Chem. Biochem.*, **29**, 229–303 (1974).

The method was first applied in studies of a bacterial polysac-charide, cholanic acid, composed of hexasaccharide repeating-units having[117] the structure **51**. The hydroxypropyl ester of this polysac-

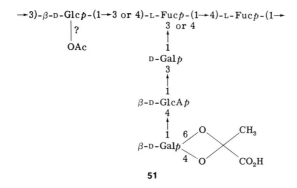

51

charide was treated with sodium methoxide in methanol in the presence of 2,2-dimethoxypropane (added to eliminate water). 4,6-O-(1-Carboxyethylidene)-D-galactose was released, demonstrating that this sugar was terminal and linked to O-4 of the adjacent D-glucuronic acid residue. Methylation analysis of the polymeric residue established that only these two residues were affected.

The degradation just discussed is subject to certain practical limitations. The esterification of the polysaccharide, by use of ethylene oxide or propylene oxide, may take several weeks. The yield in the degradation (about 50%) is moderate. A further restriction is that the uronic acid residue should be substituted at O-4. To overcome these limitations, the polysaccharide is first methylated by the Hakomori procedure,[118,119] with sodium methylsulfinyl carbanion (NaCH$_2$SOCH$_3$) and methyl iodide in dimethyl sulfoxide, whereby all hydroxyl groups are etherified and all carboxyl groups are esterified in one step. Elimination does not occur during the methylation, presumably because the strong base is rapidly decomposed by reaction with methyl iodide, which is present in excess. On treatment of the methylated product with base (either sodium methoxide in methanol, or sodium methylsulfinyl carbanion in dimethyl sulfoxide), the 4-substituent on the uronate residue (**52**) is eliminated.[120]

(117) C. J. Lawson, C. W. McCleary, H. I. Nakada, D. A. Rees, I. W. Sutherland, and J. F. Wilkinson, *Biochem. J.*, **115**, 947–958 (1969).

(118) S. Hakomori, *J. Biochem.* (Tokyo), **55**, 205–208 (1964).

(119) H. E. Conrad, *Methods Carbohyd. Chem.*, **6**, 361–364 (1972).

(120) B. Lindberg, J. Lönngren, and J. L. Thompson, *Carbohyd. Res.*, **28**, 351–357 (1973).

Anhydrous conditions (which are essential in order to avoid saponification) may be maintained by adding 2,2-dimethoxypropane and a catalytic amount of *p*-toluenesulfonic acid to the system before addition of the base. The sodium methylsulfinyl carbanion reagent also adds to the ester group, with formation of a methylsulfinylmethyl ketone,[121] but this occurs at a rate considerably lower than that of the β-elimination reaction. The unsaturated product, depicted here as an ester (**53**), is labile to acids and, on mild hydrolysis with acid, releases[120] the aglycon (R₁OH). The intermediate **54** should, by analogy with the 3-deoxyglyculosonic acids,[122] react further, and yield the furan **55** with simultaneous release of the substituents at O-2 and O-3. The reaction sequence is shown in Scheme 8.

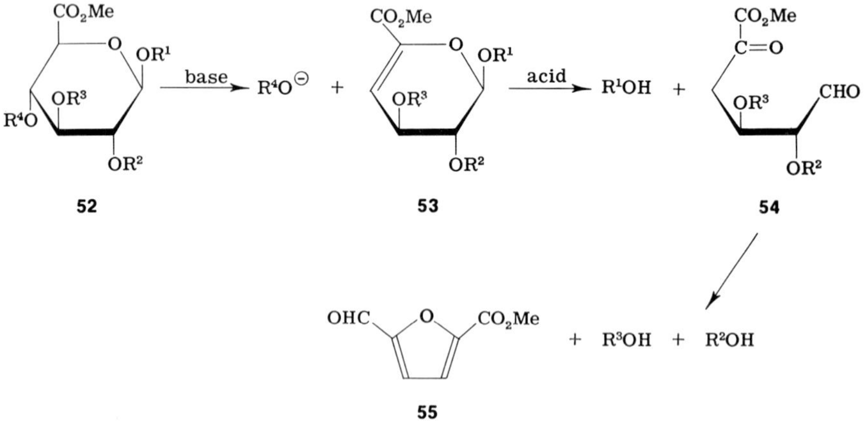

where R^1 = sugar residue; R^2, R^3, R^4 = alkyl group or sugar residue.

Scheme 8

The reactions are, however, not as simple as indicated in Scheme 8. Thus, Rees and coworkers[123] demonstrated that the unsaturated ester formed from esterified alginic acid reacts further on prolonged treatment with base. BeMiller and Kumari[124] showed that the β-elimination of esterified D-mannosiduronic acids, but not the corresponding D-glucosiduronic acids or D-galactosiduronic acids, proceeds farther, with the formation of 2,4-dienes. Some cleavage of the

(121) E. J. Corey and M. Chaykovsky, *J. Amer. Chem. Soc.*, **87**, 1345–1353 (1965).
(122) B. A. Dmitriev, L. W. Backinovski, and N. K. Kochetkov, *Dokl. Akad. Nauk SSSR*, **193**, 1304–1307 (1970).
(123) C. W. McCleary, D. A. Rees, J. W. B. Samuel, and I. W. Steele, *Carbohyd. Res.*, **5**, 492–495 (1967).
(124) J. N. BeMiller and G. V. Kumari, *Carbohyd. Res.*, **25**, 419–428 (1972).

glycosiduronic linkage has also been observed during alkaline treatment of fully methylated polysaccharides containing uronic acid residues.[125]

Elimination of a terminal uronic acid residue by successive treatments with base and mild acid produces, at the site of linkage, a hydroxyl group which may be located by etherification of the product, preferably with trideuteriomethyl or ethyl iodide, followed by acid hydrolysis, and analysis of the resulting mixture of sugars by g.l.c.–m.s.[1] The hydroxyl groups, in defined positions in an otherwise fully methylated polysaccharide, may be used as starting points for a new degradation. The method has been tested on birch xylan (partial structure 56) and *Klebsiella* type 9 capsular polysaccharide, composed of pentasaccharide repeating-units (57).[120] The expected alkyl-

$$\rightarrow 4)\text{-}\beta\text{-}D\text{-}Xyl}p\text{-}(1\rightarrow 4)\text{-}\beta\text{-}D\text{-}Xyl}p\text{-}(1\rightarrow 4)\text{-}\beta\text{-}D\text{-}Xyl}p\text{-}(1\rightarrow$$
$$\overset{\displaystyle 2}{\underset{\displaystyle 1}{\uparrow}}$$
$$4\text{-MeO-}\alpha\text{-}D\text{-GlcA}p$$

56

$$\rightarrow 3)\text{-}\alpha\text{-}D\text{-Gal}p\text{-}(1\rightarrow 3)\text{-}\alpha\text{-}L\text{-Rha}p\text{-}(1\rightarrow 3)\text{-}\alpha\text{-}L\text{-Rha}p\text{-}(1\rightarrow 2)\text{-}\alpha\text{-}L\text{-Rha}p\text{-}(1\rightarrow$$
$$\overset{\displaystyle 4}{\underset{\displaystyle 1}{\uparrow}}$$
$$\beta\text{-}D\text{-GlcA}p$$

57

ated sugars, **58** and **59**, representing the sugar residues to which the

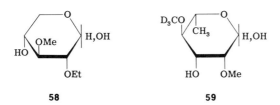

58   **59**

uronic acids had originally been linked, were obtained in the analyses of the polymeric products after ethylation or (trideuteriomethyl)ation, respectively. The yields in the degradations were[120,126] ~90%. The

(125) Unpublished results from the authors' laboratory.
(126) Unpublished results from the authors' laboratory.

terminal D-glucuronic acid residues in the *Pneumococcus* type 2 polysaccharide, composed of hexasaccharide repeating-units (**60**),

$$\rightarrow 3)\text{-}\alpha\text{-L-Rha}p\text{-}(1\rightarrow 3)\text{-}\alpha\text{-L-Rha}p\text{-}(1\rightarrow 3)\text{-}\beta\text{-L-Rha}p\text{-}(1\rightarrow 4)\text{-}\alpha\text{-D-Glc}p\text{-}(1\rightarrow$$
$$\begin{array}{c} 2 \\ \uparrow \\ 1 \\ \alpha\text{-D-Glc}p \\ 6 \\ \uparrow \\ 1 \\ \alpha\text{-D-GlcA}p \end{array}$$

60

were also eliminated by this reaction, thus producing[127] a polymer with a terminal D-glucose residue having a free hydroxyl group at C-6; this product was used for further degradations (see Section IV,4, p. 228).

The degradation becomes more complex, but also more informative, when the uronic acid residues substituted at O-4 form part of the chain. A sugar residue, R^4OH in **52** (see Scheme 8; p. 216), is then released on treatment with base, and this, having a good leaving-group at C-3, will react further by a second β-elimination. For the *Klebsiella* type 47 capsular polysaccharide, having[128] a tetrasaccharide repeating-unit (**61**), the 2,3,4-tri-O-methyl-L-rhamnose

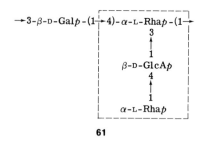

61

released is thus further degraded (and, hence, not detected in the analysis of the product). This information, together with the loss of the uronic acid residue and the appearance of 2-O-methyl-3-O-(trideuteriomethyl)-L-rhamnose (**62**) derived from the branching L-

(127) L. Kenne, B. Lindberg, and S. Svensson, *Carbohyd. Res.*, **40**, 69–75 (1975).
(128) H. Björndal, B. Lindberg, J. Lönngren, K. G. Rosell, and W. Nimmich, *Carbohyd. Res.*, **27**, 373–378 (1973).

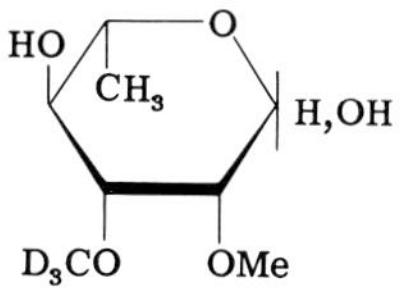

62

rhamnose residue in the analysis of the (trideuteriomethyl)ated, polymeric product, established the sequence[120] of three sugar residues, set off by dashes in **61**.

On treatment of the methylated *Klebsiella* type 52 capsular polysaccharide,[28] composed of hexasaccharide repeating-units (**63**), with a

$$\rightarrow3)\text{-D-Gal}p\text{-}(1\rightarrow2)\text{-L-Rha}p\text{-}(1\rightarrow4)\text{-D-GlcA}p\text{-}(1\rightarrow3)\text{-D-Gal}p\text{-}(1\rightarrow4)\text{-L-Rha}p\text{-}(1\rightarrow$$

$$2$$
$$\uparrow$$
$$1$$
$$\text{D-Gal}p$$

63

base, the D-glucuronic acid residues (**65**) and the L-rhamnose residues (**64**) linked to it were decomposed as depicted in Scheme 9.

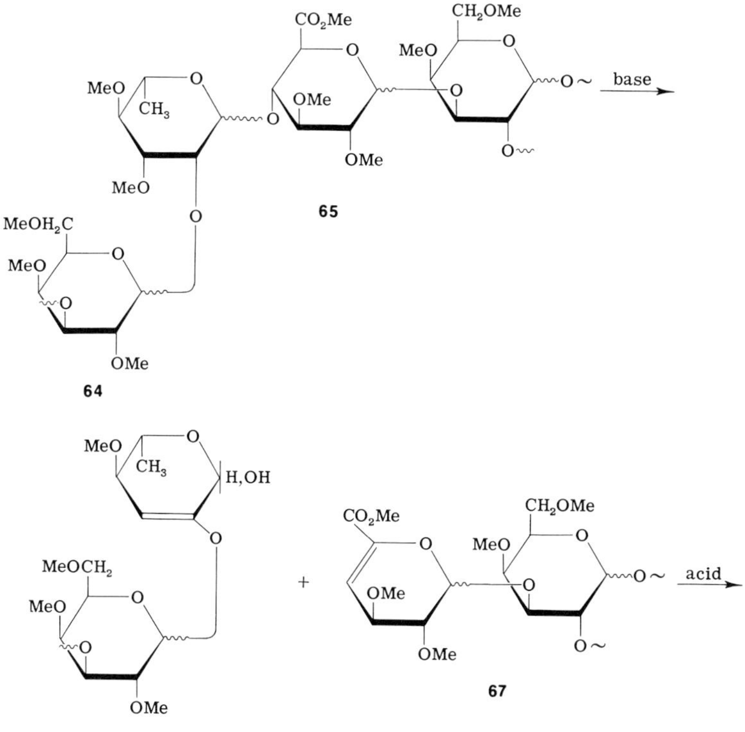

65

64

66

+

67

Scheme 9 *(Contd. on p. 220)*

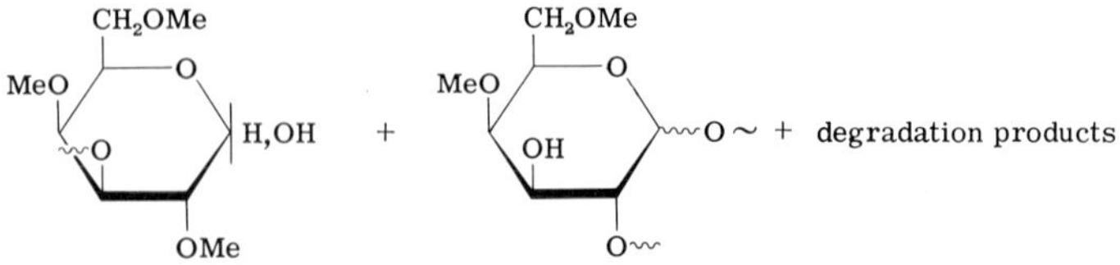

CH₂OMe ... (structures 66 and 67)

Scheme 9 (*Continued*)

The methylated D-galactose derivatives **66** and **67** were thereafter released by mild hydrolysis with acid. The product was reduced with sodium borodeuteride, the reduction product methylated with trideuteriomethyl iodide, and the ether analyzed. From the disappearance of the ethers derived from the uronic acid and the 2-substituted L-rhamnose, and the appearance of the ethers **68** and **69**, the

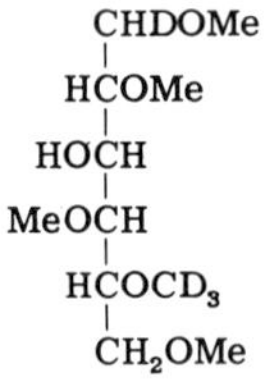

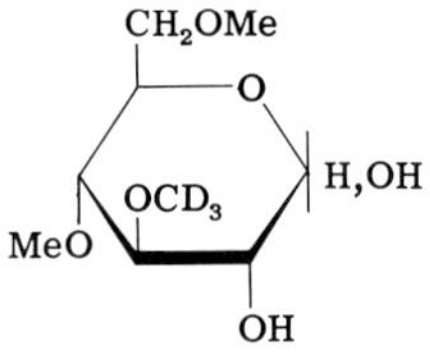

sequence of four sugar residues, set off by dashes in **63**, was deduced. It was found, however, that only ~50% of the liberated L-rhamnose residue underwent β-elimination, which was nonetheless sufficient for the structural elucidation.

It is also possible to protect hydroxyl and carboxyl groups in the polysaccharide as mixed acetals by reaction with methyl vinyl ether and an acidic catalyst in dimethyl sulfoxide.[129] The product is then degraded with base, and the degradation product subjected to mild, acid hydrolysis, which also removes the protecting groups. This method has been used to modify the *Klebsiella* type 47 capsular polysaccharide, whereby ~70% of the side chains were removed after one degradation, and 90% after a second degradation, leaving an essentially linear polysaccharide.[130] The incomplete degradation was probably attributable to the presence of traces of water during the treatment with base.

3. Degradation Preceded by Oxidation

By applying suitable, selective procedures, methylated polysaccharides having a limited number of free hydroxyl groups have been

(129) A. N. de Belder and B. Norrman, *Carbohyd. Res.*, **8**, 1–6 (1968).
(130) M. Curvall, B. Lindberg, and J. Lönngren, *Carbohyd. Res.*, **41**, 235–240 (1975).

prepared. A method has been developed by which such modified polysaccharides may be specifically degraded by oxidation of the alcohol groups to carbonyl groups, followed by alkaline β-elimination, and mild, acid hydrolysis. Although there are, as yet, only a few published examples of this degradative method, it may become of considerable applicability in structural polysaccharide chemistry. In this Section, some degradations of model compounds of low molecular weight will be treated first, followed by methods for preparing the partially methylated polysaccharides, oxidation methods, and, finally, some examples in which the degradation has been successfully employed.

Treatment of methyl 4-O-ethyl-3-O-methyl-β-D-*threo*-pentosidulose (70), a model compound having a carbonyl group at C-2, with sodium 1-propoxide in 1-propanol resulted[131] in quantitative elimination of ethanol and the formation of methyl 4-deoxy-3-O-methylpent-3-enosid-2-ulose (71). On further mild hydrolysis of this and related compounds, the substituents at C-1 and C-3 were released (see Scheme 10).[132]

On treatment with sodium ethoxide in ethanol, methyl 2,4,6-tri-O-methyl-α- (72) and -β-D-*ribo*-hexosid-3-ulose yielded 1-deoxy-2-methoxy-4,6-di-O-methyl-D-hex-1-en-3-ulose (73) and several isomeric ethyl 2,3,6-tri-O-methylhexosid-3-uloses (74) (see Scheme 10).[133] On alkaline degradation of methyl 2,3-di-O-ethyl-6-O-propyl-α-D-*xylo*-hexosid-4-ulose (75) with sodium 1-butoxide in 1-butanol, one mole of ethanol was released per mole of 75, indicating that the substituent at C-2 was preferentially eliminated. On mild, acid hydrolysis of the product, methanol, ethanol, and some 1-propanol were formed, consistent with fission at C-1, C-3, and, to some extent, C-6 (see Scheme 10).[134]

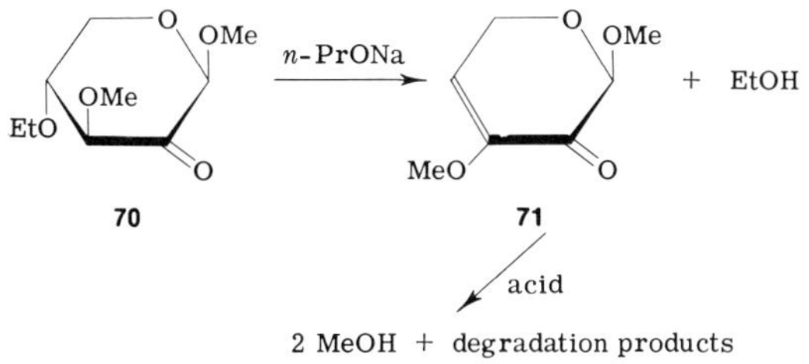

Scheme 10 *(Contd. on p. 222)*

(131) L. Kenne and S. Svensson, *Acta Chem. Scand.*, 26, 2144–2146 (1972).
(132) Unpublished results from the authors' laboratory.
(133) L. Kenne, O. Larm, and S. Svensson, *Acta Chem. Scand.*, 26, 2473–2476 (1972); 27, 2797–2801 (1973).
(134) Unpublished results from the authors' laboratory.

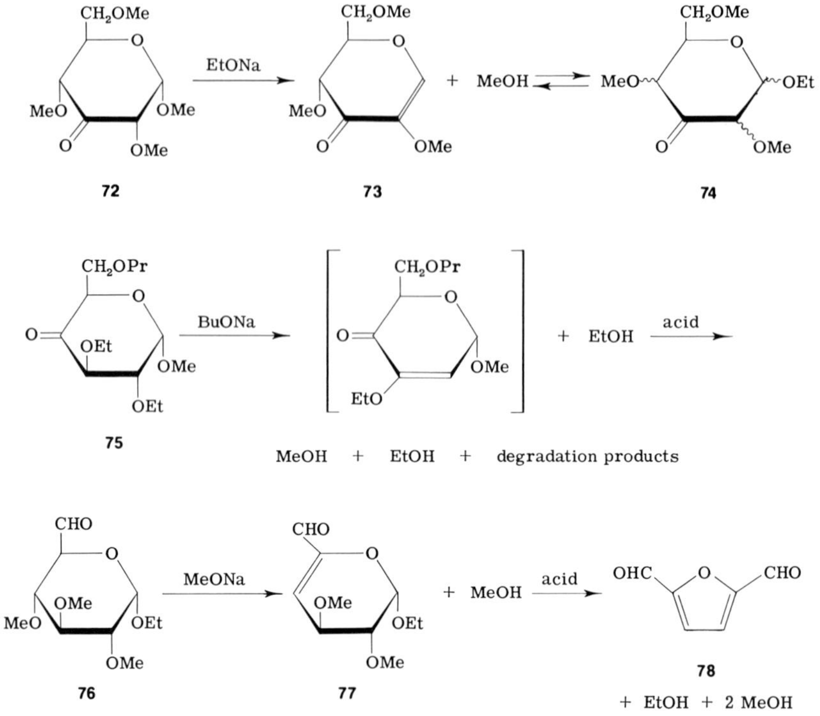

Scheme 10 (*Continued*)

On treatment of the labile ethyl 2,3,4-tri-O-methyl-α-D-*gluco*-hex-odialdo-1,5-pyranoside (**76**) with sodium methoxide, ethyl 4-deoxy-2,3-di-O-methyl-α-L-*threo*-hex-4-enodialdo-1,5-pyranoside (**77**) was formed in good yield;[135] this product seems to react further on pro-longed treatment with base. Preliminary studies indicated that com-pound **77** yields the furan **78** on mild, acid hydrolysis, with simulta-neous release of the substituents at C-1, C-2, and C-3 (see Scheme 10). These reactions are thus analogous to those of the es-terified glycosiduronic acids (see Section IV,2, p. 214).

The substituent in the β-position to the carbonyl group in a fully substituted ketopyranose residue is, therefore, readily eliminated, regardless of whether the carbonyl group is at C-2, C-3, C-4, or C-6. When the carbonyl group is at C-4, the substituent at C-2 is eliminated in preference to that at C-6, provided, of course, that the leaving groups are otherwise similar. When the reaction is performed in an al-cohol, unsaturated sugar derivatives are obtained, with the exception

(135) Unpublished results from the authors' laboratory.

of the 3-keto derivatives, which give rearranged glycosides having the alcohol residue as the aglycon; these glycosides are reasonably stable to acid hydrolysis, but the unsaturated sugars are labile. Thus, the glycosidically linked substituent is released (and, possibly, other substituents also). Although more such model experiments are needed, it is evident from the results thus far obtained that the glycosidic linkage of the oxidized sugar residue is always cleaved, either by β-elimination (carbonyl at C-3) or during the subsequent, acid hydrolysis; this should make the method valuable for the controlled degradation of polysaccharides. As discussed earlier in connection with the uronic acid elimination (see Section IV,2, p. 214), reducing sugars released during the treatment with base will be degraded further by β-elimination.

Various selective procedures that may be used for preparing methylated polysaccharides having a limited number of hydroxyl groups are presented in other Sections of this article. The most important are the following. (a) Acid hydrolysis of fully methylated polysaccharides containing sugar residues that are preferentially cleaved off (see Section II,2, p. 187); (b) Smith degradation, with methylation of the polyalcohol before application of the mild, acid hydrolysis (see Section III,3, p. 203); (c) degradation of fully methylated polysaccharides containing uronic acid residues (see Section IV,2, p. 214); (d) sulfone degradation (see Section IV,4, p. 226); (e) acid hydrolysis of fully methylated polysaccharides containing labile, acetal groups; and (f) O-acyl groups, present in some polysaccharides, that, by use of suitable protecting groups, may also provide an approach. The new hydroxyl groups released by the degradation techniques discussed should permit application of several consecutive degradations, using the same method.

There are several reagents suitable for the oxidation of an alcohol to a carbonyl compound. A number of these, based upon dimethyl sulfoxide and an activating, electrophilic reagent, are discussed in Ref. 136. In the authors' laboratory, ruthenium tetraoxide has been used for the oxidation of secondary alcohol groups to keto groups in glycosides and polysaccharides. The original reagent[137] is unsuitable, as the ruthenium dioxide precipitated seems to adsorb some polysaccharide, resulting in losses and incomplete oxidation. Instead, we use a catalytic amount of the reagent in the presence of a buffered

(136) G. H. Jones and J. G. Moffatt, *Methods Carbohyd. Chem.*, **6**, 315–322 (1972).
(137) P. J. Beynon, P. M. Collins, and W. G. Overend, *Proc. Chem. Soc.*, 342–343 (1964).

solution of periodate, as devised by Parikh and Jones.[138] The conditions for the oxidation are critical, as methoxyl groups are oxidized to formic ester groups, and β-glycosides to 5-glyculosonates, on prolonged treatment with the reagent. Chlorine–dimethyl sulfoxide[139] is the best reagent tested, and it also gives excellent yields of aldehydes from primary alcohols.[135] The β-elimination may be performed with sodium alkoxide in an alcohol (which can be diluted with dichloromethane in order to ensure complete dissolution of the methylated polysaccharide). Although anhydrous conditions are desirable, they are less important here than for the esterified glycosiduronic acids.

Two polysaccharides of known structure have been subjected to this degradation.[140] In the first example (see Scheme 11), the fully

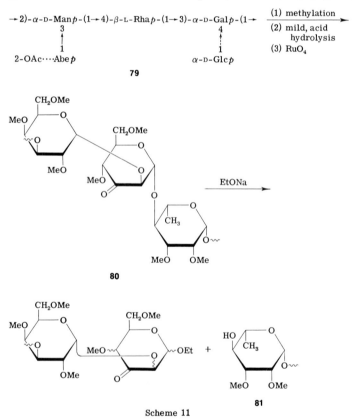

Scheme 11

(138) V. M. Parikh and J. K. N. Jones, Can. J. Chem., 43, 3452–3453 (1965).
(139) E. J. Corey and C. U. Kim, Tetrahedron Lett., 919–922 (1973).
(140) L. Kenne, J. Lönngren, and S. Svensson, Acta Chem. Scand., 27, 3692–3698 (1973).

methylated lipopolysaccharide from *Salmonella typhimurium* LT2, composed[141] of oligosaccharide repeating-units (**79**), was first subjected to mild, acid hydrolysis, whereby the terminal α-abequose (3,6-dideoxy-α-D-*xylo*-hexopyranose) residues linked to O-3 of the D-mannopyranose residues were hydrolyzed off, together with the lipid components. The partially methylated polysaccharide resulting was oxidized with ruthenium tetraoxide, and the product, composed of trisaccharide repeating-units (**80**), was treated with sodium ethoxide in ethanol–dichloromethane. (Some of the D-galactopyranose residues carry at C-4 a terminal, α-D-glucopyranose residue that, in the present treatment, may be disregarded.) From the results of the model experiments, it may be predicted that the 2,3-di-O-methyl-L-rhamnose residue (**81**) should be released by this treatment. Methylation of the degraded product with trideuteriomethyl iodide, followed by hydrolysis, yielded 2,3-di-O-methyl-4-O-(trideuteriomethyl)-L-rhamnose. All of the sugar residues, except D-mannose, in the original polysaccharide were accounted for, and, from the results of the degradation, the sequence → 3)-Man-(1 → 4)-Rha-(1 → was established.

Uronic acid degradation of the fully methylated *Klebsiella* type 47 capsular polysaccharide[120] results in a partially methylated polysaccharide having a disaccharide repeating-unit (see Section IV,2, p. 218), in which the hydroxyl groups at C-3 in the L-rhamnose residues are free. Oxidation with ruthenium tetraoxide consequently gave the product **82**, which, on treatment with base, followed by

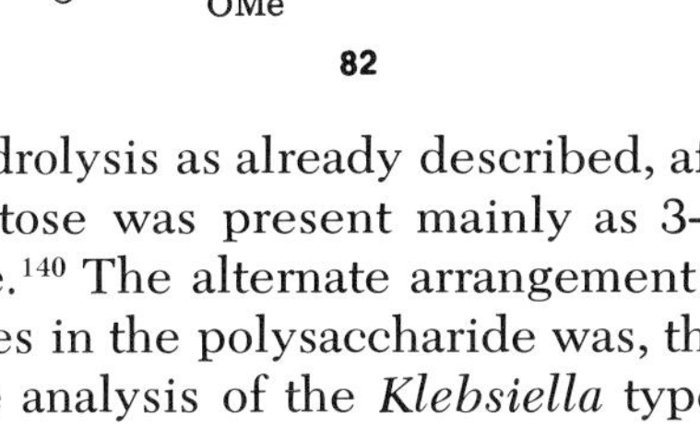

82

ethylation, and hydrolysis as already described, afforded a product in which the D-galactose was present mainly as 3-O-ethyl-2,4,6-tri-O-methyl-D-galactose.[140] The alternate arrangement of L-rhamnose and D-galactose residues in the polysaccharide was, therefore, confirmed.

In the sequence analysis of the *Klebsiella* type 59 capsular polysaccharide, the terminal D-glucuronic acid residue was eliminated

(141) C. G. Hellerqvist, B. Lindberg, S. Svensson, T. Holme, and A. A. Lindberg, *Carbohyd. Res.*, 9, 237–241 (1969).

from the fully methylated polysaccharide by treatment with base and then mild acid. The resulting, free hydroxyl group in the branching D-galactose residue was oxidized with chlorine–dimethyl sulfoxide, and the product was treated with sodium ethoxide, followed by mild, acid hydrolysis, and reduction with borohydride. Methylation analysis of the resulting mixture revealed the presence of 1,2,4,5,6-penta-O-methyl-D-glucitol, 2,3,4,6-tetra-O-methyl-D-mannose, and 2,4,6-tri-O-methyl-D-mannose; of these, only the last was present in the methylation analysis of the original polysaccharide. In this way, the sequence set off by dashes in the repeating units (83) of this polysaccharide was established.[142]

$$\rightarrow 3)\text{-}\beta\text{-}\text{D-Gal}p\text{-}(1\overset{\ulcorner}{\rightarrow} 2)\text{-}\alpha\text{-}\text{D-Man}p\text{-}(1\rightarrow 3)\text{-}\alpha\text{-}\text{D-Man}p\text{-}(1\rightarrow 3)\text{-}\beta\text{-}\text{D-Glc}p\text{-}(1\overset{\urcorner}{\rightarrow}$$

$$\underset{\substack{4 \\ \uparrow \\ 1 \\ \beta\text{-}\text{D-GlcA}p}}{}$$

83

In these examples, the chemical degradations afforded convincing evidence of the presence of an oligosaccharide repeating-unit in the original polysaccharide. Previously, the existence of such repeating units had only been demonstrated by biosynthetic studies,[143] or by degradation with bacteriophage enzymes.[144,145]

4. Degradation by Way of Sulfone Derivatives

It is possible to replace the primary hydroxyl group at C-6 in aldohexosides and related polysaccharides by an electron-attracting group by p-toluenesulfonylation, displacement with iodide, and subsequent nucleophilic substitution. Thus, a 6-deoxy-6-iodo-β-D-glucopyranoside (84) reacts with sodium p-toluenesulfinate to give the sulfone 85 in high yield.[146]

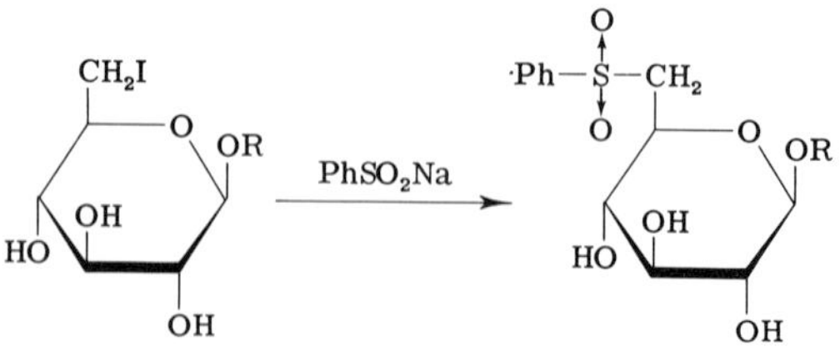

84   **85**

(142) B. Lindberg, J. Lönngren, U. Rudén, and W. Nimmich, *Carbohyd. Res.*, **42**, 83–94 (1975).

(143) F. A. Troy, F. E. Frerman, and E. C. Heath, *J. Biol. Chem.*, **246**, 118–133 (1971).

(144) I. W. Sutherland and J. F. Wilkinson, *Biochem. J.*, **110**, 749–754 (1968).

(145) E. C. Yurewicz, M. A. Ghalambor, and E. C. Heath, *J. Biol. Chem.*, **246**, 5596–5606 (1971).

(146) B. Lindberg and H. Lundström, *Acta Chem. Scand.*, **20**, 2423–2426 (1966).

6-Deoxy-6-nitro derivatives may be prepared analogously, but in lower yields, as a high percentage of nitrous ester is simultaneously formed.[147,148] On treatment with base, the glycosidic linkage of **85** is cleaved,[146] as indicated in Scheme 12. The further reactions of the unsaturated sulfone **86** have not yet been investigated, but it has been

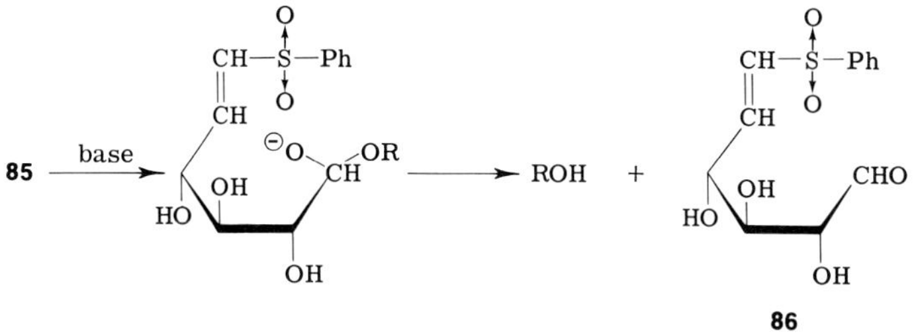

Scheme 12

shown that deoxy-nitrocyclitol derivatives are formed from analogous 6-deoxy-6-nitroglycosides.[149] By applying this sequence of reactions, certain linkages in polysaccharides may be specifically cleaved.

The method was used in studies of a fungal heterogalactan.[150] The polysaccharide was subjected to successive tritylation, methylation, detritylation, *p*-toluenesulfonylation, reaction with sodium iodide, and, finally, reaction with sodium *p*-toluenesulfinate. The product was then treated with sodium methylsulfinyl carbanion in dimethyl sulfoxide, the product remethylated, and the polysaccharide material recovered by gel chromatography. The polymer was hydrolyzed, and the sugars in the hydrolyzate were analyzed, as the alditol acetates, by g.l.c.–m.s.[1] The analysis revealed that ~60% of the hexose residues that were unsubstituted at C-6 had been eliminated. As the product was still polymeric, it was concluded that these residues had constituted a part of side chains linked to a main chain of $(1 \rightarrow 6)$-linked D-galactose residues.

This degradation has also been applied[151] to the dextran produced by *Leuconostoc mesenteroides* NRRL B-512, which is composed of $(1 \rightarrow 6)$-linked α-D-glycopyranose residues, about 5% of which carry a side chain linked to O-3 (partial structure **87**). Tritylation of

(147) J. M. Sugihara, W. J. Teerlink, R. McLeod, S. M. Dorrence, and C. H. Springer, *J. Org. Chem.* **28**, 2079–2082 (1963).
(148) B. Lindberg and S. Svensson, *Acta Chem. Scand.*, **21**, 299–300 (1967).
(149) H. H. Baer and W. Rank, *Can. J. Chem.*, **43**, 3330–3339 (1965).
(150) H. Björndal and B. Wågström, *Acta Chem. Scand.*, **23**, 3313–3320 (1969).
(151) O. Larm, B. Lindberg, and S. Svensson, *Carbohyd. Res.*, **20**, 39–48 (1971).

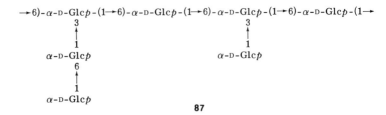

$\rightarrow$6)-α-D-Glc$p$-(1$\rightarrow$6)-α-D-Glc$p$-(1$\rightarrow$6)-α-D-Glc$p$-(1$\rightarrow$6)-α-D-Glc$p$-(1$\rightarrow$

87

this polysaccharide also resulted in some substitution at secondary hydroxyl groups. However, by applying the sequence of tritylation, methylation, detritylation, tritylation, methylation, and detritylation, a product was obtained in which virtually all secondary hydroxyl groups had been methylated, but all of the primary hydroxyl groups were free. Sulfone degradation of this material, followed by ethylation, hydrolysis, and analysis of the hydrolyzate, afforded a product whose analysis revealed that some 40% of the branches present in the original polysaccharide had disappeared, because the percentage of 2,4-di-O-methyl-D-glucose had decreased and an equivalent amount of 3-O-ethyl-2,4-di-O-methyl-D-glucose (**88**) had been formed. From the terminals of the modified polysaccharide, 6-O-ethyl-2,3,4-tri-O-methyl-D-glucose (**89**) was obtained. These results

88

89

demonstrated that the degradation had proceeded as expected, and that 40% of the branches in the original dextran consisted of a single D-glucose residue. The hydroxyl groups at C-6 that were present after the first degradation were used to initiate a second degradation. Only a small proportion ($\sim$15%) of branching points (giving 2,3-di-O-methyl-D-glucose) remained after the second degradation, demonstrating that $\sim$45% of the side chains in the original dextran contained two D-glucose residues.

The sulfone degradation has been applied to other polysaccharides, in particular to the product obtained from the *Pneumococcus* type 2 polysaccharide after methylation and uronic acid elimination (see Section IV,2, p. 218).[127] Only about one third of the terminal D-glucose residues was eliminated, and an analysis of the (trideuteriomethyl)ated product revealed the presence of an equivalent amount of 2,4-di-O-methylated L-rhamnose having a trideuteriomethyl group on O-2. These results indicated that the side chains consisted of aldobiouronic acid residues, linked to O-2 of the L-rhamnose residues, in accordance with other evidence.

The sulfone degradation is laborious, because of the many steps involved. Despite its unfavorable solubility properties, dextran offers several advantages for testing this degradation, being preponderantly (1 → 6)-linked, and available in large quantities. Nevertheless, the results obtained with this polysaccharide were not entirely satisfactory. Although some of the side chains remaining in the twice-degraded product might be attributed to incomplete reactions, there remains some doubt as to whether a proportion thereof is due to the presence of side chains of more than two D-glucose residues in the original polysaccharide. Because of the difficulties in obtaining good yields with this degradation, the method will probably be replaced by other degradations, for example, one based upon oxidation of the hydroxymethyl group to an aldehyde group, followed by treatment with base, as discussed in Section IV,3, p. 220.

V. Degradation by Oxidation with Chromium Trioxide

Angyal and James[152] showed that a fully acetylated aldopyranoside in which the aglycon is equatorially attached in the most-stable chair form (generally the β anomer) is oxidized by chromium trioxide in acetic acid. An aldohexoside yields a 5-hexulosonate in this reaction. The anomer having an axially attached aglycon group is oxidized only slowly. For some pyranosides (for example, those of dideoxy sugars) for which the energy difference between the two chair forms is small, the difference in reactivity between the anomeric forms should be less pronounced. The oxidation of acetylated furanosides proceeds independent of their anomeric configurations.

This oxidation has been used to determine the anomeric nature of sugar residues in oligosaccharides.[153] The oligosaccharide is reduced to the alditol, this is acetylated, and the ester is treated with chromium trioxide in acetic acid in the presence of an internal standard. From the sugar analysis of the product, the residues that have survived (and, consequently, are α-D-linked) may be identified. The

(152) S. J. Angyal and K. James, *Aust. J. Chem.*, **23**, 1209–1215 (1970).
(153) J. Hoffman, B. Lindberg, and S. Svensson, *Acta Chem. Scand.*, **26**, 661–666 (1972).

reaction is illustrated in Scheme 13 for cellobiitol nonaacetate (**90**) and the virtually inert maltitol nonaacetate (**91**).

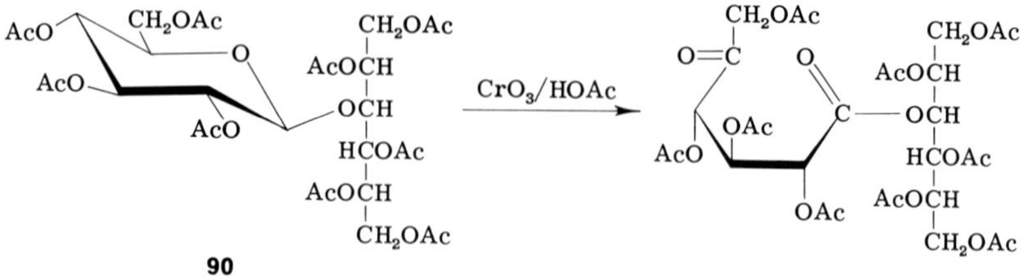

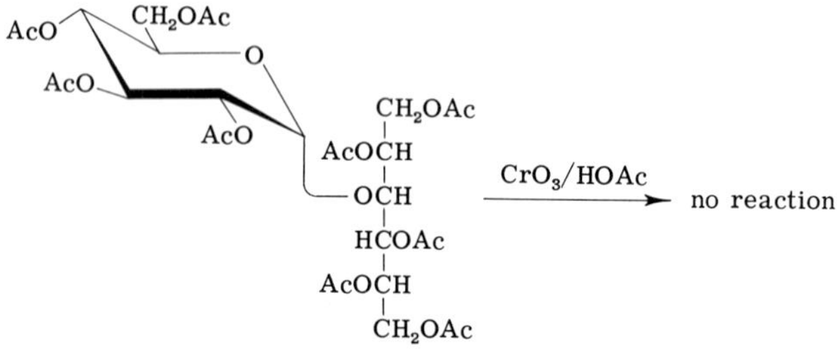

Scheme 13

The method has also been applied to acetylated polysaccharides. Thus, whereas the D-mannopyranose residue in the acetylated lipopolysaccharide from *Salmonella typhi* was found to be inert, that in *Salmonella strasbourg* was oxidized, demonstrating that the former is α-D-linked and the latter β-D-linked.[154] The D-mannose residues in the acetylated lipopolysaccharide from *Klebsiella* O-group 3 were found resistant to oxidation, and are consequently α-D-linked.[99]

Ester linkages formed during the oxidation are cleaved during a subsequent methylation by the Hakomori procedure.[118,119] Hence, by comparing the original methylation analysis of the polysaccharide with that of the oxidized sample, the sequence of sugar residues may be determined. Oxidation of the acetylated lipopolysaccharide from *Salmonella kentucky* revealed that the L-rhamnose and D-galactose residues are β-linked and the D-mannose residues α-linked. On methylation analysis of the oxidized product, comparable amounts of 2,3,4,6-tetra-O-methyl-D-mannose and 3,4,6-tri-O-methyl-D-mannose were obtained. As the two D-mannose residues are the only α-linked sugars in the chain, it was concluded[155] that they are adjacent, as

(154) C. G. Hellerqvist, J. Hoffman, B. Lindberg, Å. Pilotti, and A. A. Lindberg, *Acta Chem. Scand.*, **25**, 1512–1513 (1971).

(155) C. G. Hellerqvist, J. Hoffman, A. A. Lindberg, B. Lindberg, and S. Svensson, *Acta Chem. Scand.*, **26**, 3282–3286 (1972).

shown in the structure of the oligosaccharide repeating-unit (**92**).

α-Abe$p$ 2-AcO---D-Glc$p$
1   1
$\downarrow$ $\downarrow$
3   4
$\rightarrow$4)-β-L-Rha$p$-(1$\rightarrow$2)-α-D-Man$p$-(1$\rightarrow$2)-α-D-Man$p$-(1$\rightarrow$3)-β-D-Gal$p$-(1$\rightarrow$

92

Similar results were obtained with the related, *Salmonella newport* lipopolysaccharide.[155]

Oxidation of the carboxyl-reduced and acetylated *Pneumococcus* type 2 capsular polysaccharide revealed that only one L-rhamnose residue in the hexasaccharide repeating-unit, later demonstrated to have the structure **60**, was oxidized and, consequently, β-L-linked.[156] Replacement of 2,3,6-tri-*O*-methyl-D-glucose in the methylation analysis of the original polysaccharide by 2,3,4,6-tetra-*O*-methyl-D-glucose in that of the oxidized polysaccharide established that this L-rhamnose residue is linked to O-4 of a D-glucose residue. The analysis also showed that it was an L-rhamnose residue in the chain (and not the branching L-rhamnose residue) that was β-linked.

Trisaccharide **93**, isolated from the urine of patients having the

α-D-Man$p$-(1$\rightarrow$3)-β-D-Man$p$-(1$\rightarrow$4)-D-GlcNAc

93

genetic disease mannosidosis, contains[157] two D-mannose and one 2-acetamido-2-deoxy-D-glucose residue. Oxidation of the acetylated trisaccharide alditol revealed that one D-mannose residue was α-linked and the other β-linked. On methylation analysis of the oxidized product, 2,3,4,6-tetra-*O*-methyl-D-mannose and 2-deoxy-1,3,4,5,6-penta-*O*-methyl-2-(*N*-methylacetamido)-D-glucitol (**94**) were

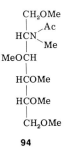

94

(156) O. Larm, B. Lindberg, and S. Svensson, *Carbohyd. Res.*, **31**, 120–126 (1973).
(157) N. E. Nordén, A. Lundblad, S. Svensson, P. A. Öckerman, and S. Autio, *J. Biol. Chem.*, **248**, 6210–6215 (1973).

obtained, demonstrating that the oxidizable β-D-mannose residue is linked to the 2-acetamido-2-deoxy-D-glucose; this result was confirmed by enzymic studies.

A problem that may be encountered when this method is applied is the difficulty in preparing fully acetylated polysaccharides. Sugar residues containing free hydroxyl groups will be oxidized with chromium trioxide independent of their anomeric configuration, and this will obscure the results. For this reason, the method has failed completely in some instances.[158] Furthermore, it is not feasible to protect the hydroxyl groups by methylation, as methoxyl groups are oxidized to formyl groups[159] at a rate comparable to that at which β-glycosides are oxidized.

Deslongchamps and Moreau[160] showed that ozone oxidizes acetylated β-glycopyranosides, but not the corresponding α-glycopyranosides. Ruthenium tetraoxide seems to show a similar selectivity.[161] These oxidation agents have not, however, been systematically investigated, and it is not known whether they offer any advantages over chromium trioxide.

VI. Degradation by Deamination

The most common 2-amino-2-deoxyaldoses found in polysaccharides and glycoconjugates are 2-amino-2-deoxy-D-glucose and 2-amino-2-deoxy-D-galactose. They generally occur as the N-acetyl derivatives, and have to be N-deacetylated, as already discussed (see Section II,2, p. 194), before deamination by reaction with nitrous acid. N-Sulfated amino sugar residues, which occur in heparin, can be deaminated without desulfation. Deamination of 2-amino-2-deoxyhexoses or their hexopyranosides, in which the amino group is equatorially attached in the most stable chair form, yields 2,5-anhydrohexose residues. These reactions have been discussed in this Series by Shafizadeh,[162] Defaye,[163] and Williams.[163a] The rearrangement results from the attack upon the intermediate diazonium ion by the ring-oxygen atom, which is in a *trans* and antiparallel disposition. The glycosidic linkage is simultaneously cleaved,

(158) Unpublished results from the authors' laboratory.
(159) S. J. Angyal and K. James, *Carbohyd. Res.*, 12, 147–149 (1970).
(160) P. Deslongchamps and C. Moreau, *Can. J. Chem.*, 49, 2465–2467 (1971).
(161) Unpublished results from the authors' laboratory.
(162) F. Shafizadeh, *Advan. Carbohyd. Chem.*, 13, 9–61 (1958).
(163) J. Defaye, *Advan. Carbohyd. Chem. Biochem.*, 25, 181–228 (1970).
(163a) J. M. Williams, *Advan. Carbohyd. Chem. Biochem.*, 31, 9–79 (1975).

thereby rendering the reaction useful in structural studies. The ions from a 2-amino-2-deoxy-D-glucoside (95) and 2-amino-2-deoxy-D-galactoside yield 2,5-anhydro-D-mannose (96) and 2,5-anhydro-D-talose, respectively. Amino sugars incorporated in a polysaccharide chain react similarly.

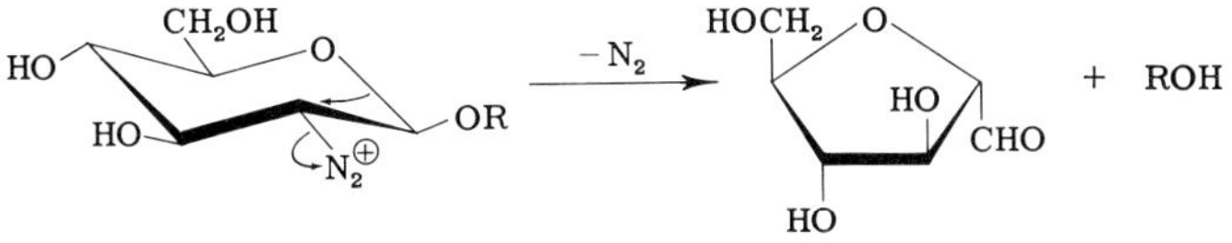

Studies have revealed that the deamination of 2-amino-2-deoxy-D-glucosides is more complicated than previously supposed. It has been reported that methyl 2-amino-2-deoxy-α- and -β-D-glucopyranoside give, in addition to 2,5-anhydro-D-mannose, methyl 2-deoxy-2-C-formyl-α- and -β-D-ribofuranoside (for example, 97 to 98) in yields of ~25 and ~19%, respectively, under the experi-

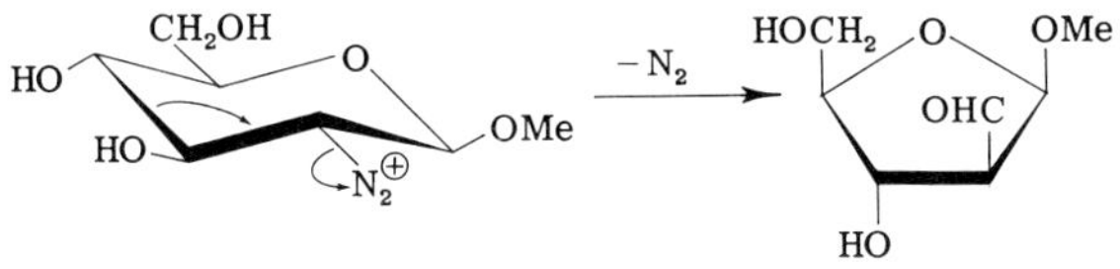

mental conditions used.[164] (The D-*arabino* configuration was erroneously assigned to this compound in the original publication.) This rearrangement, which is not unexpected, results from an attack of the other atom, namely, C-4, which is also *trans* and antiparallel to the diazonium ion. The extent of this side reaction for 2-amino-2-deoxy-D-galactosides, and its implications for the deamination of amino sugar residues in polysaccharides, have not yet been investigated. The glycosidic linkages of a 2-deoxy-2-C-formylpentofuranose residue, or of the corresponding 2-(hydroxymethyl) residue obtained by reduction, should be sensitive to acid hydrolysis, and cleaved under conditions that do not affect ordinary glycosidic linkages. Substituents at C-3 of the amino sugar should be released during this rearrangement.

(164) B. Erbing, B. Lindberg, and S. Svensson, *Acta Chem. Scand.*, **27**, 3699–3704 (1973); see also D. Horton and K. D. Philips, *Carbohyd. Res.*, **30**, 367–374 (1973).

Kochetkov and coworkers studied the deamination of two model disaccharide derivatives. Benzyl 2-amino-2-deoxy-3-O-β-D-galactopyranosyl-α-D-glucopyranoside (99) yielded 2,5-anhydro-3-O-β-D-galactopyranosyl-D-mannose (100), which was reduced, the product

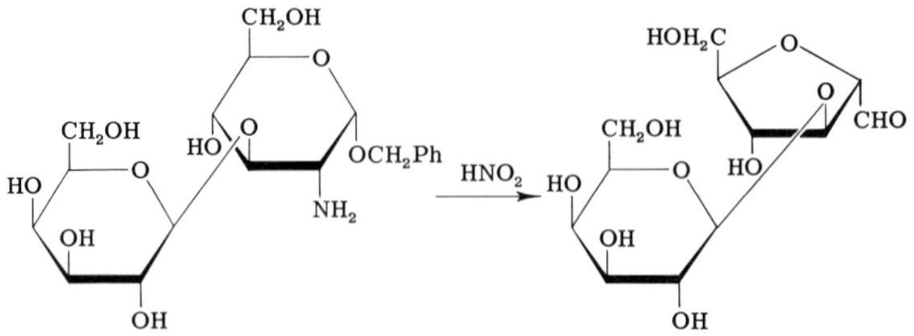

acetylated, and the peracetate characterized by m.s.[48] Some galactitol was also formed; it was most probably derived from the D-galactose released in the side reaction discussed.

Benzyl 2-amino-2-deoxy-6-O-α-D-mannopyranosyl-α-D-glucopyranoside yielded 2,5-anhydro-6-O-α-D-mannopyranosyl-D-mannose,[48a] and was characterized as already described. Also formed was a byproduct which, in the light of the results just discussed, may be benzyl 2-deoxy-2-C-formyl-5-O-α-D-mannopyranosyl-D-ribofuranoside.

The deamination of 2-amino-2-deoxyaldopyranosides having an axially attached amino group has not yet been extensively studied. Llewellyn and J. M. Williams[165] reported that the deamination of methyl 2-amino-2-deoxy-α-D-mannopyranoside (101) yielded a mixture of methyl 2-deoxy-α-D-*erythro*-hexopyranosid-3-ulose (102) and 2-O-methyl-D-glucose (103). On deamination of 6-O-(2-amino-

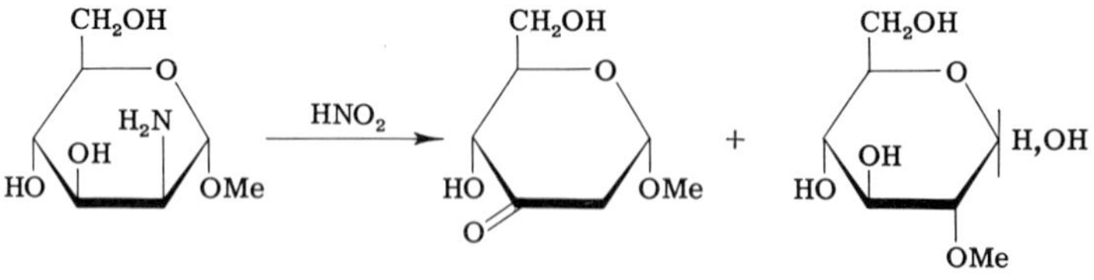

2-deoxy-β-D-mannopyranosyl)-D-glucose (104) and its alditol, substitution to the corresponding β-glycosides (for example, 105) was

(165) J. W. Llewellyn and J. M. Williams, *J. Chem. Soc. Perkin I*, 1997–2000 (1973).

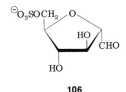

104   **105**

observed, together with rearranged products, possibly 2-deoxy-β-D-*erythro*-hexopyranosid-3-uloses.[166,167] Deamination of 2-amino-2-deoxy-D-mannopyranose residues in polysaccharides will, therefore, probably give complicated mixtures. It is conceivable, however, that reduction of the deaminated polysaccharide, and acid hydrolysis of the highly reactive 2-deoxyhexopyranosidic linkages, followed by characterization of the product, may furnish valuable structural evidence.

There are several applications of the deamination reaction in structural studies of polysaccharides and glycoconjugates, and some examples will be given. One of the first polysaccharides to be studied was chitin, which was *N*-deacetylated with hydrazine, and the product deaminated to give 2,5-anhydromannose.[168] Stacey and coworkers[169] studied the deamination of heparin, and obtained mono- and di-sulfuric esters of a 2,5-anhydro-*O*-(glycosyluronic acid)-D-mannose.[170] Wolfrom and coworkers[171] subjected heparin to a Smith degradation, followed by deamination of the product, and isolated 2,5-anhydro-D-mannose 6-sulfate (**106**) from the reaction mixture;

106

(166) S. Hase and Y. Matsushima, *J. Biochem.* (Tokyo), **72**, 1117–1128 (1972).

(167) S. Hase, Y. Tsuji, and Y. Matsushima, *J. Biochem.* (Tokyo), **72**, 1549–1555 (1972).

(168) Y. Matsushima and N. Fujii, *Bull. Chem. Soc. Jap.*, **30**, 48–50 (1957).

(169) A. B. Foster, E. F. Martlew, and M. Stacey, *Chem. Ind.* (London), 825 (1953).

(170) A. B. Foster, R. Harrison, T. D. Inch, M. Stacey, and J. M. Webber, *J. Chem. Soc.*, 2279–2287 (1963).

(171) M. L. Wolfrom, P. Y. Wang, and S. Honda, *Carbohyd. Res.*, **11**, 179–185 (1969).

this result demonstrated that the O-6 atoms of the 2-amino-2-deoxy-
D-glucose residues are sulfated in the polysaccharide. It has now
been demonstrated that the main fragment obtained on deamination
of heparin is 2,5-anhydro-4-O-(L-idopyranosyluronic acid 2-sulfate)-
D-mannose 6-sulfate[172,173] (107) which, according to Perlin and co-

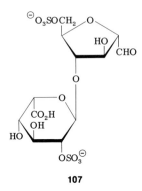

107

workers,[173] has the α-L configuration. From a consideration of the
products obtained by the reaction of heparitin sulfate with nitrous
acid, it was proposed that this glycosaminoglycan (mucopolysac-
charide) is composed of alternating sequences of 2-acetamido-2-
deoxy-O-(glycosyluronic acid)-D-glucose or 2-deoxy-O-(glycosyluronic
acid)-2-sulfamino-D-glucose units.[174]

On using the same reaction in studies on the human plasma α_1-acid
glycoprotein, Isemura and Schmid obtained 2,5-anhydro-4-O-β-D-
galactopyranosyl-D-mannose (108) and three derivatives of this compound containing a modified sialic acid residue linked to different
positions in the D-galactose moiety.[175]

Bayard investigated a number of glycoproteins (α_1-acid glycoprotein, fetuin, lactotransferrin, transferrin, and ovomucoid) by
N-deacetylation followed by deamination.[176] They all gave the oligosaccharides 108 and 109, demonstrating the presence of common
structural elements in these compounds.

(172) U. Lindahl and O. Axelsson, *J. Biol. Chem.*, **246**, 74–82 (1971).
(173) A. S. Perlin, N. M. K. Ng Ying Kin, S. S. Bhattacharjee, and L. F. Johnson, *Can. J. Chem.*, **50**, 2437–2441 (1972).
(174) J. A. Cifonelli, *Carbohyd. Res.*, **8**, 233–242 (1968).
(175) M. Isemura and K. Schmid, *Biochem. J.*, **124**, 591–604 (1971).
(176) B. Bayard, personal communication.

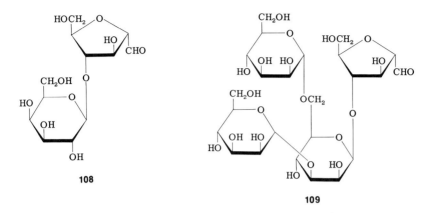

108

109

Baddiley and coworkers used N-deacetylation–deamination in structural studies on two oligosaccharides containing 2-amino-2-deoxy-D-galactose that had been obtained by dephosphorylation of the capsular substances from *Pneumococcus* types 10A (Ref. 177) and 29 (Ref. 13), respectively. The former (**110**) yielded 2-*O*-α-D-galac-

$$\text{D-Gal}f\text{-}(1\rightarrow3)\text{-D-Gal}p\text{-}(1\rightarrow4)\text{-D-Gal}p\,\text{NAc-}(1\rightarrow3)\text{-D-Gal}p\text{-}(1\rightarrow2)\text{-D-Ribitol}$$
$$6$$
$$\uparrow$$
$$1$$
$$\text{D-Gal}f$$

110

topyranosyl-D-ribitol and a product that, on borohydride reduction and acid hydrolysis, gave 2,5-anhydro-D-talose and D-galactose. The second oligosaccharide (**19**; see p. 196) afforded 2,5-anhydro-D-talose and a tetrasaccharide.

VII. DEGRADATION BY THE WEERMANN AND LOSSEN REARRANGEMENTS

The use of the Weermann and Lossen rearrangements for degrading periodate-oxidized polysaccharides has already been discussed. These reactions may also be applied directly to polysaccharides and glycoconjugates containing carboxyl groups. Kochetkov and coworkers[178] applied the Weermann degradation to hexopyranosiduronamides (**111**). The intermediate **112** was not isolated, but was

(177) E. V. Rao, J. G. Buchanan, and J. Baddiley, *Biochem. J.*, **100**, 801–810 (1966).
(178) N. K. Kochetkov, O. S. Chizhov, and A. F. Svirdov, *Carbohyd. Res.*, **14**, 277–285 (1970).

subjected to mild hydrolysis with acid, releasing the aglycon and
a pentodialdose (**113**) (see Scheme 14). The method was applied to

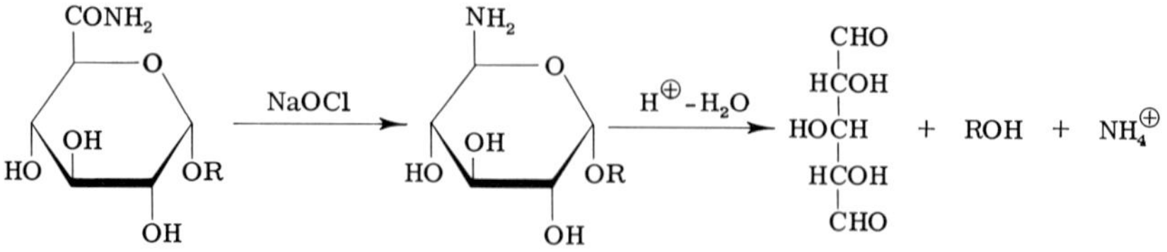

Scheme 14

birch xylan (**56**; see p. 217). Esterification was achieved by treating
the acetylated polysaccharide with diazomethane. On subsequent
treatment with liquid ammonia, the product was converted into the
amide, with concomitant deacetylation. When this product was
treated with hypochlorite, followed by mild hydrolysis with acid,
94% of the original uronic acid residues were removed, leaving an
essentially linear polysaccharide.

The lipopolysaccharide from *Mycobacterium phlei* contains glyceric acid, glycosidically linked to a D-glucosyl group as shown in
114. The glyceric acid was selectively removed by a Lossen rearrangement performed under very mild conditions by using a water-
soluble carbodiimide (see Scheme 15).[179]

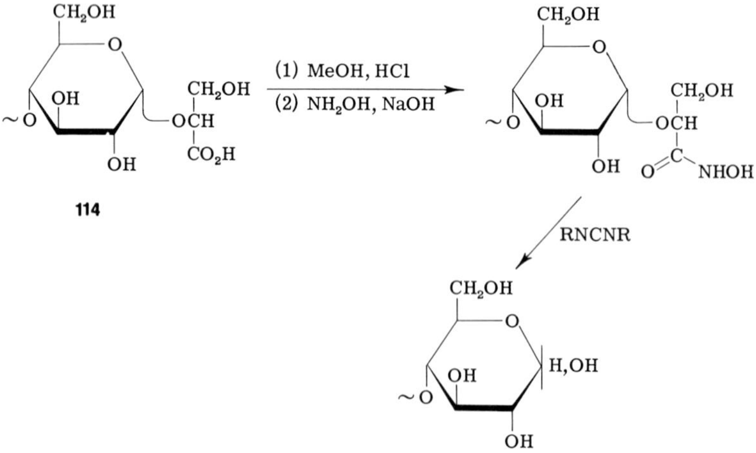

Scheme 15

(179) M. H. Saier, Jr., and C. E. Ballou, *J. Biol. Chem.*, **243**, 992–1005 (1968).

VIII. Conclusion

The results discussed in this Chapter show that a range of conceivable methods for the specific degradation of polysaccharides already exists. In principle, any reaction that can be performed at the monomer level should be applicable to polysaccharides, and rapid development in this field will probably be seen. Application of consecutive degradations, each of which should ideally be quantitative, should facilitate determination of complicated structures with moderate effort. G.l.c. is an established method for the separation of the low-molecular weight products formed, and high-pressure, liquid chromatography will most probably become equally important. Mass spectrometry and nuclear magnetic resonance spectroscopy, especially when combined with pulse Fourier transformation (PFT-n.m.r.), are becoming increasingly important for the characterization of the degradation products.

IX. Addendum

Selective cleavage of the 2-acetamido-2-deoxyaldosidic linkage in peracetylated benzyl 2-acetamido-2-deoxy-4-O-β-D-galactopyranosyl-β-D-glucoside by treatment with ferric chloride, to afford the corresponding oxazoline, has been reported.[180]

When guaran is oxidized by periodate, the protection of D-mannose residues is essentially due to hemiacetal formation with oxidized D-mannose residues that carry an α-D-galactopyranosyl group linked to O-6. Methylation analysis of oxidized, borohydride-reduced guaran therefore gave information on the distribution of these α-D-galactopyranosyl groups, and the results were consistent with an irregular (rather than a block) arrangement.[181]

Further examples of modified Smith degradation, in which the polyalcohol is methylated prior to hydrolysis under mild conditions, uronic acid degradation, and β-elimination preceded by oxidation have been reported in connection with structural studies on the *Klebsiella* type 28 (Ref. 182), 57 (Ref. 183), 59 (Ref. 142), and 81 (Ref. 184) capsular polysaccharides.

(180) B. A. Dmitriev, Yu. A. Knirel, and N. K. Kochetkov, *Izv. Akad. Nauk SSSR, Ser. Khim.*, 411–416 (1974); *Chem. Abstr.*, **81**, 63,917 (1974).

(181) J. Hoffman, B. Lindberg, and T. Painter, *Acta Chem. Scand.*, **B29**, 137 (1975).

(182) M. Curvall, B. Lindberg, J. Lönngren, and W. Nimmich, *Carbohyd. Res.*, **42**, 95–106 (1975).

(183) J. P. Kamerling, B. Lindberg, J. Lönngren, and W. Nimmich, *Acta Chem. Scand.*, in press.

(184) M. Curvall, B. Lindberg, J. Lönngren, and W. Nimmich, *Carbohyd. Res.*, **42**, 73–82 (1975).

The β-elimination reaction with esterified uronic acid residues has been performed by using the nonionic base 1,5-diazabicyclo[5.4.0]-undec-5-ene[185] plus acetic anhydride in benzene.[186] By performing the reaction under acetylating conditions, eventual further degradation, as already discussed, is prevented.

(185) G. O. Aspinall and P. E. Barron, *Can. J. Chem.*, **50**, 2203–2210 (1972).

(186) G. O. Aspinall, Reported at the VIIth International Carbohydrate Symposium, Bratislava, 1974.

CHEMISTRY AND INTERACTIONS OF SEED GALACTOMANNANS

By Iain C. M. Dea and Anthony Morrison

*Unilever Research Laboratories, Colworth House,
Sharnbrook, Bedfordshire MK44 1LQ, England*

I. Introduction

1. Scope of Article

Galactomannans have attracted considerable academic and industrial attention because of the central role they play in the area of polysaccharide interactions. On cooling even concentrated solutions of galactomannans, true gels do not form, although, when mixed with certain gelling polysaccharides, they can considerably increase the gel strength and modify the gel structure. In these synergistic interactions, those galactomannans in the lower range of D-galactose contents are more effective than those in the higher range. Early studies indicated that even small additions of the galactomannan from *Ceratonia siliqua* (locust-bean gum) to carrageenan and agar gels improve their mechanical properties.[1]

Mixtures of locust-bean gum with the non-gelling polysaccharide from *Xanthomonas campestris* have since been shown to interact synergistically to give firm, rubbery gels, whereas the use of the galactomannan from *Cyamopsis tetragonolobus* (guar gum) results only in viscosity enhancement.[2] These interactions are important in many of the industrial applications of galactomannans, and a study of them may also help provide an understanding of how associations between polysaccharide chains contribute to biological cohesion and texture. A large proportion of this article has, therefore, been devoted to the current knowledge and understanding of the basis of the interaction of galactomannans with other polysaccharides. It has also been proposed that galactomannans can interact with milk proteins,[3] plant lectins,[4] and protein antibodies,[5] but, as these interactions have received little attention, they will not be discussed here.

The general chemistry of the galactomannans has been reviewed by Whistler and Smart[6] and others,[7,8] and papers frequently appear that discuss the uses of guar and locust-bean gums in various industrial applications, such as in cosmetics[9] and food.[10,11] Most of this article deals with investigations of the structural chemistry of galac-

(1) G. L. Baker, J. W. Carrow, and C. W. Woodmansee, *Food Ind.*, **21**, 617–619 (1949).

(2) Kelco Co., Brit. Pat. 1,108,376 (1968); *Chem. Abstr.*, **29**, 2255 (1935).

(3) A. S. Ambrose, U. S. Pat. 1,991,189 (1935); *Chem. Abstr.*, **29**, 2255 (1935).

(4) J. P. van Wauwe, F. G. Loontiens, and C. K. de Bruyne, *Biochim. Biophys. Acta*, **313**, 99–105 (1973).

(5) M. Heidelberger, *J. Amer. Chem. Soc.*, **77**, 4308–4311 (1958).

(6) R. L. Whistler and C. L. Smart, "Polysaccharide Chemistry," Academic Press, New York, 1953, p. 291.

(7) B. N. Stepanenko, *Bull. Soc. Chim. Biol.*, **42**, 1519–1536 (1960).

(8) F. Smith and R. Montgomery, "Chemistry of Plant Gums and Mucilages," Reinhold, New York, 1959.

(9) R. J. Chudzikowski, *J. Soc. Cosmet. Chem.* **22**, 43–60 (1971).

(10) M. Glicksman, "Gum Technology in the Food Industry," Academic Press, New York, 1969, p. 130.

(11) W. A. Carlson, E. M. Ziegenfuss, and J. D. Overton, *Food Technol.*, **16** (10), 50–54 (1962).

tomannans, especially as revealed by enzymic studies, and discusses the significance of this information in understanding the interactions of these polysaccharides with themselves and with other polysaccharides. The article commences with a short discussion of the botanical significance of galactomannans in the *Leguminoseae,* and then describes the occurrence, isolation, and purification of the gums. To complete the account, the applications of galactomannans in industry are briefly treated.

In describing and interpreting some of the more important properties of plant galactomannans, comparisons will be made with structurally similar polysaccharides, including the closely related glucomannans and galactoglucomannans, and those based on $(1 \rightarrow 4)$-β-D-xylan main-chains (for example, the arabinoxylans) and $(1 \rightarrow 4)$-β-D-glucan main-chains [for example, the amyloids and sodium *O*-(carboxymethyl)cellulose].

2. Historical Background

The two main groups of galactomannan polysaccharides are those derived from (*a*) the endosperm of plant seeds, the vast majority of which originate in the *Leguminoseae,* and (*b*) microbial sources, in particular, the yeasts[12,13] and other fungi.[12,14,15] D-Mannose and D-galactose are also found in numerous other plant polysaccharides, for example, glucomannans,[7] mannans, and galactans.[8]

In the two main groups of galactomannans, those obtained from the seed mucilage of the *Leguminoseae* are usually based on a β-D-$(1 \rightarrow 4)$-linked backbone of β-D-mannopyranosyl residues having side stubs linked α-D-$(1 \rightarrow 6)$ and consisting of single α-D-galactopyranosyl groups. These polysaccharides are normally edible, and some are used in the food industry as thickening agents.

On the other hand, a variety of linkages has been detected in microbial galactomannans,[12,13,16,17] and, in addition, the sugar residues are frequently present in the furanoid form.[12-14] These polysaccharides are produced by organisms responsible for a number of skin diseases, including, amongst others, the dermatophytes;[17] the galac-

(12) P. A. J. Gorin and J. F. T. Spencer, *Advan. Carbohyd. Chem.,* **23**, 367 (1968).
(13) See "Carbohydrate Chemistry," Vol. 3. R. D. Guthrie, ed., Specialist Report, The Chemical Society, London, 1970, p. 236.
(14) K. O. Lloyd, *Biochemistry,* **9**, 3446–3453 (1970).
(15) O. Sakaguchi, K. Yokota, and M. Suzuki, *Jap. J. Microbiol.,* **13**, 1 (1969).
(16) P. A. J. Gorin and J. F. T. Spencer, *Can. J. Chem.,* **46**, 2299–2304 (1968).
(17) C. T. Bishop, M. B. Perry, F. Blank, and F. P. Cooper, *Can. J. Chem.,* **43**, 30–39 (1965).

tomannans may even be causative agents.[18] They are believed to be immunologically reactive[15,19] and provoke hypersensitive responses.[20] The nuclear magnetic resonance (n.m.r.) spectra of the yeast mannans and galactomannans have been used to aid the taxonomic classification of yeasts.[16,21-23] There have also been reports of the isolation of peptido-galactomannans from yeasts.[12,14,18,20] Microbial galactomannans have been reviewed[12,13] and will not be discussed further here. The hemicellulose galactoglucomannans, often considered together with plant galactomannans, will be only briefly treated.

A few of the plants that produce a mucilage containing galactomannan polysaccharides have been known and cultivated for many hundreds, and even thousands, of years. The best known example is the locust bean[24] or carob tree (*Ceratonia siliqua* L.), which was originally native to Southern Europe and the Near East, but which has since been transported to many other parts of the world, including[25] Australia and the U. S. A. Carob pods have a long history of use as both animal and human foodstuff,[24] although, as the latter, more use was made of them in times of desperate need than as a welcome, staple diet. The synonym "St. John's Bread" reveals the Biblical associations of the carob, as the "locusts" (Matthew 3:4) eaten by John the Baptist in the wilderness are believed to have been locust-bean pods. One of the shortcomings of the carob tree is the very long time it takes to mature; it fruits only after 5 years and is not fully grown until it is 25 years old.[25] The carob tree is of considerable economic importance to Cyprus, where it is cultivated on very poor soil.[26]

(18) S. A. Barker, O. Basarab, and C. N. D. Cruickshank, *Carbohyd. Res.*, **3**, 325–332 (1967).

(19) L. Borecký, V. Lackovič, D. Blaškovič, L. Masler, and D. Šikl, *Acta Virol.*, **11**, 264–266 (1967).

(20) S. A. Barker, C. N. D. Cruickshank, J. H. Morris, and S. R. Wood, *Immunology*, **5**, 627–632 (1962).

(21) P. A. J. Gorin, J. F. T. Spencer, and R. J. Magus, *Can. J. Chem.*, **47**, 3569–3576 (1969).

(22) J. F. T. Spencer, P. A. J. Gorin, and L. J. Wickerham, *Can. J. Microbiol.*, **16**, 445–448 (1970).

(23) J. F. T. Spencer and P. A. J. Gorin, *J. Bacteriol.*, **96**, 180–183 (1968).

(24) J. E. Coit, *Econ. Bot.*, **5**, 82 (1951).

(25) R. J. Binder, J. E. Coit, K. T. Williams, and J. E. Brekke, *Food Technol.*, **13**, 213–216 (1959).

(26) W. N. L. Davies, *Econ. Bot.*, **24**, 460–470 (1970).

In World War II, the supply of locust beans from Mediterranean countries to the U. S. A. was greatly reduced, and a search was initiated for a suitable replacement for the galactomannan[27] extracted from locust-bean gum. Further studies with similar objectives were initiated later.[28-31] The plant that eventually emerged as a rival to the locust was guar (*Cyamopsis tetragonolobus*).[32] Like the locust bean, guar seeds yield a high level of galactomannan from the endosperm (locust bean, 48%; guar bean, 42%), and from both sources, the yield of soluble galactomannan is[27] ~80%.

The guar plant is native to N. W. India and Pakistan,[33] where it is of considerable economic importance.[34] It was introduced into the U. S. A. at the beginning of this century (1903) and became a commercial commodity[9] in 1940. Guar gum was produced commercially by General Mills Inc., Minneapolis, Minnesota, in 1942. Unlike the locust-bean tree, which grows to height of ~8 m, the guar plant is only about 0.6 m high and grown as an annual crop.[9] Like the carob pod, the guar pod and seeds have been used as both human food and animal feedstuffs, although it is certainly not a complete nutrient in itself.[35] The family *Leguminoseae* embraces a wide range of plants, from such full-sized trees as the locust-bean tree to such small, herbaceous plants as clover. Many of these plants are native to the Near East, where they have frequently been coveted for their medicinal properties, for example, *Cassia fistula* and *Cassia absus*.[36,37]

Galactomannans were first examined chemically in 1897, when

(27) E. Anderson, *Ind. Eng. Chem.*, **41**, 2887–2890 (1949).

(28) J. Y. Morimoto, I. C. J. Unrau, and A. M. Unrau, *J. Agr. Food Chem.*, **10**, 134–137 (1962).

(29) H. L. Tookey, R. L. Lohmar, I. A. Wolff, and Q. Jones, *J. Agr. Food Chem.*, **10**, 131–133 (1962).

(30) H. L. Tookey, V. F. Pfeiffer, and C. R. Martin, *J. Agr. Food Chem.*, **11**, 317–321 (1963).

(31) H. L. Tookey and Q. Jones., *Econ. Bot.*, **19**, 165–174 (1965).

(32) F. J. Poats, *Econ. Bot.*, **14**, 241–246 (1960).

(33) S. P. Mital, V. Swarup, M. M. Kohli, and H. B. Singh, *Indian J. Genetics Plant Breeding*, **29**, 98–103 (1969).

(34) V. K. Saxena, *Res. Ind.* (New Delhi), **10**, 101–106 (1965).

(35) M. L. Nagpal, O. P. Agrawal, and I. S. Bhatia, *Indian J. Anim. Sci.*, **41**, 283–293 (1971).

(36) R. N. Choppa, S. L. Nayar, and I. C. Choppa, "Glossary of Indian Medical Plants," Council of Scientific and Industrial Research, New Delhi, 1956, pp. 53 and 54.

(37) K. R. Kirtikar and B. D. Basu, "Indian Medicinal Plants," Basu, Allahabad, India, 2nd Edition, 1936, Vol. II.

Effront[38] investigated the mucilaginous carbohydrate from the locust bean. Two years later, Bourquelot and Hérissey[39] reported that the gum is composed of 83.5% of D-mannose and 16.5% of D-galactose. This D-mannose:D-galactose ratio of 5.1:1 is in fair agreement with values obtained by modern techniques, which range from 3.35:1 (Ref. 40) to 4.50:1 (Ref. 41). Since then, the ratios of D-mannose:D-galactose have been determined for over 70 plant galactomannans, and range from 1:1 for *Medicago sativa* galactomannan[42] and 1.04:1 for *Trifolium resupinatum* galactomannan[29] to 5.26:1 for *Sophora japonica* galactomannan.[43] However, if the reserve polysaccharides from ripe seeds of species in the family *Palmae*, which contain only small proportions of D-galactose, are considered to be galactomannans, then much higher proportions of D-mannose are obtained; thus, mannan A and mannan B from the ivory nut (*Phytelephas macrocarpa*) have D-mannose:D-galactose ratios of 50:1 and 90:1, respectively.[44-46]

The unique, thickening properties of locust-bean gum and guar gum have found extensive application in industry. The initial incentive for finding a substitute for locust-bean gum came from the paper industry, but industries now[34] using either or both of these gums also include food, mining, pharmaceuticals, explosives, and petroleum technology. The viscosity and thickening properties of the other galactomannans are quite similar to those of locust-bean gum and guar gum, although most of them are obtained in much lower yield. At present, only locust-bean gum and guar gum are used extensively. However, as the interesting and novel physical properties of such galactomannans as tara gum, which is extracted from the seeds of *Caesalpinia spinosa*,[27] become appreciated, industrial exploitation may eventually prove economical and practical.[19]

(38) J. Effront, *Compt. Rend.*, **125**, 38–40 (1897).
(39) E. Bourquelot and H. Hérissey, *Compt. Rend.*, **129**, 228, 391–395 (1899).
(40) I. C. M. Dea, A. A. McKinnon, and D. A. Rees, *J. Mol. Biol.*, **68**, 153–172 (1972).
(41) P. A. Hui and H. Neukom, *Tappi*, **47**, 39–42 (1964).
(42) J. E. Courtois, C. Anagnostopoulos, and F. Petek, *Bull. Soc. Chim. Biol.*, **40**, 1277–1285 (1958).
(43) P. Kooiman, *Carbohyd. Res.*, **20**, 329–337 (1971).
(44) G. O. Aspinall, E. L. Hirst, E. G. V. Percival, and J. R. Williamson, *J. Chem. Soc.*, 3184–3188 (1953).
(45) F. Klages, *Ann.*, **509**, 159–181 (1934).
(46) F. Klages, *Ann.*, **512**, 185–194 (1935).

II. Occurrence and Biological Function

1. Botanical Sources

As mentioned earlier (see p. 243), the vast majority of galactomannan polysaccharides extracted from higher plants have originated in the *Leguminoseae* family. Of 163 species of legume seeds examined by E. Anderson,[27] 119 contained endosperm mucilage, and, of these 119, galactomannan was identified in 13. Most of the remainder were not examined, but might be expected to be composed of galactomannan. In all, galactomannans have been identified in about 70 species of the *Leguminoseae* (see Table I). Except for *Gymnocladus dioica*,[64] in which the mucilage is extracted from the inner side of the seed coat, and *Glycine max.*,[71] in which it occurs in the hulls, the sole source of the galactomannan has been the endosperm. This is the translucent, white region of the seed which lies between the outer husk and the innermost seed-germ, and which functions as a food reserve for the germinating embryo.

(47) V. P. Kapoor and S. Mukherjee, *Curr. Sci.* (India), **38**, 38 (1969).
(48) V. P. Kapoor and S. Mukherjee, *Phytochemistry*, **10**, 655–659 (1971).
(49) V. P. Kapoor and S. Mukherjee, *Indian J. Chem.*, **10**, 155–158 (1972).
(50) J. Y. Morimoto and A. M. Unrau, *Hawaii Farm Sci.*, **11**, 6–8 (1962).
(51) P. S. Kelkar and S. Mukherjee, *Indian J. Chem.*, **9**, 1085–1087 (1971).
(52) S. A. I. Rizvi, P. C. Gupta, and R. K. Kaul, *Planta Medica*, **20**, 24–32 (1971).
(53) D. S. Gupta and S. Mukherjee, forthcoming publication.
(54) R. S. Saxena and S. Mukherjee, forthcoming publication.
(55) J. E. Courtois and P. Le Dizet, *Carbohyd. Res.*, **3**, 141–151 (1966).
(56) R. D. Jones and A. Morrison unpublished results.
(57) F. Smith, *J. Amer. Chem. Soc.*, **70**, 3249–3253 (1948).
(58) E. L. Hirst and J. K. N. Jones, *J. Chem. Soc.*, 1278–1282 (1948).
(59) A. M. Unrau and Y. M. Choy, *Carbohyd. Res.*, **14**, 151–158 (1970).
(60) V. P. Kapoor, *Phytochemistry*, **11**, 1129–1132 (1972).
(61) A. S. Cerezo, *J. Org. Chem.*, **30**, 924–927 (1965).
(62) J. E. Courtois and P. Le Dizet, *Bull. Soc. Chim. Biol.*, **45**, 731–741 (1963).
(63) C. Leschziner and A. S. Cerezo, *Carbohyd. Res.*, **15**, 291–299 (1970).
(64) E. B. Larson and F. Smith, *J. Amer. Chem. Soc.*, **77**, 429–432 (1955).
(65) A. M. Unrau, *J. Org. Chem.*, **26**, 3097–3101 (1961).
(66) Z. F. Ahmed and A. M. Rizk, *J. Chem. U.A.R.*, **6**, 217–226 (1963).
(67) A. M. Unrau and Y. M. Choy, *Can. J. Chem.*, **48**, 1123–1128 (1970).
(68) E. Heyne and R. L. Whistler, *J. Amer. Chem. Soc.*, **70**, 2249–2252 (1958).
(69) H. C. Srivastava, P. P. Singh, and P. V. Subba Rao, *Carbohyd. Res.*, **6**, 361–366 (1968).
(70) R. L. Whistler and J. Saarnio, *J. Amer. Chem. Soc.*, **79**, 6055–6057 (1957).
(71) G. O. Aspinall and J. N. C. Whyte, *J. Chem. Soc.*, 5058–5063 (1964).

TABLE I

Galactomannans of Leguminoseae Species

Taxon	Botanical name / Species	Yield of gum (%)	D-Mannose to D-galactose ratio	Specific rotation at 589 nm (degrees)	Comments	References
CAESALPINIACEAE (=Caesalpinioideae)						
Caesalpinioideae Group 2	Cassia absus		3.00	+24.2 (water)	permethyl ether, $[\alpha]_D$ +28.3° (chloroform)	47–49
	emarginata L.	27.4	2.70	+21 (M NaOH)		29
	fistula L.	27	3.0			28,50
		1.5	4.0			51
	leptocarpa Benth.	19.3	3.05		contains 10% of pentosan	27
	marylandica L.	27.5	3.76	+27 (M NaOH)		29
						30
	nodosa	27.5	2.7–3.5	+70.5 (M NaOH)	permethyl ether, $[\alpha]_D$ +44° (chloroform)	52
	occidentalis		3.0	+22.4		53
	tora		3.0	+21.8		54
Group 4	Ceratonia siliqua L. (locust-bean gum)	38	3.75			27
			4.5			41
			1.2	+19 (water)	cold-water soluble fraction	41
			3.3	+4 (water)	hot-water soluble fraction	41
			5.2		hot-water insoluble fraction	41
			3.88		extracted at 20°C	55
			3.55		purified, commercial gum	56
			4.0			57
			5.25		permethyl ether, $[\alpha]_D$ −12° (water)	58
Group 5	Caesalpinia cacalaco Humb and Bonpl. (Huizache)	23	2.50			27
	pulcherima (L.) Sw.	31	3.0			50,59
	spinosa Kuntz (tara)	24	2.7	+6 (0.1 M NaOH)		27

			$[\alpha]_D$		Ref.
Cercidium torreyanum Sarg. (Palo Verde)		3.38			27
Delonix regia (Boj.) Raf. (Flame tree)	20	4.28		permethyl ether, $[\alpha]_D$ +46.5° (chloroform)	27
	18.6	2.0			60
Gleditsia amorphoides Taub *ferox* Desf.	28–30	2.5	+22.4 (water)	immature seeds	61
		3.73	−12	immature seeds	62
		3.82	−12.2	mature seeds extracted with cold water	62
		3.90	−12.2	mature seeds extracted with	62
		3.86	−12.2	mature seeds extracted with hot water	62
triacanthos L. (Honey locust)	15–20	3.2	+25	permethyl ether, $[\alpha]_D$ +30.1° (chloroform)	63
	27	2.71		extracted with water at 20°C	27
		3.82		extracted with water at 50°C	55
		3.76			55
Gymnocladus dioica C. Koch (Kentucky coffee-bean)	15	2.71	+22 (NaOH)	permethyl ether, $[\alpha]_D$ ±0° (acetone)	27
		4.0	+29 (water)		64
			−10 (M NaOH)		64
Parkinsonia aculeata L.	28.1	2.70			29
	14				27
MIMOSACEAE (=*Mimosoideae*)					
Mimoseae tribe 4 *Desmanthus illinoensis* Mac. Mill	25	2.69			27
Leucaena glauca (Willd.) Benth. (Koa Hoale)		1.33			28,50,65
FABACEAE (=*Lotoideae*)					
Sophoreae tribe 3 *Sophora japonica* L. (Japanese pagoda tree)	20	5.19	−9 (water)	permethyl ether, $[\alpha]_D$ −25° (benzene)	27
		5.26			43
Genisteae tribe 9 *Genista raetam* (*Retama raetam*)	5	4.14	+59 (water)		66
scoparia	3.3	1.59	+46.5	extracted with water at 20°C	62
		1.66	+36.9	extracted with water at 50°C	62

(Continued)

TABLE I (Continued)

Taxon	Botanical name / Species	Yield of gum (%)	D-Mannose to D-galactose ratio	Specific rotation at 589 nm (degrees)	Comments	References
Crotalarieae	tribe 13					
	Crotalaria incana L.	17.5	2.70	+23 (M NaOH)		29
	intermedia Kotschy	23.5	2.29			27,30
	juncea L.		59% of mannose		no figure for galactose	27
	lanceolata E. Mey	18.5	2.57	+22 (M NaOH)		29
	mucronata		3.0	+6–10 (0.1 M NaOH)		67
		23	2.28			50
			2.27			28
	retusa L.	18	2.85	+29 (M NaOH)		29
	spectabilis Roth.	20.1	2.85	+20 (M NaOH)		29
	striata		60% of mannose		no figure for galactose	27
Indigofereae	tribe 25					
	Cyamopsis tetragonoloba (L.) Taub. (guar)	35	1.54			27
		23.5	1.70	+53 (M NaOH)		29,30
			1.77	+54.5 (M NaOH)		68
			1.37		commercial gum	56
			2.0	+77 (water)		41
			1.3		extracted with cold water	41
			1.7		extracted with hot water	41
			7.0		insoluble in hot water	41
	Indigofera hirsutum		3.14			27
Sesbanieae	tribe 26					
	Sesbania grandiflora (Poir)		1.8	+50 (water)	permethyl ether, $[\alpha]_D$ +42° (chloroform)	69
Astragaleae	tribe 30					
	Astragalus cicer L.	15.8	1.33	+64 (M NaOH)		29
	glycyphyllos L.	16	1.22	+72 (M NaOH)		29
	nuttallianus DC.	20.5	1.38	+60 (M NaOH)		29
	sinicus L.	19	1.63	+74 (M NaOH)		29
	tenellus Bunge	19.2	1.38	+80 (M NaOH)		29

Glycineae	tribe 38	Centrosema plumieri				92% of galactose	50
		Glycine max. (L.) Merr	2	0.09	+65 (water)	permethyl ether, $[\alpha]_D$ +58° (chloroform)	70
		var. Lindarin		1.5	+68 (water)	permethyl ether, $[\alpha]_D$ +56° (chloroform)	71
				1.4	+26.5 (water)	permethyl ether, $[\alpha]_D$ +12° (chloroform)	71
				2.35		extracted with alkaline borate	71
Trifolieae	tribe 42	Medicago hispida Gaertner	19.4	11.5	+76 (water)		29
		lupulina L.	4.2	1.22	+85 (M NaOH)		29,72
			17.7	1.13	+55 (M NaOH)		29,73
		orbicularis (L.) Bartal	18.6	1.17	+89		29
		sativa L. (alfalfa, lucerne)	5.5	1.56	+118		74
				0.5	+90 (M KOH)		75
				1.25			76
				1.08			73
				1.20			42
		Melilotus albus		1.00	+89 (water)		73
		indica (L.) All.	22.1	1.13	+89 (M NaOH)		29
		officinalis		1.04			73
		Trifolium alexandrinum		1.20	+72 (water)		77
		dubium		1.17			73
		hirtum All.	19.9	1.04	+88 (M NaOH)		29
		incarnatum	3.5	1.17	+78 (water)		73
		pratense L.	1.6	1.28	+77.8 (water)		75
				1.3			78
		repens L.	5.25	1.07	+58.8	mature seeds extracted with water at 20°C	62
				1.04	+59.1	mature seeds extracted with hot water	62
		resupinatum L.	15.6	1.04	+84 (M NaOH)		29
		Trigonella caerulea L. Ser.		1.17			73
		calliceras Fisch.		1.13			73
		corniculata L.		1.17			73

(Continued)

TABLE I (Continued)

Taxon	Botanical name Species	Yield of gum (%)	D-Mannose to D-galactose ratio	Specific rotation at 589 nm (degrees)	Comments	References
	cretica L.		1.56–1.67		4 different seed sources	73
	foenum-graecum L. (fenugreek)		1.2	+70 (water)	permethyl ether, $[\alpha]_D$ +50° (chloroform)	79
			1.1			80
	hamosa L.		1.13			73
	monspeliaca L.		1.17			73
	polycerata Bieb.		1.08			73
	radiata (L.) Boiss or		1.13			73
	Medicago radiata L. Heyn		1.17			73
Loteae tribe 43	Anthyllis vulneraria L.	1.8	1.33	+80		81
	Lotus corniculatus L.	1.8	1.25	+87 (water)		82
	pedunculatus Cav.		1.04	+82.5		83
	scoparius Ottley	15.2	1.13	+79 (M NaOH)		29
Desmodieae tribe 48	Alysicarpus vaginalis DC	15.8	1.14	+57 (M NaOH)		29
	Desmodium pulchellum	1	2.00	+60 (0.5 M NaOH)	permethyl ether, $[\alpha]_D$ +42.2° (chloroform)	84

True galactomannans, as defined by Aspinall[85] (that is, those mannans containing more than 5% of D-galactose) have also been extracted from members of the *Annonaceae*,[43] the *Convolvulaceae*,[43,86] and the *Palmae* families[43,87-90] (see Table II). In addition, polymers containing D-galactose and D-mannose have been identified in the *Ebenaceae*[91] and *Loganiaceae*,[91] although figures are not yet available for the relative proportions of the sugars. The *Palmae* family also produce gums of the mannan variety that frequently contain a low level of galactose; for example, ivory nut,[44-46] date palm,[93,94] and doum palm[92] (see Table II).

From *Centrosema plumari* has been isolated[50] an unusual polysaccharide containing 92% of D-galactose and only 8% of D-mannose; the structure of this mannogalactan obviously cannot conform to that of the normal, seed galactomannan.

(72) R. Sømme, *Acta Chem. Scand.*, **22**, 870–876 (1968).

(73) J. S. G. Reid and H. Meier, *Z. Pflanzenphysiol.*, **62**, 89–92 (1970).

(74) E. L. Hirst, J. K. N. Jones, and W. O. Wolder, *J. Chem. Soc.*, 1443–1446 (1947).

(75) P. Andrews, L. Hough, and J. K. N. Jones, *J. Amer. Chem. Soc.*, **74**, 4029–4032 (1952).

(76) R. J. McCredie, *Dissertation Abstr.*, **19**, 432 (1958).

(77) N. R. Krishnaswamy, T. R. Seshadri, and B. R. Sharma, *Curr. Sci.* (India), **35**, 11–12 (1966).

(78) K. F. Horvei and A. Wickstrøm, *Acta Chem. Scand.*, **18**, 833–835 (1964).

(79) P. Andrews, L. Hough, and J. K. N. Jones, *J. Chem. Soc.*, 2744–2750 (1952).

(80) K. M. Daoud, *Biochem. J.*, **26**, 255–263 (1932).

(81) R. Sømme, *Acta Chem. Scand.*, **21**, 685–690 (1967).

(82) R. Sømme, *Acta Chem. Scand.*, **20**, 589–590 (1966).

(83) E. L. Richards, R. J. Beveridge, and M. R. Grimmett, *Aust. J. Chem.*, **21**, 2107–2113 (1968).

(84) M. P. Sinha and R. D. Tiwari, *Phytochemistry*, **9**, 1881–1883 (1970).

(85) G. O. Aspinall, *Advan. Carbohyd. Chem.*, **14**, 429–468 (1959).

(86) S. N. Khanna and P. C. Gupta, *Phytochemistry*, **6**, 605–609 (1967).

(87) A. K. Mukherjee, D. Choudhury, and P. Bagchi, *Can. J. Chem.*, **39**, 1408–1418 (1961).

(88) V. Subrahamnyan, G. Bains, C. Natarijan, and D. Bhatia, *Arch. Biochem. Biophys.*, **60**, 27–34 (1956).

(89) C. V. N. Rao and A. K. Mukherjee, *J. Indian Chem. Soc.*, **39**, 711–716 (1962).

(90) C. V. N. Rao, D. Choudhury, and P. Bagchi, *Can. J. Chem.*, **39**, 375–381 (1961).

(91) P. Kooiman and D. R. Kreger, *Koninkl. Ned. Akad. Wettenschap. Proc. Ser. C*, **63**, 634–645 (1960).

(92) H. S. El Khadem and M. A. E. Sallam, *Carbohyd. Res.*, **4**, 387–391 (1967).

(93) V. K. Jindal and S. Mukherjee, *Curr. Sci.* (India), **38**, 459–460 (1969).

(94) V. K. Jindal and S. Mukherjee, *Indian J. Chem.*, **8**, 417–419 (1970).

TABLE II

Galactomannans of Non-leguminous Plants

Botanical name	D-Mannose to D-galactose ratio	Specific rotation at 589 nm (degrees)	Specific rotation of methyl ether (degrees)	References
ANNONACEAE				
Annona muricata	4.46	+3.4 (water)	+4.7 (chloroform)	43
CONVOLVULACEAE				
Convolvulus tricolor	1.75	+44 (water)	+41 (chloroform)	43
Ipomoea muricata	1.8		+40 (chloroform)	86
EBENACEAE				
Diospyros virginiana L. (Persimmon)	mannose and galactose identified, but no ratios given			91
LOGANIACEAE				
Strychnos nux-vomica	mannose and galactose identified, but no ratios given			91
PALMAE				
Borassoideae				
Borassus flabellifer (unripe) (Palmyra palm nut)	2.4	+8.5 (0.1 M NaOH)		87,89
	2.83			88
Cocosoideae				
Cocos nucifera (unripe) (Coconut)	2.57	+27 (0.1 M NaOH)	+14 (chloroform)	43
	2.0	−85 (0.1 M NaOH)	+25 (chloroform)	90
Caryotoideae				
Arenga saccharifera (unripe)	2.26	+50 (water) +35 (0.1 M NaOH)	+33.5 (chloroform)	43
Phytelephantoideae				
Phytelephas macrocarpa A (Ivory nut)	50.0	−44 (water)		44,45
B	90.0	−38.2 (M NaOH)		44,46
Hyphaene thebaica (Doum palm)	19.0	−28.7 (acetic acid)		92
Phoenicoideae				
Phoenix dactylifera (Date palm)	10.0		−11.8 (chloroform) fractionation −60.6 (chloroform) of permethyl −26.1 (chloroform) ether	93,94

2. Role of Galactomannan in Seeds of *Leguminoseae* Species

For many years, it has been known that (*a*) during germination of leguminous seeds, both the mucilage and the endosperm are degraded,[95,96] and (*b*) the endosperm may be composed almost entirely of galactomannan polysaccharide, and a serious study of the function and metabolism of these gums in the seed has now been made. It has been generally accepted[78,97] that the galactomannan is a reserve polysaccharide which is utilized during germination. In a careful examination of the seeds of *Trigonella foenum-graecum* (fenugreek) of various states of maturity, Reid and Meier[98] found that, although the galactomannan level increases steadily with maturity to a maximum and then decreases slightly, the ratio of sugars remains constant. They suggested that, if the decrease in galactomannan is caused by metabolic degradation, a variable D-mannose:D-galactose ratio would be likely. They were unable to detect any variation in the proportions of sugars, and concluded that the decrease in yield of galactomannan is due to insolubilization of the gum on drying out in the seed. They considered that the galactomannan is formed by the concomitant laying down of D-galactosyl and D-mannosyl groups by a specific mechanism, and not by the random attachment of D-galactosyl groups to a preformed mannan chain, as proposed earlier.[75]

In fenugreek gum (D-mannose:D-galactose ratio, 1.05:1), in which almost every D-mannosyl residue is substituted with a D-galactosyl group, it is impossible for there to be a wide spread of D-mannose:D-galactose ratios between chains. This is not so with locust-bean gum (D-mannose:D-galactose ratio, ~3.5:1), the measured ratio of which probably represents the average, sugar composition over a wide range. Courtois and Le Dizet[62] fractionated galactomannan from immature and mature seeds of *Gleditsia ferox* and *Trifolium repens* by extraction with hot and cold water, and obtained products having almost constant D-mannose:D-galactose ratios; they varied from 3.73:1 to 3.90:1 for *Gleditsia ferox*, and from 1.07:1 to 1.04:1 for *Trifolium repens*, indicating that the galactomannans were highly homogeneous with regard to sugar composition. Interestingly, although the *Gleditsia ferox* galactomannan and locust-bean gum have similar D-mannose:D-galactose ratios, the actual distribution of D-galactose is

(95) A. Tschirch, "Angewandte Pflanzenanatomie," Urgan and Schwarzenberg, Vienna, 1889.
(96) E. Schulze, *Landwirtsch. Jahrb. Schweiz*, **23**, 1 (1894).
(97) J. W. Hylin and K. Sawai, *J. Biol. Chem.*, **239**, 990–992 (1964).
(98) J. S. G. Reid and H. Meier, *Phytochemistry*, **9**, 513–520 (1970).

fundamentally different (this point will be amply demonstrated in Section IV,2; p. 271). We believe that, although the mechanism for the biosynthesis of galactomannans is probably that suggested by Reid and Meier,[98] there must be fundamental differences in either the enzymic-control mechanisms or the chemical activities of the enzymes responsible (in order to account for the diversity of galactomannans noted here).

The biodegradative hydrolysis of a galactomannan requires the presence of at least three enzymes[99] in the germinating seeds: α-D-galactosidase for removal of the $(1 \rightarrow 6)$-α-D-galactose side-chains, β-D-mannanase for fission of the $(1 \rightarrow 4)$-β-D-mannan backbone into oligosaccharides, and β-D-mannosidase for complete hydrolysis of the D-manno-oligosaccharides to D-mannose. α-D-Galactosidase has been reported[100] in the seeds of a number of galactomannan-yielding Leguminoseae, including Gleditsia ferox, Cyamopsis tetragonolobus, Medicago species, and Trigonella foenum-graecum. It has also been detected in Coffea species,[100] which are a non-leguminous source of galactomannan (for a review of α-D-galactosidases, see Ref. 101). β-D-Mannanases have also been reported in the seeds of leguminous plants, including Cyamopsis tetragonolobus,[102] Gleditsia ferox,[100] Leucaena glauca,[97] and Trigonella foenum-graecum.[103] α-D-Galactosidases and β-D-mannanases have been used in a study of the fine structure of galactomannans (see Section IV,2; p. 272). β-D-Mannosidase had not been detected in galactomannan-containing seeds until it was reported to occur in germinating seeds of Trigonella foenum-graecum.[104]

The products of the enzymic hydrolysis of galactomannans include D-galactose, D-mannose, and D-manno-oligosaccharides; these have been detected in several Leguminoseae species. Courtois and Le Dizet[62] showed that, during germination of seeds of Gleditsia ferox and Gleditsia triacanthos, there is a preliminary removal of D-galactosyl groups; this is followed by scission of the mannan chain, and concomitant release of D-manno-oligosaccharides. D-Mannobiose, D-mannotriose, and D-mannotetraose were isolated, in addition to D-mannose and D-galactose, and a galactomannan having an increased

(99) E. T. Reese and Y. Shibata, Can. J. Microbiol., 11, 167–183 (1965).
(100) J. E. Courtois and P. Le Dizet, Bull. Soc. Chim. Biol., 45, 743–760 (1963).
(101) P. M. Dey and J. B. Pridham, Advan. Enzymol., 36, 91–130 (1972).
(102) R. L. Whistler and C. G. Smith, J. Amer. Chem. Soc., 74, 3795–3796 (1952).
(103) S. Clermont-Beaugiraud and F. Percheron, Bull. Soc. Chim. Biol., 50, 633–639 (1968).
(104) J. S. G. Reid and H. Meier, Planta, 112, 301–303 (1973).

content of D-mannose was identified. Similar results were obtained with germinated seeds of *Trigonella foenum-graecum*.[105]

In further studies on the germinating seeds of *Trigonella foenum-graecum* by use of light microscopy and electron microscopy, Reid[106] differentiated three distinct phases in the germination and metabolism of the seeds. During the first phase (lasting 18 hours; zero time defined as the penetration of the seed coat by the radicle), there was, apparently, no change in the galactomannan. D-Galactose appeared in the endosperm, but starch was not present in the cotyledons. Raffinose decreased in both the endosperm and the cotyledons.

During the following 24 hours (phase 2), the content of galactomannan decreased to zero, and the endosperm began to liquefy, starting in the region of the aleurone layer (the outer layer of the endosperm, farthest from the embryo). Starch granules appeared in that part of the cotyledon bordering on the endosperm, and sucrose levels increased. The concentrations of D-galactose and D-mannose also built up. In phase 3, the endosperm virtually ceased to exist, and the concentrations of starch and sucrose started to decrease in the cotyledons.

Reid and Meier[107] were able to demonstrate, with endosperm separated from the embryo, and by the use of chemical inhibition of protein synthesis and oxidative phosphorylation, that the enzymes responsible for the degradation of galactomannan are synthesized in, and secreted from, the aleurone layer. The same authors[104] assayed, as a function of time, the levels of α-D-galactosidase and β-D-mannosidase in the endosperm of germinating seeds of *Trigonella feonum-graecum*, and showed that the maximal amounts of these enzymes coincided with the release of D-galactose and the breakdown of galactomannan.

Contrary to the results of Reid and Meier,[104,107] Sømme[108] was unable to detect either D-mannose or β-D-mannosidase in germinating seeds of *Trifolium repens, Trifolium pratense, Medicago sativa, Anthyllis vulneraria*, or *Lotus corniculatus*. This failure to detect β-D-mannosidase led her to suggest phosphorolysis as an alternative to hydrolysis for the mechanism of the degradation of galactomannan.[108] A report[109] that crude homogenates of germinating

(105) A. Sioufi, F. Percheron, and J. E. Courtois, *Phytochemistry*, 9, 991–999 (1970).
(106) J. S. G. Reid, *Planta*, 100, 131–142 (1971).
(107) J. S. G. Reid and H. Meier, *Planta*, 106, 44–60 (1972).
(108) R. Sømme, *Acta Chem. Scand.*, 24, 72–76 (1970).
(109) M. J. Foglietti and F. Percheron, *Compt. Rend., D*, 274, 130–132 (1972).

seeds of *Trigonella foenum-graecum* possess phosphorolytic activity and are able to catalyze the reversible phosphorolysis of $(1 \rightarrow 4)$-β-D-manno-oligosaccharides, appears to support this suggestion. Reid and Meier,[104] however, pointed out that the enzymic degradation of galactomannan just reported occurs in the storage tissue (which is not living tissue), and that the enzymes are released into this tissue from the aleurone layer. Thus,[107–109] the enzymes are extracellular, and yet no extracellular phosphorylases have so far been reported in any biodegradative system.

3. Chemotaxonomy of *Leguminoseae*

Several authors[73,110,111] have attempted to discuss the taxonomy of the *Leguminoseae* in terms of either the yield of galactomannan derived from the endosperm or of the D-mannose : D-galactose ratio of the galactomannan. As far as the yields of gum from the endosperms are concerned, the data available are insufficient to permit the drawing of meaningful conclusions. The values in the literature are very variable (see Table I) and, presumably, depend on the methods employed for extraction and purification of the gum. Many *Leguminoseae* do not yield a mucilage, and, consequently, do not afford a galactomannan.[27]

There is still considerable disagreement[112] regarding the classification of the *Leguminoseae*, and, in Table I, we have followed Bailey[110] and classified the *Leguminoseae* according to Hutchinson (see Ref. 112). Although the D-mannose:D-galactose ratios of the galactomannans may sometimes be dependent on the extraction techniques used, they are possibly more reliable guides to taxonomy than are the yields. For example, the *Astragaleae* yield galactomannans having D-mannose:D-galactose ratios of 1.2 : 1 to 1.8 : 1 (see Table I), and an investigation of the tribe *Trifolieae*[73] showed a remarkable consistency of the D-mannose:D-galactose ratio (1 : 1 to 1.2 : 1) throughout the various species; the only exception was *Trigonella cretica*, which had a D-mannose:D-galactose ratio of 1.6 : 1. Seeds of *Trigonella cretica* from four different sources were examined, and all

(110) R. W. Bailey, in "Chemotaxonomy of Leguminoseae," J. B. Harborne, D. Boulter, and B. L. Turner, eds., Academic Press, London, 1971, p. 519.

(111) P. Kooiman, *Carbohyd. Res.*, **25**, 1 (1972).

(112) V. H. Heywood, in "Chemotaxonomy of Leguminoseae," J. B. Harborne, D. Boulter, and B. L. Turner, eds., Academic Press, London, 1971, p. 1; J. Hutchinson, "The Genera of Flowering Plants," Oxford University Press, London, 1964.

were found to have this "anomalous" D-mannose:D-galactose ratio. Examination of eight other *Trigonella* species showed them to conform to the general pattern for *Trifolieae*. There is, therefore, a reason for re-examining the classification of *Trigonella cretica*.

It is clear that only such careful examination as that already described[73] will lead to useful chemotaxonomic progress. Observations based solely on collations of D-mannose : D-galactose ratios and yields from the literature should be treated with caution.

III. Extraction and Preliminary Characterization

1. Extraction and Purification

The extraction on a commercial scale of galactomannan gums from carob and guar seeds usually involves a preliminary removal of the tough seed-coat by some form of milling or grinding. For research samples, the whole seed is normally ground to a fine powder, and the gum is extracted by prolonged stirring with hot or cold water, or dilute, aqueous sodium hydroxide solution; sometimes, 1% acetic acid has been used. The soluble portion of the gum is then separated by centrifugation or filtration, and the polysaccharide is recovered by precipitation with ethanol. Methanol[113] has also been used for the recovery stage, and isopropyl alcohol is the alcohol of choice for industrial processes. One novel approach, used to extract gum from Doum-palm kernels,[92] involved the acetylation of the dried kernels, followed by regeneration of the polymer. The specific rotation (−31.50°) of the product, a polysaccharide low in D-galactose, was very similar to that (−28.75°) of the gum extracted by use of 20% sodium hydroxide solution.

Purification of the crude galactomannan is achieved either by repeated precipitations with ethanol, or by precipitation of the copper or barium complex, regeneration of the galactomannan, and, finally, reprecipitation with ethanol. The purification step may be repeated several times. Fehling solution[63,75,79,92] is the reagent commonly used, although cupric acetate[59,71,114] and barium hydroxide[60,115] have also been employed. It has been suggested that galactomannans may be degraded in the presence of metal ions, but all of the evidence available indicates that purification by way of the copper or

(113) C. M. Rafique and F. Smith, *J. Amer. Chem. Soc.*, **72**, 4634–4636 (1950).
(114) A. J. Erskine and J. K. N. Jones, *Can. J. Chem.*, **34**, 821–826 (1956).
(115) H. Meier, *Methods Carbohyd. Chem.*, **5**, 45–46 (1965).

barium complex does not alter the chemical composition. Thus, for *Crotalaria mucronata*,[67] *Caesalpinia pulcherima*,[59] and *Cocos nucifera*,[90] the sugar ratios remained almost constant throughout two purifications with Fehling solution[90] or three with cupric acetate.[59,67] Furthermore, purifications by way of copper complexes, or fractional precipitation with ethanol, gave, from *Gleditsia triacanthos*,[63] galactomannan samples having identical optical rotations. Acetylation and regeneration of the galactomannan apparently also causes no change in the chemical composition, as shown for those from *Gleditsia amorphoides*[61] and *Gymnocladus dioica*.[64] For the former, the optical rotation of the regenerated gum (+23.6°) differed little from that of the ethanol-precipitated product (+22.4°).

It is, however, possible that purification by way of copper complexes or peracetyl derivatives may lead to chain cleavage, and, thus, to smaller molecular weights (this has been demonstrated for Konjac mannan[116]), but such cleavage would not significantly alter either the specific rotation or the D-mannose:D-galactose ratio of the sample. The relatively low molecular weights[43] of a number of galactomannans purified by way of the copper complexes may be the result of chain scission (see Table V, p. 279). However, it has been found that, for a guar gum, the number-average molecular weight determined by membrane osmometry was unchanged before and after acetylation (for the method, see Ref. 61), provided that a correction was made for the acetyl groups.[117]

In only one instance, namely, *Borassus flabellifer*, has there been any report of decomposition leading to altered D-mannose:D-galactose ratios caused by purification through the copper complexes, and it seems likely that this was due to fortuitous fractionation.[37]

A method for purifying certain neutral polysaccharides by precipitation in borate buffer with cetyltrimethylammonium bromide was developed by Stacey and coworkers,[118] but it does not appear to have been used for the isolation of galactomannan gums. It has, however, been claimed[119] that guar gum and locust-bean gum can be separated by carefully controlled precipitation of the cetylpyridinium bromide complexes by 0.1 M sodium hydroxide. Guar gum is precipitated between pH ~7.0 and 8.3, and locust-bean gum between 8.6 and 9.3.

(116) N. Sugiyama, H. Shimahara, T. Andoh, M. Takemoto, and T. Kamata, *Agr. Biol. Chem.* (Tokyo), **36**, 1381–1387 (1972).
(117) R. A. Jones and A. Morrison, unpublished results.
(118) S. A. Barker, M. Stacey, and G. Zweifel, *Chem. Ind.* (London), 330 (1957).
(119) R. G. Morley, Ph. D. Thesis, University of Salford (1972).

a. Identification of Sugar Residues, and the D-Mannose to D-Galactose Ratio. — A particularly convenient method favored by the authors is that developed by Albersheim and coworkers,[120] in which the polysaccharides are hydrolyzed with 2 *M* trifluoroacetic acid, and the liberated hexoses converted into alditol acetates, which are then separated by gas–liquid chromatography (g.l.c.). *myo*-Inositol may be added as an internal standard. Suitable column-packings include ECNSS-M and OV-225. Identification of the residues can be achieved unambiguously (except for chirality), as all six hexitol acetates derived from the eight aldohexoses can be separated and identified by g.l.c.–mass spectrometry. Fortunately, galactitol and mannitol are uniquely derived from galactose and mannose, respectively, whereas glucitol is produced by reduction of both glucose and gulose, and altritol from both altrose and talose.

2. Homogeneity

The subject of the homogeneity of polysaccharides can lead to considerable confusion, and a definition of the term, as used in this article, will now be given. In a discussion of molecular weights of proteins and polysaccharides, Tanford[121] attributed "a high degree of homogeneity" to those polymers that have almost identical weight-average ($\overline{M}_w$) and number-average ($\overline{M}_n$) molecular weights. No polysaccharide has as yet been shown to be homogeneous in molecular weight, and as the ratio of $\overline{M}_w$ to $\overline{M}_n$ increases, so does the polydispersity increase also.[122] For galactomannan polysaccharides, not only does the possibility of polydispersity of molecular weights arise, but there is also the likelihood of the existence of a range of molecules of differing D-mannose:D-galactose ratios. The ratios may vary about a mean, which is the average D-mannose:D-galactose ratio as determined experimentally on the sample. Such a collection of polysaccharides will be regarded as homogeneous; if the distribution shows two or more maxima of D-mannose:D-galactose ratios, the sample will be described as heterogeneous.

As mentioned previously (see Section II,2; p. 255), those galactomannans having D-mannose:D-galactose ratios approaching unity must, of necessity, possess, at the most, a narrow range of D-man-

(120) P. Albersheim, D. J. Nevins, P. D. English, and A. Karr, *Carbohyd. Res.*, **5**, 340–345 (1967).
(121) C. Tanford, "Physical Chemistry of Macromolecules," Wiley, New York, 1961, pp. 145 and 291.
(122) C. T. Greenwood, *Advan. Carbohyd. Chem.*, **7**, 289–332 (1952).

nose: D-galactose ratios if they have the usual structure (see Section IV,1; p. 263); for example, the *Trifolieae* (see Table I; p. 251). For galactomannans having larger ratios, there is at least the possibility of a wide range of ratios. Locust-bean gum and guar gum can be fractionated simply by stirring with cold water, centrifuging, stirring the residue with hot water, and re-centrifuging. Measurement of the D-mannose: D-galactose ratios of the three fractions, namely, cold-water soluble, hot-water soluble, and hot-water insoluble, indicates a wide spread of sugar compositions. The fractionation, originally conducted by Hui and Neukom,[41] has been repeated in our laboratory. Fractionation was, indeed, achieved, but the D-galactose ratios were different, probably because our sample was a commercial preparation that had already been partially fractionated.[56] Hui and Neukom's fractionation[41] also separated fractions of differing molecular weight. Attempts to fractionate *Gleditsia ferox* galactomannan[62] (D-mannose: D-galactose ratio, 3.8:1) failed, suggesting that this gum exhibits a narrow range of distribution of sugar compositions.

The gums extracted from *Cassia absus*[47] were fractionated by means of their copper complexes. One fraction was soluble in dilute acetic acid, and the other was insoluble therein. As, on regeneration, both gave polysaccharides having a D-mannose:D-galactose ratio of 3:1, the different solubility was, presumably, caused by differences in molecular weight, or, possibly, by differences in the distribution of D-galactosyl groups along the D-mannan chain (see Section IV, p. 269). Two copper complexes were also isolated from soybean-hull gums.[71] One was insoluble in water, and the other was precipitated from aqueous solution by ethanol. The regenerated materials from the second complex had a D-mannose: D-galactose ratio of 1.4:1. Further extraction of the soybean hulls yielded two more copper complexes, one of which gave a polysaccharide having a D-mannose: D-galactose ratio of 2.35:1. Unfortunately, D-mannose:D-galactose ratios were not given for the other products. The lower solubility of the gum having less substitution by D-galactosyl groups is a feature typical of the galactomannans. Thus, locust-bean gum having a D-mannose: D-galactose ratio of 3.3–4.5:1 can be dispersed in water only after heating the suspension to ~80° for 30 minutes, whereas guar gum (D-mannose: D-galactose ratio, 1.5:1) is readily dispersed at room temperature.

When gum is extracted from seeds in which the mucilage exhibits a broad range of species differing in D-mannose and D-galactose composition, the D-mannose:D-galactose ratio of the product obtained depends very much on the method of extraction used. Thus, if cold water alone is used, the only species to be efficiently recovered will have a high content of D-galactose, whereas hot-water extraction will yield more of the polysaccharide that has a low content of D-galactose. To summarize, galactomannan extracts of some plant mucilages may contain glycans exhibiting polydispersity of molecular weight and variations in the content of D-galactose. It will be shown later that variations in the distribution of D-galactosyl groups along the D-mannan chain also occur.

IV. Structure of Galactomannans and Related Polysaccharides

1. Main Covalent Features

a. Galactomannans. — Pioneering work, long regarded as classical, in the structural chemistry of plant-seed galactomannans was conducted in the late 1940's and early 1950's; these studies were mainly concerned with two commercially important galactomannans, guar gum and locust-bean gum, and indicated that these polysaccharides contain a β-D-$(1 \rightarrow 4)$-linked D-mannan backbone to which are attached single α-D-galactosyl stubs at O-6 of certain of the D-mannosyl residues. Guar gum contains approximately twice as many α-D-galactosyl stubs as does locust-bean gum. These conclusions were based mainly on the results of methylation analysis[57,58,113,123] and partial-hydrolysis experiments.[124,125] Thus, methylation and subsequent hydrolysis yielded mixtures of 2,3,4,6-tetra-O-methyl-D-galactose, 2,3,6-tri-O-methyl-D-mannose, and 2,3-di-O-methyl-D-mannose, whereas partial hydrolysis yielded the disaccharides 6-O-α-D-galactopyranosyl-D-mannose and 4-O-β-D-mannopyranosyl-D-mannose. The results of periodate oxidation[113,126] supported these results. Formulas 1 and 2 show possible repeating units for locust-bean gum and guar gum, respectively.

$$\alpha\text{-D-Gal}p$$
$$1$$
$$\downarrow$$
$$6$$
$$\beta\text{-D-Man}p\text{-}(1 \rightarrow 4)\text{-}\beta\text{-D-Man}p\text{-}(1 \rightarrow 4)\text{-}\beta\text{-D-Man}p\text{-}(1 \rightarrow 4)\text{-}\beta\text{-D-Man}p\text{-}(1 \rightarrow 4)\text{-}$$

1

(123) Z. F. Ahmed and R. L. Whistler, *J. Amer. Chem. Soc.*, **72**, 2524–2525 (1950).
(124) R. L. Whistler and J. Z. Stein, *J. Amer. Chem. Soc.*, **73**, 4187–4188 (1951).
(125) R. L. Whistler and D. F. Durso, *J. Amer. Chem. Soc.*, **73**, 4189–4190 (1951).
(126) O. E. Moe, S. E. Miller, and M. H. Iwen, *J. Amer. Chem. Soc.*, **69**, 2621–2625 (1947).

α-D-Gal$p$ α-D-Gal$p$

1   1

$\downarrow$ $\downarrow$

6   6

β-D-Man$p$-(1$\rightarrow$4)-β-D-Man$p$-(1$\rightarrow$4)-β-D-Man$p$-(1$\rightarrow$4)-β-D-Man$p$-(1$\rightarrow$4)-

2

In the wake of these investigations, numerous other plant-seed galactomannans have been examined by essentially the same techniques, and most have been found to have the same fundamental structure, although the degree of substitution of the β-D-mannan backbone by D-galactosyl groups varies widely. Galactomannans reported to have structures of this type include those from the seeds of *Gleditsia amorphoides*,[61] *Gleditsia triacanthos*,[63] *Cassia nodosa*,[52] *Cassia fistula*,[51] *Cassia tora*,[54] *Cassia occidentalis*,[53] *Lotus corniculatus*,[82] *Lotus pedunculatus*,[83] *Anthyllis vulneraria*,[81] *Trigonella foenum-graecum*,[75] *Medicago sativa*,[75] *Medicago lupulina*,[72] and *Desmodium pulchellum*[84] (*Leguminoseae* family); *Arenga saccharifera*,[43] *Cocos nucifera*,[43] and *Borassus flabellifer*[87] (*Palmae* family); *Convolvulus tricolor*[43] and *Ipomoea muricata*[86] (*Convolvulaceae* family); and *Annona muricata*[43] (*Annonaceae* family).

However, a few galactomannans have been shown to have structures that deviate from this classical type; this first became apparent from Unrau and Choy's examination of the galactomannans from *Crotalaria mucronata*[67] and *Caesalpinia pulcherima*.[59] *Crotalaria mucronata* galactomannan[67] has a D-mannose : D-galactose ratio of 3 : 1. Methylation and hydrolysis yielded 2,3,4,6-tetra-*O*-methyl-D-mannose, 2,3,4,6-tetra-*O*-methyl-D-galactose, 2,3,6-tri-*O*-methyl-D-mannose, 2,4,6-tri-*O*-methyl-D-mannose, 3,4,6-tri-*O*-methyl-D-mannose, and 2,3-di-*O*-methyl-D-mannose in the molar ratios of 1 : 23 : 28 : 11 : 4 : 26. The presence of (1 $\rightarrow$ 3)-linkages in the D-mannan main-chain was further supported by the finding that one in every six D-mannose residues is resistant to periodate oxidation. Smith degradation yielded equimolar amounts of 2-*O*-β-D-mannopyranosyl-D-erythritol and *O*-β-D-mannopyranosyl-(1 $\rightarrow$ 3)-2-*O*-β-D-mannopyranosyl-D-erythritol, indicating that there are significant amounts of isolated (**3**) and consecutive (**4**) β-D-(1 $\rightarrow$ 3)- linkages in the D-mannan backbone.

β-D-Man$p$-(1$\rightarrow$4)-β-D-Man$p$-(1$\rightarrow$3)-β-D-Man$p$-(1$\rightarrow$4)-β-D-Man$p$-

3

β-D-Man$p$-(1$\rightarrow$4)-β-D-Man$p$-(1$\rightarrow$3)-β-D-Man$p$-(1$\rightarrow$3)-β-D-Man$p$-(1$\rightarrow$4)-β-D-Man$p$-

4

A further conclusion from the methylation results was that approximately 6% of the residues in the mannan backbone are $(1 \rightarrow 2)$-linked. Single D-galactosyl stubs are linked to the main chain by $(1 \rightarrow 6)$-bonds.

In a similar series of experiments, Unrau and Choy[59] found that the structure of *Caesalpinia pulcherima* galactomannan differs from the normal type of structure, but to a much lesser extent; ~4% of the D-mannose units in the main chain are β-D-$(1 \rightarrow 3)$-linked, and ~2% are ($1 \rightarrow 2$)-linked. Smith degradation again indicated that isolated and consecutive β-D-$(1 \rightarrow 3)$-linkages were present in the main chain.

Kapoor and Mukherjee showed that the structure of the galactomannan from *Cassia absus* seeds differs from the "classical" structure of galactomannans not only in the D-mannan backbone but also in the mode of linkage of the D-galactosyl groups to the main chain.[48,49,127] *Cassia absus* galactomannan has a D-mannose:D-galactose ratio of 3:1. Methylation analysis[127] yielded 2,3,4,6-tetra-*O*-methyl-D-galactose, 2,3,6-tri-*O*-methyl-D-mannose, 2,3-di-*O*-methyl-D-mannose, and 4,6-di-*O*-methyl-D-mannose in the molar ratios of 2:42:1:1. Partial hydrolysis[48] with acid yielded the disaccharides 2-*O*-α-D-galactopyranosyl-D-mannose, 6-*O*-α-D-galactopyranosyl-D-mannose, and 4-*O*-β-D-mannopyranosyl-D-mannose, and the trisaccharides 4-*O*-(6-*O*-α-D-galactopyranosyl-β-D-mannopyranosyl)-β-D-mannose and 4-*O*-(4-*O*-β-D-mannopyranosyl-β-D-mannopyranosyl)-β-D-mannose.

This was the first report of 2-*O*-α-D-galactopyranosyl-D-mannose as a fragmentation product of any seed galactomannan, although it had previously been isolated from the galactomannan of the yeast *Trichosporon fermentans*.[16] From these results, it was deduced that ~85% of the D-mannose residues in the main chain are β-D-$(1 \rightarrow 4)$-linked and the rest are $(1 \rightarrow 3)$-linked. All of the $(1 \rightarrow 3)$-linked D-mannose residues are substituted at O-2 by an α-D-galactosyl group, accounting for ~50% of the D-galactosyl, side-chain stubs. The rest are attached through α-D-$(1 \rightarrow 6)$-linkages to β-D-$(1 \rightarrow 4)$-linked D-mannosyl residues. The average repeating-structure proposed for the galactomannan is shown in formula 5. Smith-degradation experiments were not carried out, and so it is not known whether the $(1 \rightarrow 3)$-linkages in the main chain are isolated and/or consecutive in this case.

(127) V. P. Kapoor and S. Mukherjee, *Can. J. Chem.*, **47**, 2883–2887 (1969).

α-D-Gal$p$
1

↓

6
-[β-D-Man$p$-(1→4)]$_2$-β-D-Man$p$-(1→4)-β-D-Man$p$-(1→4)-β-D-Man$p$-(1→3)-β-D-Man$p$-(1→4)-

2

↑

1

α-D-Gal$p$

5

Another possible variation in the chemical structure of galactomannans is the presence of infrequent, short chains of α-D-(1 → 6)-
linked D-galactosyl residues attached to the D-mannan backbone, in
addition to the single α-D-galactosyl stubs. Thus, after methylation
and subsequent hydrolysis of the galactomannans of *Gleditsia
ferox*,[55,62] *Gleditsia amorphoides*,[61] and *Trifolium repens*,[55] small proportions of 2,3,4-tri-*O*-methyl-D-galactose were obtained. As this
methylated derivative does not seem to be an artefact of partial
demethylation or undermethylation, it appears likely that, in these
galactomannans, there are infrequent, short chains of α-D-(1 → 6)-
linked D-galactosyl residues attached to the main chain, as depicted
in formula **6.**

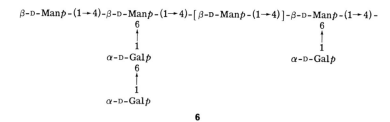

β-D-Man$p$-(1→4)-β-D-Man$p$-(1→4)-[β-D-Man$p$-(1→4)]-β-D-Man$p$-(1→4)-
6
↑
1
α-D-Gal$p$
6
↑
1
α-D-Gal$p$

6
↑
1
α-D-Gal$p$

6

The possibility of there being branches in the mannan main-chain
should not be excluded. Estimates from a number of methylation
analyses of galactomannans have shown a slightly higher molar
quantity of 2,3-di-*O*-methyl-D-mannose than 2,3,4,6-tetra-*O*-
methyl-D-galactose[55,78,79]. However, if branching of the mannan main-
chain is a feature of galactomannan structure, its extent seems to be
fairly low.

Although most of the galactomannans studied are reported to have
structures based on a completely β-D-(1 → 4)-linked backbone having
single α-D-galactosyl stubs attached to this backbone by (1 → 6)-
linkages, the reports just discussed indicate that deviations, sometimes substantial, can occur. Indeed, it is possible that deviations
from the "classical" type of structure, to the extent shown by *Caesal-*

pinia pulcherima galactomannan, may not have been observable before the development of modern techniques of structure determination, and that the presence of β-D-(1 → 3)-linkages in the D-mannan backbone may be more widespread in galactomannans than previously supposed. However, a careful search for (1 → 3)- and (1 → 2)-linkages, by methylation analysis, in guar gum, locust-bean gum, tara gum (from *Caesalpinia spinosa*), and fenugreek mucilage (from *Trigonella foenum-graecum*) proved negative,[128] and so the classical type of structure probably represents the structure of many galactomannans accurately.

Kooiman[111] and Leschziner and Cerezo[129] used the fact that, to a first approximation, the specific rotation of a polysaccharide depends on the relative proportions of α-D- and β-D-linkages to develop empirical equations relating the sugar compositions of galactomannans with their specific rotations. Only 12 of the 75 galactomannans considered were found to deviate seriously from predicted values. The authors suggested that the lack of correspondence between the calculated and experimental values in these instances might be due to experimental error in the measurements of optical rotations attributable to opacity of solutions, to errors in determining sugar compositions, or to minor structural deviations of the galactomannan from the normal type.

b. Related Polysaccharides.—The general structure of galactomannans indicated by these experiments bears a close resemblance to that of many other common, plant polysaccharides. For instance, the structures of the galactomannans show an obvious similarity to those of the hemicellulosic galactoglucomannans. These cell-wall polysaccharides consist of a main chain of β-D-(1 → 4)-linked D-mannosyl and D-glucosyl residues, some of which are substituted at O-6 with a single α-D-galactosyl stub. The ratio of D-glucose to D-mannose is usually in the range of 1:3; and the proportion of D-galactose can range from 0.1 to >1 per D-glucosyl residue.[130] Those polysaccharides that contain little D-galactose are soluble in aqueous alkali, whereas those rich in D-galactose are water-soluble.[130]

Analogies may be drawn between galactomannans and xyloglucans. The latter polysaccharides, often referred to as "amyloids," are found in seed endosperms of plants from many different fami-

(128) I. C. M. Dea and R. Moorhouse, unpublished results.
(129) C. Leschziner and A. S. Cerezo, *Carbohyd. Res.*, **11**, 113–118 (1969).
(130) G. C. Hoffmann and T. E. Timell, *Tappi*, **53**, 1896–1899 (1970).

lies,[131] where, like galactomannans, they function as food reserves.[132] Very similar polysaccharides are also found extracellular to,[133] and in the primary cell-wall[134] of, sycamore cells grown in culture. Like that of the galactomannans, the typical structure of seed xyloglucans comprises a main chain of β-D-(1 $\rightarrow$ 4)-linked D-glucosyl residues to which are attached 2-O-β-D-galactopyranosyl-D-xylopyranosyl disaccharide stubs by α-D-(1 $\rightarrow$ 6)-linkages to certain of the D-glucosyl residues.[135,136] The degree of substitution by disaccharide stubs varies with the source of the polysaccharide. Certain of the seed xyloglucans (see, for example, Ref. 132) and the primary, cell-wall xyloglucan[134] deviate slightly from this structure.

A further related group comprises the arabinoxylans and 4-O-methylglucuronoxylans. These hemicelluloses occur widely in the cell walls of higher plants, and as the "pentosans" in the endosperm of wheat and other cereals, where they function as food reserves. The arabinoxylans consist of a linear chain of β-D-(1 $\rightarrow$ 4)-linked D-xylosyl residues to which single α-L-arabinofuranosyl groups are attached by (1 $\rightarrow$ 2) and (1 $\rightarrow$ 3)-linkages. The extent of substitution by L-arabinose differs with the source of the polysaccharide. In the 4-O-methylglucuronoxylans, single α-D-glucosyluronic acid groups are attached by (1 $\rightarrow$ 2)-linkages to the same main-chain; in softwood xylans, L-arabinosyl stubs also occur, and, in hardwoods, the xylan chain is partially acetylated (for reviews, see Ref. 137). A related group of polysaccharides occurs as food reserves in the corm sacs of *Watsonia* species; here, the degree of substitution of the xylan chain is unusually high.[138]

Considering the close resemblance in general structure and food-reserve function of these polysaccharides, a certain similarity in properties might be expected, and this similarity will be discussed in later Sections. Also, the parent polysaccharides cellulose, mannan, and esparto xylan are all intrinsically water-insoluble; the progressive introduction of the monosaccharide and disaccharide stubs, if

(131) P. Kooiman, *Acta Botan. Neerl.*, **9**, 208–219 (1960).
(132) S. E. B. Gould, D. A. Rees, and N. J. Wight, *Biochem. J.*, **124**, 47 (1971).
(133) G. O. Aspinall, J. A. Malloy, and J. W. T. Craig, *Can. J. Biochem.*, **47**, 1063–1070 (1969).
(134) W. D. Bauer, K. W. Talmadge, K. Keegstra, and P. Albersheim, *Plant Physiol.*, **51**, 174–187 (1973).
(135) P. Kooiman, *Rec. Trav. Chim.*, **80**, 849–865 (1961).
(136) P. Kooiman, *Phytochemistry*, **6**, 1665–1675 (1967).
(137) T. E. Timell, *Advan. Carbohyd. Chem.*, **19**, 247–302 (1964); **20**, 409–483 (1965).
(138) D. H. Shaw and A. M. Stephen, *Carbohyd. Res.*, **1**, 400–413 (1966).

present in large enough proportion, renders the molecules more water-soluble. A controlled degree of acetylation in main chains of this type has a similar effect. Thus, β-D-$(1 \rightarrow 4)$-linked glucomannan is insoluble in water, but the glucomannans from the tubers of *Amorphophallus konjac*[139] (Konjac mannan) and *Tubera salep*,[140] which contain approximately one acetyl group per ten glycosyl residues, are moderately soluble in water. These acetylated polysaccharides (which *also* function as food reserves) have some properties in common with galactomannans.

2. Fine Structure

Although experiments of the type outlined in the previous Section have led to a good understanding of galactomannan structure, a knowledge of the distribution of α-D-galactosyl stubs along the main chain is necessary before a correlation can be made between structure and properties. Formulas **7–9** indicate three possible extremes of structures for a galactomannan having a D-mannose (M): D-galactose (G) ratio of $2:1$.

```
-M-M-M-M-M-M-M-M-M-M-M-M-M-M-M-M-
 |   |   |   |   |   |   |   |
 G   G   G   G   G   G   G   G
```

7

```
-M-M-M-M-M-M-M-M-M-M-M-M-M-M-M-M-M-
 |   | |     |     | | |       | |
 G   G G     G     G G G       G G
```

8

```
-M-M-M-M-M-M-M-M-M-M-M-M-M-M-M-M-M-M-M-M-
 | | | | | |               | | | | | | |
 G G G G G G               G G G G G G G
```

9

Formula **7** represents a regular arrangement of side chains; **8**, a random distribution of side chains; and **9**, a structure in which the side chains occur in blocks. Intermediates between these structures are, of course, possible. The fact that conclusions from structural investigations of galactomannans are commonly expressed in an average repeating-structure, such as **1** to **5**, does not necessarily mean that a regular arrangement of side chains was assumed.[57,67]

(139) K. Kato and K. Matsuda, *Agr. Biol. Chem.* (Tokyo), **33**, 1446–1453 (1969).
(140) H. Bittiger and E. Hüsemann, *J. Polym. Sci., Part B*, **10**, 367–371 (1972).

Similar structural problems exist for the distribution of side chains in galactoglucomannans, xyloglucans, arabinoxylans, and 4-*O*-methylglucuronoxylans, and even for the distribution of acetyl groups in glucomannans and for related features in many other polysaccharides.

For arabinoxylans, this problem of side-chain distribution can to a large extent be tackled chemically. Thus, Aspinall and Ross,[141] using the Smith degradation, indicated that, in rye-flour arabinoxylan, no more than three, adjacent, D-xylosyl residues ever carry L-arabinosyl side-chains, but that, otherwise, the distribution is random. The basis of this approach is that the L-arabinose side-chains are attached to the D-xylosyl residues in the main chain by $(1 \rightarrow 2)$- and $(1 \rightarrow 3)$-linkages, so protecting each such D-xylosyl residue from attack by periodate. Complementary information has been obtained by enzymic methods (see p. 276). Unfortunately, the use of periodate is not applicable to galactomannans, as the α-D-$(1 \rightarrow 6)$-linked D-galactosyl stubs do not have this protecting effect.

The determination of the fine structure of galactomannans should, however, be quite amenable to enzymic study, as shown by the results of Courtois and coworkers. α-D-Galactosidases (α-D-galactoside galactohydrolases, EC 3.a.1.22) have proved particularly useful in this context for the selective cleavage of α-D-$(1 \rightarrow 6)$-linked D-galactosyl stubs. α-D-Galactosidases are found in a variety of microorganisms, and in seeds, including those rich in galactomannans. In seeds, the enzyme may exist in multiple forms. Thus, coffee-bean α-D-galactosidase can be separated into two forms (I and II) that have different pH optima;[142] multiple forms of α-D-galactosidase have also been noted from many other species.[143] The specificity of α-D-galactosidases is found to vary with the source and the method of purification. Thus, the purified enzymes from *Vicia sativa*[144] and *Mostierella vinacea*[145] cause hydrolysis of phenyl α-D-galactosides, but not of galactomannans, whereas the purified enzymes from *Phaseolus vulgaris*[146] and coffee beans (*Coffea* sp.)[147] catalyze hydrolysis of both substrates.

(141) G. O. Aspinall and K. M. Ross, *J. Chem. Soc.*, 1681–1686 (1963).
(142) F. Petek and T. Dong, *Enzymologia*, **23**, 133–142 (1961).
(143) D. Barham, P. M. Dey, D. Griffiths, and J. B. Pridham, *Phytochemistry*, **10**, 1759–1763 (1971).
(144) F. Petek, E. Villarroya, and J. E. Courtois, *Eur. J. Biochem.*, **8**, 395–402 (1969).
(145) H. Suzuki, S. C. Li, and Y. T. Li, *J. Biol. Chem.*, **245**, 781–786 (1970).
(146) K. M. L. Agrawal and O. P. Bahl, *J. Biol. Chem.*, **243**, 103–111 (1968).
(147) J. E. Courtois and F. Petek, *Methods Enzymol., Complex Carbohydrates*, 565 (1965).

Most of the structural investigations of galactomannans have been carried out by using the coffee-bean enzyme. Release of D-galactose from galactomannans by this enzyme takes place readily at first, but then slows down until an enzyme-resistant product is formed. The initial rates of release of D-galactose have proved to be indistinguishable for the galactomannans from *Trifolium repens* (D-mannose:D-galactose ratio, 1.07:1), *Genista scoparia* (ratio 1.59:1), *Gleditsia ferox* (ratio 3.9:1) and *Gleditsia triacanthos* (ratio 3.82:1). This insensitivity with regard to the content of D-galactose was confirmed by Hui[148] for the galactomannans from *Medicago sativa* and *Trigonella foenum-graecum* (D-mannose:D-galactose ratio, 1:1), and *Caesalpinia spinosa* (ratio 3:1).[148] However, the results of prolonged action by α-D-galactosidase differ quite significantly for galactomannans from different species. Thus, the action of α-D-galactosidase on *Trifolium repens* galactomannan for 6 days raised the D-mannose:D-galactose ratio from 1.07:1 to 1.54:1; a further 20 days of treatment with the enzyme raised the ratio[55] to 1.72:1; however, similar treatment of *Medicago sativa* galactomannan raised[149] the ratio from 1.1:1 to 3.04:1. This large difference in the products of enzymic treatment of galactomannans having an almost identical D-mannose:D-galactose ratio (and in which the D-mannosyl residues and D-galactosyl groups are joined by the same linkages) presumably indicates that the arrangement of D-galactosyl stubs along the D-mannan backbone must differ markedly.

A comparison of *Gleditsia ferox* galactomannan (D-mannose:D-galactose ratio, 3.9:1) and locust-bean gum (ratio 3.88:1) revealed an even more striking difference.[55] On treating *Gleditsia ferox* galactomannan with α-D-galactosidase, a precipitate appeared after 2 days; this was separated by centrifugation, and washed in ethanol to remove free D-galactose. The D-mannose:D-galactose ratio determined for this precipitate was 13.8:1. This precipitate was treated with the enzyme for a further 48 hours, to yield an insoluble material having a D-mannose:D-galactose ratio of 17:1; a further treatment with the enzyme yielded an insoluble galactomannan of D-mannose:D-galactose ratio 28:1. The fraction remaining in the first solution (D-mannose:D-galactose ratio, 7.8:1) was also treated with the enzyme, and yielded an insoluble product having a D-mannose:D-galactose ratio of 30:1. The ratio increased steadily with the number of successive enzyme treatments, but it did not prove possible to detach all of the D-galactosyl groups from the insoluble fractions.

(148) P. A. Hui, Ph. D. Thesis, Juris, Zürich (1962).

(149) J. E. Courtois and P. Le Dizet, *Bull. Soc. Chim. Biol.*, **52**, 15–22 (1970).

In contrast, treatment of locust-bean gum with α-D-galactosidase for 5 days yielded only a small proportion of a precipitate that had a D-mannose : D-galactose ratio of 4.57 : 1. A further 5 days of treatment with the enzyme caused the insoluble fraction to have a slightly increased ratio of 5.27 : 1. In a number of other experiments, Courtois and coworkers were unable to isolate from locust-bean gum an insoluble fraction having an appreciably increased D-mannose : D-galactose ratio. However, by using different samples of locust-bean gum and coffee-bean α-D-galactosidase, both soluble and insoluble products having D-mannose : D-galactose ratios as high as 11.6 : 1 and 19 : 1, respectively, have since been obtained.[150] This result probably indicates batch-to-batch variation in the fine structure of locust-bean gum preparations, resulting from different extraction and purification procedures. Notwithstanding, the results obtained for *Gleditsia ferox* galactomannan are different from those for locust-bean gum. As *Gleditsia ferox* galactomannan and locust-bean gum have very similar D-mannose:D-galactose ratios, it seems that they have markedly different distributions of the α-D-$(1 \rightarrow 6)$-linked D-galactosyl stubs along the D-mannan backbone.

Courtois and Le Dizet[149] also used β-D-mannanase and a combination of β-D-mannanase and α-D-galactosidase to probe the fine structure of galactomannans. Similar results were obtained for the β-D-mannanases from *Bacillus subtilis*, germinated *Medicago sativa* grains, and germinated *Leucaena glauca* grains.[149] These β-D-mannanases,[149] like that from *Trigonella foenum-graecum* seeds,[103] were found to be endo-enzymes.

Reaction with β-D-mannanase was conducted with various natural galactomannans and α-D-galactosidase-treated galactomannans. (β-D-Mannanase has practically no effect on galactomannans having a D-mannose : D-galactose ratio close to unity, as shown viscometrically,[151] but acts only on galactomannans in which some of the D-mannosyl residues in the main chain do not carry D-galactosyl stubs.) Treatment with β-D-mannanase was carried out at pH 6.8 for 4 days at 37°; the enzyme was then inactivated, and the nondialyzable product isolated in 10–20% yield. Examples of typical results are shown in Table III. Both of the other β-D-mannanases gave similar results. Thus, for all of the galactomannans treated with β-D-mannanase, a residual, nondialyzable galactomannan product was obtained that had an appreciably higher content of D-galactose than the starting

(150) I. C. M. Dea, C. Hitchcock, S. Hall, and A. Morrison, unpublished results.
(151) J. E. Courtois and P. Le Dizet, *Bull. Soc. Chim. Biol.*, **50**, 1695 (1968).

TABLE III

D-Mannose to D-Galactose Ratio[151] of Residual Nondialyzable Fractions
after Treatment with *Bacillus subtilis* β-D-Mannanase

Substrate galactomannan	D-Mannose to D-galactose ratio	D-Mannose to D-galactose ratio after enzyme action
Locust-bean gum	3.88 : 1	1.08 : 1
Gleditsia ferox galactomannan	3.90 : 1	1.13 : 1
α-D-Galactosidase-treated locust-bean gum	5.27 : 1	1.1 : 1
α-D-Galactosidase-treated *Gleditsia ferox* galactomannan	4.60 : 1	1.1 : 1
α-D-Galactosidase-treated *Medicago sativa* galactomannan	3.04 : 1	1.1 : 1
α-D-Galactosidase-treated *Trifolium repens* galactomannan	1.72 : 1	1.18 : 1

material. Paper-chromatographic analysis of the dialyzed fraction obtained by use of *Bacillus subtilis* β-D-mannanase showed that it contained 4-O-β-mannopyranosylmannose and its higher oligosaccharides, but no galactose.

The effect of treating locust-bean gum with β-D-mannanase for a short time was also investigated.[149] On treatment with *Bacillus subtilis* β-D-mannanase, locust-bean gum shows an abrupt drop in viscosity; there is a 95% drop in viscosity while 2.5% of the glycosidic linkages are cleaved, as shown by estimation of the free reducing sugars resulting. The changes in the D-mannose : D-galactose ratio of the nondialyzable fraction were monitored as a function of the diminution in viscosity. Sequential treatments of locust-bean gum with β-D-mannanase gave nondialyzable products having the properties shown in Table IV.

TABLE IV

Progressive Action of β-D-Mannanase on Locust-bean Gum[149]

Property	Number of successive enzyme treatments		
	1	2	3
Viscosity (as a percentage of that of locust-bean gum)	64.3	29.5	5.0
D-Mannose to D-galactose ratio of the non-dialyzable galactomannan	3.04 : 1	3.12 : 1	2.9 : 1

The large lowering of the viscosity that accompanyies the cleavage of a small proportion of the β-D-(1 → 4)-linkages in the D-mannan backbone, and the lessening (~25%) in the D-mannose:D-galactose ratio of the nondialyzable fraction implies that the β-D-mannanase splits the locust-bean gum into large fragments. Together with the resistance of highly substituted galactomannans to the action of β-D-mannanase, this suggests that the point of attack is at the D-mannosyl residues not substituted with D-galactosyl groups. The distribution of D-galactosyl stubs along the main chain in galactomannans having a D-mannose:D-galactose ratio much greater than 1:1 does not appear to be regular, and Courtois and Le Dizet[149,151] proposed that their structure may be represented schematically as **10**. Zones *b* and

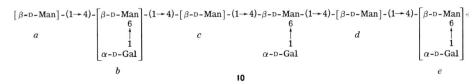

e represent portions of the galactomannan in which blocks of the mannan backbone are totally, or almost totally, substituted with galactose stubs. Along the chain of unsubstituted D-mannosyl residues are interspersed some D-mannosyl residues substituted with D-galactosyl groups (between *c* and *d* in **10**). This partial, block structure therefore departs from the ideal structure illustrated in formula **9**.

Courtois and Le Dizet proposed[149] that coffee-bean α-D-galactosidase does not hydrolyze off α-D-galactosyl stubs at random along the chain, but hydrolyses those next to D-mannosyl residues not substituted with D-galactosyl groups (that is, D-galactosyl stubs between *c* and *d* in **10**). After hydrolysis of these isolated D-galactosyl groups, the α-D-galactosidase is presumed to act at the extremities of zones *b* and *e* in **10**, but it was found impossible to remove all of the D-galactosyl stubs. Galactomannans having a D-mannose:D-galactose ratio close to 1:1 are practically unaffected by β-D-mannanase, whereas galactomannans having ratios greater than 1:1 are susceptible, because the hydrolysis occurs at the sequences of unsubstituted D-mannosyl residues, leaving residual galactomannans having D-mannose:D-galactose ratios of 1:1–1.2:1.

The galactomannans from only six species have been examined by these methods, and so, caution must be exercised in attributing a block type of structure to all seed galactomannans. Indeed, the large differences found between galactomannans of very similar D-mannose:D-galactose ratio suggest that galactomannans having totally regular (7) or random (8) structures (see p. 269) may well exist. Final judgment regarding these possibilities must wait until many more galactomannans have been examined by these techniques.

The enzymic examination of locust-bean gum and *Gleditsia ferox* galactomannan indicated that the nature of their block structures is quite different. Compared with *Gleditsia ferox* galactomannan, relatively little D-galactose is released from locust-bean gum by α-D-galactosidase, indicating that fewer, isolated D-galactosyl stubs are interspersed in the smooth regions on locust-bean gum than in those of *Gleditsia ferox* galactomannan. However, treatment of *Gleditsia ferox* galactomannan with α-D-galactosidase leads to a much greater proportion of insoluble product compared with that from locust-bean gum; this indicated that the D-galactose-rich regions are shorter, and the D-galactose-poor regions longer, in the *Gleditsia* galactomannan, because association between blocks of unsubstituted D-mannan backbone, leading to insolubility, would be expected to increase markedly with increase in block length.[152] Similarly, as regards *Medicago sativa* galactomannan and *Trifolium repens* galactomannan, the former has more, isolated stubs in the D-galactose-poor regions; here, however, it is not possible to state whether the galactomannans differ in block length. It is, therefore, apparent that the relative degree of substitution in the D-galactose-poor region should be distinguished from the block length of these regions.

Variation in the block length of the D-galactose-poor regions would be expected to affect many of the properties of galactomannans and α-D-galactosidase-treated galactomannans. A comparison of the block length of α-D-galactosidase-treated galactomannans having similar D-mannose:D-galactose ratios could be made were the average molecular weight of the respective, nondialyzable products from sequential treatment with α-D-galactosidase and β-D-mannanase estimated. The galactomannan yielding the product of higher molecular weight would have the greater block length of smooth region, as depicted in 11. Indeed, a comparison of the block lengths of galactomannans

```
M−M−M−M−M−M−M−M−M−M−M−M−M−M−M−M−M−M−M−M−M−M−
  |  |  |  |              |  |  |  |              |  |  |  |
  G  G  G  G              G  G  G  G              G  G  G  G        )

  M−M−M−M−M−M−M−M−M−M−M−M−M−M−M−M−M −M−M−M−M−M−M−
  |  |  |  |  |  |        |  |  |  |  |  |
  G  G  G  G  G  G        G  G  G  G  G  G
```

11

(152) W. N. Haworth and H. Machemer, *J. Chem. Soc.*, 2372–2374 (1932).

having quite different D-mannose:D-galactose ratios could be made in the same way, so long as the D-mannose (M):D-galactose (G) ratio of the α-D-galactosidase-treated products was estimated. So far as we are aware, experiments of this type have not yet been attempted.

Of the related polysaccharides mentioned earlier (see p. 267), only arabinoxylans have been examined in any detail by enzymic methods. Goldschmid and Perlin[153] studied the fine structure of wheat arabinoxylan by using the β-D-xylanase from *Streptomyces* QMB 814. Their results indicated that arabinoxylan molecules are mainly constituted of highly branched regions in which isolated and paired L-arabinosyl (A) branches are separated by single D-xylosyl (X) residues, as shown in 12, but that, at unequal intervals (averaging

$$\text{X–X–X–X–X–X–X–X–X–X–X–X–X–X–X–}$$
$$\quad|\quad\,|\;\,|\quad\;\,|\quad\;|\;\,|\quad\;\,|\;\,|\quad\;\,|$$
$$\quad\text{A}\quad\;\text{A A}\quad\;\text{A}\quad\;\text{A A}\quad\;\text{A A}\quad\;\text{A}$$

12

20–25 main-chain residues), this type of sequence is interrupted by L-arabinose-free regions. It was supposed that these L-arabinose-free regions may sometimes be as large as those depicted in **13**. They considered that the ready solubility of wheat arabinoxylan, as contrasted with the low solubility of xylans containing relatively low proportions of L-arabinosyl stubs,[154] suggested that few of its molecules are likely to have arabinose-free regions much longer than that shown in **13**.

$$\text{X–X–X–X–X–X–X–X–X–}$$
$$|\qquad\qquad\qquad\qquad\quad|$$
$$\text{A}\qquad\qquad\qquad\qquad\text{A}$$

13

Similar work by Tsujisaka and coworkers[155] on the enzymic degradation of rice-straw arabinoxylan with *Aspergillus niger* hemicellulase also suggested that portions of the molecule occur as open regions. It is well known that the group of arabinoxylans comprises a series of polysaccharides in which the L-arabinose:D-xylose ratio varies substantially.[154,156] Therefore, some molecules probably contain a relatively large proportion of the highly branched type of

(153) H. R. Goldschmid and A. S. Perlin, *Can. J. Chem.*, **41**, 2272–2277 (1963).
(154) A. S. Perlin, *Cereal Chem.*, **28**, 382–393 (1951).
(155) Y. Tsujisaka, S. Takenishi, and J. Fukumoto, *Nippon Nogei Kagaku Kaishi*, **45**, 253–259 (1971).
(156) A. S. Perlin, *Cereal Chem.*, **28**, 370–381 (1951).

sequence, while others contain more of the L-arabinose-free regions. Comparison with the Smith-degradation results of Aspinall and Ross[141] on rye arabinoxylan (mentioned on p. 270) is difficult, as Smith degradation of the highly substituted regions postulated by Goldschmid and Perlin[153] could give essentially similar results.

In similar experiments on birch 4-O-methylglucuronoxylan, it was shown that blocks in which all of the D-xylosyl residues carry an acidic sugar residue do not occur.[157] The fact that substantial yields of oligomers containing one acidic sugar group were obtained suggested that a block type of structure of the same form as postulated by Goldschmid and Perlin[153] for wheat arabinoxylans may exist (that is, relatively highly substituted regions separated by more-open regions). Although more work is obviously required on these other groups of polysaccharides before firm conclusions can be made, it could well turn out that block structures similar to those suggested for galactomannans are important in these related, natural polysaccharides.

3. Analysis by Use of Physicochemical Methods

Two fundamental properties determine much of the behavior of polymers: the molecular weight and the conformation (molecular shape). Because of the very high viscosities of even dilute solutions of galactomannans, determinations of the molecular weight are technically difficult.

a. Molecular Weights. — Galactomannans, like many other natural polysaccharides, tend to be polydisperse and, consequently, cannot be accurately described by a single molecular weight. Instead, average molecular weights are obtained, the values of which are *dependent on the techniques used.* Thus, light-scattering methods give a value which is exaggerated by aggregation and which tends to overemphasize the species of high molecular weight in the sample: this is the weight-average molecular weight, $\overline{M}_w$. Chemical methods and osmometry give the number-average molecular weight, $\overline{M}_n$.

A number of problems arise in connection with the measurement of molecular weights of galactomannans. Firstly, it is very difficult to obtain true solutions of many of the gums, especially of locust-bean gum and guar gum; consequently, the solutions must be filtered or centrifuged to remove undissolved or partially hydrated gum. This immediately raises the problem of fractionation of the sample, as the

(157) T. E. Timell, *Svensk Papperstidn.*, **65**, 435–447 (1962).

result may not mean very much if a large portion of the sample is removed before analysis. Secondly, the values depend upon the rigorous evaluation of the molecular shape, particularly those obtained by light-scattering and viscosity measurements. Deb and Mukherjee[158] used a dissymmetry method, and assumed for molecules of guar gum a spherical shape, which is highly unlikely; the molecular weight derived was 1,720,000, a value that seems unreasonably large but may, however, be due to aggregation.

A third problem, pertaining to chemical methods, arises because of the very high molecular weights of some galactomannans. A gum having a molecular weight of 250,000 has only one reducing endgroup for every fifteen hundred or so sugar residues; thus, in methods involving end-group analysis, a very high level of accuracy is required; in addition, highly purified samples are essential. Periodate-oxidation methods can be complicated by over-oxidation.[159] A further problem is the need to assume the chemical nature of the end-groups that are undergoing reaction; this characterization is difficult to check, especially for polymers of such high molecular weight. A method of Richards and Whelan[160] that uses scintillation counting to detect end-groups reduced with sodium borohydride-t has been published; the method has not yet been applied to galactomannans, but appears to be promising, although problems may arise because of gelling caused by borate ions. It is well known that borate ions cause galactomannans to gel by interaction with the cis-diol groupings in the sugar residues.[161]

The fullest examination made to date of the molecular weight of a galactomannan was that by Koleske and Kurath,[162] who used carefully fractionated samples of peracetylated guar gum. Light-scattering, sedimentation-velocity, viscosity, and osmotic-pressure measurements were employed in a detailed examination of the hydrodynamic properties of the fractions. The polymolecularity was measured, and the ratio of the mean-square radius of gyration to the molecular weight was found to decrease with increase in molecular weight. The dependence of radius of gyration on chain length was found to be identical to that observed for O-(hydroxymethyl)cellulose. The authors concluded that the D-galactose side-

(158) S. K. Deb and S. N. Mukherjee, *Indian J. Chem.*, **1**, 413–414 (1963).
(159) G. W. Hay, B. A. Lewis, F. Smith, and A. M. Unrau, *Methods Carbohyd. Chem.*, **5**, 251 (1965).
(160) G. N. Richards and W. J. Whelan, *Carbohyd. Res.*, **27**, 185–191 (1973).
(161) A. O. Pittet, *Methods Carbohyd. Chem.*, **5**, 3 (1965).
(162) J. V. Koleske and S. F. Kurath, *J. Polym. Sci.*, Part A, **2**, 4123–4149 (1964).

chains have little, if any, effect on the conformation of the main chain.

By the application of membrane osmometry to a commercial sample of guar gum, a number-average molecular weight of $\overline{M}_n = $ 240,000 for both the gum and its peracetate was obtained, the value for the latter being corrected for the acetyl groups.[117] Light-scattering and viscosity measurements gave a weight-average molecular weight of $\overline{M}_w = 950,000$. The value of $\overline{M}_w$ was calculated from viscosity data by using constants derived by Koleske and Kurath.[162] The large difference between $\overline{M}_n$ and $\overline{M}_w$ indicated the presence of species having a broad spectrum of molecular weight in commercial guar gum.

The molecular weights published for a number of galactomannans are given in Table V.

b. **Conformational Analysis and Molecular Shape.**—As in the whole field of polymer science, perhaps some of the most formidable problems involve the satisfactory assignment of molecular shape and

TABLE V

Molecular Weights of Galactomannans

Sample	Molecular weight	Degree of polymerization	Method	References
Annona muricata	8,700		chemical	43
Arenga saccharifera	17,000		chemical	43
Borassus flabellifer	139,000		light-scattering	87
Caesalpinia pulcherima	60,000		sedimentation analysis	59
Ceratonia siliqua	300,000 ($\overline{M}_w$)			163
	300,000 ($\overline{M}_n$)	1,500	chemical	41,148
(hot-water soluble fraction)	650,000		chemical	41,148
(cold-water soluble fraction)	150,000		chemical	41,148
Cocos nucifera	7,200		chemical	43
Convolvulus tricolor	11,000		chemical	43
Crotalaria mucronata	50,000		sedimentation analysis	67
Cyamopsis tetragonoloba	250,000 ($\overline{M}_n$)	900	chemical	41
	1,900,000 ($\overline{M}_w$)		sedimentation analysis	41
	1,720,000		light-scattering	158
	240,000 ($\overline{M}_n$)		osmometry	117
	950,000 ($\overline{M}_w$)		viscosity and light-scattering	117
Gleditsia amorphoides		116	chemical	61
triacanthos		110–120	chemical	63
Glycine max.	32,000		osmometry	70
Hyphaene thebaica		14	chemical	92
		23	chemical	92
Sophora japonica	6,000		chemical	43

(163) J. V. Kubal and N. Gralén, *J. Colloid Sci.*, 3, 457–471 (1948).

dimensions to galactomannans. Very little progress has been made on the shape of galactomannan molecules themselves, but interesting comparisons can be made with related polysaccharides, for example, $(1 \rightarrow 4)$-β-D-mannan and cellulose. Two main approaches to the problem have been made: (i) through X-ray diffraction measurements, and (ii) by making interaction-energy calculations.

i. X-Ray Crystallography. The results of X-ray diffraction studies of galactomannans indicated a close similarity with the known structures of the β-D-$(1 \rightarrow 4)$-linked D-glycans, namely, cellulose,[164] D-mannans,[165] and D-xylans.[166] They all show two- or three-fold fiber structures having a rise per residue of >5.1 Å. The fiber structure of ordered films of guar gum containing 16.5% of moisture was examined by Palmer and Ballantyne[167] in 1950. These authors suggested that the material had an orthorhombic, unit cell with $a = 15.5$ Å, $b = 10.3$ Å (fiber axis), and $c = 8.65$ Å. The presence of meridional reflections on alternate layer-lines suggested a 2-fold screw-axis parallel to the galactomannan chains and lying between adjacent sheets. A structure was proposed in which the chains are arranged in sheets, with the side chains and the planes of the pyranose rings lying in the planes of the sheets, and the side chains of adjacent sheets pointing in opposite directions. A comparison with the X-ray diffraction pattern of mannans I and II and cellulose is interesting.

The galactomannan from *Gleditsia amorphoides* has been studied by X-ray diffraction, and an orthorhombic cell was proposed[168] having $a = 12.2$ Å, $b = 10.3$ A (fiber repeat), and $c = 8.8$ Å. This molecule, also, displays two-fold symmetry. The results are consistent with those for guar gum. Mannan I has[165] a two-fold, screw axis and an orthorhombic, unit cell with $a = 7.21$ Å, $b = 10.27$ Å (fiber-repeat distance), and $c = 8.82$ Å. However, mannan II, obtained by treatment of mannan I with alkali, is monoclinic,[165] with $a = 18.8$ Å, $b = 10.2$ Å (fiber repeat), $c = 18.7$ Å, and $\beta = 57.5°$. Cellulose[169] and cellulose triacetate[164] also have a two-fold screw-axis, and the former has a fiber-repeat distance of 10.3 Å. Mannan triacetate and β-D-$(1 \rightarrow 4)$-linked D-xylans,[166] on the other hand, have three-fold screw-axes and fiber repeats of 15.24 Å (Ref. 170) and 14.8 Å, respectively. A trend

(164) W. J. Dulmage, *J. Polym. Sci.*, **26**, 277 (1957).
(165) E. Frei and R. D. Preston, *Proc. Roy. Soc., Ser. B.*, **169**, 127–145 (1968).
(166) R. H. Marchessault and C. Y. Liang, *J. Polym. Sci.*, **59**, 357–378 (1962).
(167) K. J. Palmer and M. Ballantyne, *J. Amer. Chem. Soc.*, **72**, 736–741 (1950).
(168) E. Aisenberg, E. E. Smolko, and A. S. Cerezo, to be published.
(169) K. H. Meyer and L. Misch, *Helv. Chim. Acta*, **20**, 232–244 (1937).
(170) H. Bittiger and R. H. Marchessault, *Carbohyd. Res.*, **18**, 469–470 (1971).

towards formation of a crystalline structure as D-galactosyl groups are removed from either locust-bean gum or guar gum was demonstrated by Dugal and Swanson[171] by using α-D-galactosidase to hydrolyze off the D-galactosyl groups, and X-ray diffraction to study the products. Although the X-ray powder diffractograms of the starting materials were structureless, and characteristic of the amorphous state, the end products gave diffractograms characteristic of crystalline material and very similar to those obtained from ivory-nut mannan. It has been noted[168] that, for mannan I, *Gleditsia amorphoides* galactomannan, and guar gum, the *a* dimension increases with increasing substitution by D-galactosyl groups.

ii. Conformational Analysis. The common sugar residues in polysaccharides normally exist in the $^4C_1$ (D) [that is, the $^1C_4$ (L)] conformation and, by using this conformation and the standard values assumed for bond angles and bond lengths in carbohydrates, Rees and Skerrett[172] calculated the energetically most favorable conformations for cellulose and $(1 \rightarrow 4)$-β-D xylan. A number of simplifying assumptions were made, including treatment of the hydroxymethyl group on C-5, or C-4, as equivalent to a methyl group (on the grounds that the hydroxyl group could probably rotate away from regions of steric overcrowding), and the assumption that polar forces and hydrogen bonding are effectively quenched in aqueous solution. They concluded that the Hermans[172a] conformation for cellulose is more likely than that of Meyer and Misch.[172b] Favorable energetic stability is achieved both by hydrogen bonding (O-5–HO-3') and by the possibility of crystal packing that results in a two-fold screw-axis. These features were demanded by the experimental data obtained by infrared dichroism studies and X-ray crystallography, respectively.

Sundararajan and Rao[173] also computed nonbonded interaction-energies, but with respect to the $(1 \rightarrow 4)$-β-D-mannan system, and observed for mannan a larger region of allowed conformations than for cellulose. The monomer units of three β-D-$(1 \rightarrow 4)$-linked glycans, namely, mannan, cellulose, and xylan, had increasing rotational freedom in the order: cellulose < mannan < xylan. Like Rees and Skerrett,[172] these workers[173] neglected the effects of the hydroxymethyl groups on the C-5 atoms. They proposed an extended-

(171) H. S. Dugal and J. W. Swanson, *Tappi*, **55**, 1362–1367 (1972).

(172) D. A. Rees and R. J. Skerrett, *Carbohyd. Res.*, **7**, 334–348 (1968).

(172a) P. H. Hermans, "Physics and Chemistry of Cellulose Fibres," Elsevier, New York, 1949.

(172b) K. H. Meyer and L. Misch, *Helv. Chim. Acta.*, **20**, 232–244 (1937).

(173) P. R. Sundararajan and V. S. R. Rao, *Biopolymers*, **9**, 1239–1247 (1970).

ribbon conformation having a hydrogen bond between O-5 and O-3′ similar to that suggested for cellulose. In a third examination of the conformations of mannan and cellulose, Atkins and coworkers[174] allowed for the hydroxymethyl group on C-5, and found that the conformational flexibility of mannan is more restricted than had previously been calculated, although the conformation is still more flexible than that of cellulose.

Although direct evidence for the shape of galactomannan molecules in the solid state exists only in the form of the X-ray fiber diffraction patterns of guar gum and *Gleditsia amorphoides* galactomannan, these computer, model-building calculations strongly supported an extended, ribbon-like conformation as the ordered structure favored by galactomannans, and indicated that the D-galactose side-chains may not necessarily affect the conformation of the main chain. However, conformational studies indicated that $(1 \rightarrow 4)$-β-D-galactan, which is equatorial–axial-linked (instead of equatorial–equatorial-linked), should exist in a helical form[173] similar to that of amylose.

c. **Solution Properties.**—Much of the interest in galactomannans has resulted directly from their ability to give aqueous solutions of very high viscosity. Most of the investigations of their rheological properties have been commercially motivated and, consequently, the commercially important gums (locust-bean and guar gums) have been studied in much greater detail than any others. In aqueous solution, galactomannans can be confidently expected to exist in the random-coil conformation,[40] a state that would only revert to a more-ordered form were conditions favorable for aggregation or interaction with other species (see Section V, p. 284).

It has been shown that the D-galactose side-chains probably have little effect on the conformation of the D-mannan backbone in the solid state; however, they play a very important role in determining the ease with which galactomannans can be dissolved. Fractionation of locust-bean gum by use of its solubility in water gives a cold-water-soluble fraction that has a high content of D-galactose, and a hot-water-soluble fraction having a low content thereof. An insoluble fraction containing even less D-galactose may remain[41] [see Sections III,2 (p. 262) and V,3 (p. 296)]. The difficulty in dispersing locust-bean gum is, presumably, due to the strong interactions between adjacent chains.

(174) E. D. T. Atkins, E. D. A. Hopper, and D. H. Isaac, *Carbohyd. Res.*, **27**, 29–37 (1973).

Locust-bean gum has been implicated as a promoter of the retrogradation of amylose; possibly, it interacts strongly through
regions of the backbone sparsely substituted with D-galactosyl
groups [see Sections IV,2 (p. 269) and V,4 (p. 305)] to cause precipitation. Guar gum, having more side-chains, disperses more readily
than locust-bean gum, presumably because the side groups hold the
main mannan chains far enough apart to render ineffective the non-
covalent interactions. An interesting comparison can be made with
O-(carboxymethyl)cellulose, O-methylcellulose, amyloids, arabinoxylans, and acetylated glucomannans, the substituents of which
render soluble the insoluble parent polysaccharides (see Section
IV,1; p. 267). Usually, locust-bean gum is completely dispersed in
water after heating at ~80° for 20–30 minutes, although high viscosity is rapidly attained[56] if the gum is dissolved in 6 M urea solution at
20°. Presumably, the urea assists in destroying the non-covalent
bonds between adjacent chains.

When completely dispersed in water at the 1% level, guar gum
and locust-bean gum give highly viscous solutions that are non-New-
tonian and pseudoplastic. At lower concentrations (for example,
<0.3%), the pseudoplasticity is less evident, and the behavior
approximates to Newtonian. The maximum viscosity and the stability
of the viscous solution in storage depend very much on the way in
which the gum was dispersed, and, in particular, on the dispersion
time and temperature. Heating for long periods at temperatures
above 60° tends to give high initial viscosity, but prolonged heating
leads to breakdown and poor stability of the solutions.[11] The optimal
conditions for achieving maximum viscosity of solutions of guar gum
involve heating at 25–40° for ~2 hours. In contrast, a large proportion of locust-bean gum is not dispersed at room temperature, and
the resulting "solution" has a very low viscosity.

For a given range of galactomannans having identical D-man-
nose:D-galactose ratios, the viscosities are proportional to the molecular weights. However, there is some evidence that gums having a
high content of D-galactose have viscosities higher than those of
gums with lower D-galactose contents. Thus, it was found[148] that a
sample of locust-bean gum with $\overline{M}_w$ 650,000 had a lower viscosity
than a sample of guar gum with $\overline{M}_w$ 250,000. The most likely explanation for this behavior is the presence of aggregates in the sample of
locust-bean gum. It is probable that the extent of the D-galactose-free
mannan backbone is more important in this respect than the D-man-
nose:D-galactose ratio.

As galactomannans are non-ionic polysaccharides, the properties of their solutions are only slightly affected by moderate changes in pH. Thus, the viscosity tends to remain constant over the pH range of 1–10.5, although degradation soon occurs in highly acidic and alkaline solutions, especially if they are heated. Above pH 11, the rate of dissolution is depressed. When mixed with the dry gum and dispersed, sucrose[175] and some salts (for example, sodium sulfate, but not sodium chloride) compete for water, and the rate of dissolution of the gum is lowered.[11] This procedure gives better control over the dispersion of the polysaccharides, and results in a more homogeneous solution and the elimination of clumping.

V. INTERACTION OF GALACTOMANNANS WITH CARRAGEENAN, AGAR, AND OTHER POLYSACCHARIDES

1. Introduction

It has long been realized in industry that galactomannans interact with certain gelling polysaccharides to give beneficial results. For example, Baker and coworkers[1] and Deuel and coworkers,[176] on investigating the effect of locust-bean gum on the mechanical properties of carrageenan and agar gels, found that even small additions of locust-bean gum make the gels firmer, less brittle, and more elastic. Similarly, measurements of rigidity indicated that locust-bean gum, guar gum, and fenugreek-seed mucilage interact with agarose to increase the gel strength.[177] Early work indicated that it is mainly the hot-water-soluble fraction of locust-bean gum that is responsible for strengthening agar gels;[148] this fraction of locust-bean gum is depleted in D-galactose.[41,178] The industrial importance of these interactions lies in the fact that locust-bean gum and guar gum are much less expensive than carrageenan and agar. Thus, gels of a specified strength can be made with a mixture of a galactomannan and less carrageenan or agarose than would be required if either of the last two was used alone, at a significant savings in cost. In addition, the resulting gels, although of the required gel strength, have textural properties different from those of the pure carrageenan and agarose gels, and may be preferable in some applications.

The interaction of galactomannans (particularly locust-bean gum) with the exopolysaccharide from the plant pathogen *Xanthomonas*

(175) W. A. Carlson and E. M. Ziegenfuss, *Food Technol.*, **19**, 954–958 (1965).
(176) J. Deuel, G. Huber, and J. Solms, *Experientia*, **6**, 138, (1950).
(177) J. Boyd, to be published.
(178) I. C. M. Dea, A. A. McKinnon, and D. A. Rees, in preparation.

campestris has become important industrially.[179,180] Both locust-bean gum and the *Xanthomonas* polysaccharide form viscous solutions, but not gels, at 0.5% concentration; however, a mixture comprising 0.5% of each polysaccharide forms a firm, rubbery gel having a distinctive texture and mouth-feel.[181] The fact that non-gelling galactomannans can increase the firmness of agar and carrageenan gels, and cause gel-formation in the presence of *Xanthomonas* polysaccharide, indicates that they take part in the formation of the gel framework.

2. Interactions with Carrageenans

The fundamental, chemical repeat of the carrageenans[182] is shown in formula **14**. Different types of carrageenan occur naturally. Thus,

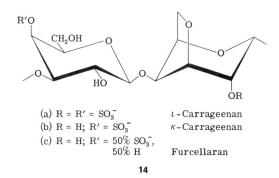

(a) R = R' = SO$_3^-$ ι-Carrageenan
(b) R = H; R' = SO$_3^-$ κ-Carrageenan
(c) R = H; R' = 50% SO$_3^-$,
              50% H   Furcellaran

14

Idealized, Chemical Repeat of the Carrageenans

ideally, ι-carrageenan bears a sulfate group on each sugar residue; κ-carrageenan, a sulfate group on each D-galactose residue; and furcellaran, a sulfate group on about 50% of the D-galactose residues. As the sulfate content of the carrageenans decreases, the concentration required for gelation decreases. Thus ι-carrageenan gels at ~2.5%, κ-carrageenan at ~1.5%, and furcellaran at ~0.4%.

Gelation and liquefaction of these polysaccharides may be readily monitored by observation of the optical rotation.[183] A typical graph of

(179) "Xanthan Gum, a Natural Biopolysaccharide for Scientific Water Control," Kelco Co., San Diego, California, 1972.
(180) J. K. Rocks, *Food Technol.*, **25**, 476–483 (1971).
(181) *Federal Register*, U. S. Government Printing Office, Washington, D. C., 1969, Xanthan Gum, Section 121.1224 5376–5377, March 19.
(182) D. A. Rees, *Advan. Carbohyd. Chem. Biochem.*, **24**, 267 (1969).
(183) D. A. Rees, I. W. Steele, and F. B. Williamson, *J. Polym. Sci., Part C*, **28**, 261–276 (1969).

optical rotation *versus* temperature for κ-carrageenan is shown in Fig. 1. Both the heating and cooling curves have a sigmoidal form, suggesting a cooperative conformational change. ι-Carrageenan and furcellaran show optical rotational behavior similar to that of κ-carrageenan, and differ only in the absence of hysteresis for ι-carrageenan and the slightly greater hysteresis for furcellaran.

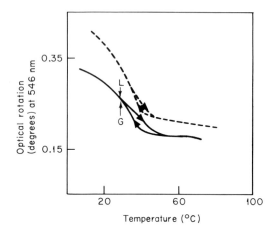

FIG. 1.—Comparison of the Variation of Optical Rotation with Temperature for Native κ-Carrageenan (—) and Segmented κ-Carrageenan (----). [Heating and cooling curves are distinguished by the arrows. Measurements were made at 546 nm with 3% (w/v) solutions. The gel point (G) and liquefaction point (L) are shown.]

Detailed X-ray analysis of ι-carrageenan indicated that its ordered conformation is a parallel, exactly staggered, right-handed, double helix having three disaccharide residues per turn and a pitch of 26 Å (Na$^+$ salt)[184] and 26.6 Å (Ca^{2+} salt).[185] Such an ordered conformation may be regarded as a tertiary structure. The ordered conformations of κ-carrageenan and furcellaran are less well defined, but are believed to be parallel, 3-fold, right-handed, double helices slightly displaced from the exactly staggered position; the pitches in these are slightly less[184,186] at ~25 Å. In all three of the ordered conformations, the sulfate groups lie on the outside of the double helix.

(184) N. S. Anderson, J. W. Campbell, M. M. Harding, D. A. Rees, and J. W. B. Samuel, *J. Mol. Biol.*, **45**, 85–99 (1969).

(185) S. Arnott and W. E. Scott, in preparation.

(186) I. C. M. Dea and W. E. Scott, unpublished work.

For the carrageenans, the changes in optical rotation calculated by assuming double-helix formation are closely similar to the changes in optical rotation experimentally observed on gelation. This result strongly supports the thesis that the chain associations necessary for gel formation of the carrageenans occur through such double-helical junction-zones; this is indicated schematically in Fig. 2. In ι-carrageenan gels, these double-helical junctions are non-aggregated (see later). In κ-carrageenan and furcellaran gels, the decrease in sulfate substitution leads to a decrease in electrostatic repulsion between double helices; this decrease results in aggregation of the double-helical junction-zones, and cloudiness of the gels.[182] It is, therefore, apparent that the degree of sulfation has a large effect on the gelling ability and gel properties of the carrageenans. It will be shown later (see p. 288) that the degree of sulfation similarly affects the extent of interaction with galactomannans.

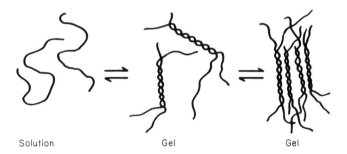

Solution   Gel   Gel

FIG. 2.—Schematic Form of the Proposed Mechanism for Gelation of ι- and κ-Carrageenan. [In ι-carrageenan gels, aggregation of double-helix junction-zones does not occur.]

The carrageenans are not totally the regular, alternating structures indicated in formula **14**. Along the chain, 3,6-anhydro-D-galactosyl residues are occasionally replaced by D-galactosyl residues or their 6-sulfate.[182] These residues in the chain may be cleaved by Smith degradation to yield segments having low molecular weight and a degree of polymerization[187,188] of ~30–60. On heating and cooling, these segments still undergo helix–coil transitions, as shown by op-

(187) A. A. McKinnon, D. A. Rees, and F. B. Williamson, *Chem. Commun.*, 701–702 (1969).
(188) R. A. Jones, E. J. Staples, and A. Penman, *J. Chem. Soc. Perkin II*, 1608–1612 (1973).

tical rotation measurements, but they do not form gels, even at high concentration.[187] The graph of optical rotation *versus* temperature for segmented κ-carrageenan shows a hysteresis loop, although it is much smaller than that of the native polymer (see Fig. 1).[40,189] Double helices of segmented ι-carrageenan, like those of the undergraded polysaccharide, do not aggregate; this is shown by determinations of molecular weight,[188] and their formation follows an equilibrium law that depends on the second power of the concentration of the polysaccharide.[190]

When diluted to below a gelling concentration, native κ-carrageenan gels if mixed with locust-bean gum. Similarly, the non-gelling, segmented κ-carrageenan forms firm, rubbery gels on admixture with locust-bean gum.[40] These striking interactions are not observed for ι-carrageenan or segmented ι-carrageenan, but appear to be stronger for furcellaran.[178] For both κ-carrageenan and furcellaran, the interaction diminishes from one galactomannan to another, in the direction of increasing content of D-galactose.[40]

On heating and cooling, mixtures of κ-carrageenan and segmented κ-carrageenan with galactomannans show the type of optical rotation changes normally obtained for κ-carrageenan systems.[40] In both cases, the bulk changes of liquefaction and gelation occur at much lower helix contents (as indicated by the magnitude of the optical rotation change) than they do in the absence of locust-bean gum (compare Figs. 1 and 3). Careful comparison of the graphs of optical rotation *versus* temperature for segmented κ-carrageenan in the presence and absence of locust-bean gum indicated that the presence of locust-bean gum broadened the hysteresis loop and shifted it to higher temperature; this indicated that the interaction between κ-carrageenan and locust-bean gum is accompanied by double-helix formation of the κ-carrageenan. These double-helical junction-zones are more stable in the presence than in the absence of locust-bean gum.

The ability of certain galactomannans to form gel structures when mixed with these non-gelling polysaccharides implies that a network is formed by the interaction of unlike chains. Fiber diffraction evidence, and indications from computer model-building[191,192] (dis-

(189) A. A. McKinnon, Ph. D. Thesis, University of Edinburgh (1973).
(190) T. A. Bryce, D. A. Rees, and D. S. Reid, to be published.
(191) E. D. T. Atkins, W. Mackie, K. D. Parker, and E. E. Smolko, *J. Polym. Sci., Part B*, **9**, 311–316 (1971).
(192) D. A. Rees and W. E. Scott, *J. Chem. Soc. (B)*, 469–479 (1971).

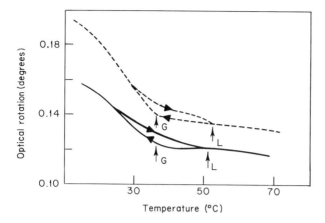

FIG. 3.—Comparison of the Variation of Optical Rotation with Temperature for Native κ-Carrageenan (—) and Segmented κ-Carrageenan (----) in the Presence of Locust-bean Gum. [Heating and cooling curves are distinguished by the arrows. Measurements were made at 546 nm with solutions that were 1% with respect to locust-bean gum and 2% with respect to carrageenan. The gel point (G) and liquefaction point (L) are shown.]

cussed earlier for guar gum; see p. 281), indicated that any ordered, secondary structure of a galactomannan must involve the backbone in an extended, ribbon-like form. The enzymic evidence cited earlier (see p. 274) indicated that this extended backbone would be, alternately, rather densely and rather lightly substituted with D-galactosyl groups.

The mixed carrageenan–galactomannan gels show sharp melting and setting behavior consistent with cooperative transitions and, hence, with networks that are cross-linked by ordered, non-covalent associations. The fact that galactomannans containing less D-galactose (and, presumably, more-sparsely substituted regions) interact best with κ-carrageenan and furcellaran indicates that the D-galactose-poor regions are the most effective in any binding. Experimental evidence already cited also indicates that the κ-carrageenan and furcellaran chains enter the mixed cross-links in their double-helical form, and, therefore, that ordered binding occurs between the ribbon-like, sparsely substituted regions of the galactomannan chains and the double-helical regions of the carrageenan (see Fig. 4). For non-gelling concentrations of κ-carrageenan and furcellaran, optical rotation evidence[40,178] indicated that double-helix formation may still occur. Presumably, a continuous network does not form throughout the solution; instead, the chains are associated by double-helical junction-zones, to give soluble, gel islands. By the binding of the sparsely substituted regions along galactomannan molecules onto double helices, as indicated in Fig. 4, a continuous network leading to gel formation occurs. For segmented κ-carrageenan–galactomannan gels, the double helices must either bind more than one D-galactose-poor region of galactomannan or obtain the necessary, extra cross-linking through aggregation of the double helices. A combination of these mechanisms is, of course, possible. Because of the extra, mixed-junction zones in the mixed carrageenan–galactomannan gels, it is not surprising that less carrageenan double-helix content is required for gelation to occur, as is indicated in Figs. 1 and 3.

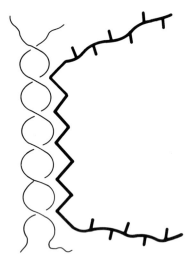

FIG. 4.—Model Proposed for the Interaction Between Chains of κ-Carrageenan and the regions of Galactomannan Sparsely Substituted with D-Galactosyl Groups.

The model proposed for mixed-junction zones is also supported by changes in the hysteresis behavior, as shown by the optical rotation (see Figs. 1 and 3). Thus, the shift in the cooling transition to higher temperature shows that the critical nucleus for helix formation of κ-carrageenan and furcellaran corresponds to a lower free-energy in the presence of galactomannan, such as could be caused by spontaneous, non-covalent and ordered binding of the galactomannan to the growing, double helix (that is, ligand-induced formation of the nucleus[193]). The effect on helix melting (see Figs. 1 and 3) is substantially larger, and is likewise consistent with stabilization by binding.

The exact geometry of the proposed mixed-junction zone, as, indeed, of the pure κ-carrageenan and furcellaran double-helices, has yet to be defined. It is, however, apparent that the ordered binding of the mannan backbone to the carrageenan double-helix is hindered by substitution with D-galactosyl groups, presumably because of steric hindrance. The increase in binding of galactomannan with decrease in sulfate content from κ-carrageenan to furcellaran, together with the complete lack of interaction between ι-carrageenan and galactomannan, suggests that an increase of sulfuric ester groups on the exterior of the carrageenan double-helix also sterically hinders binding. The prediction may safely be made that any carrageenan variant containing less sulfuric ester than furcellaran should interact extremely well with galactomannans. The disposition of sulfuric ester in different varieties of furcellaran might also be expected to affect the interaction with galactomannans materially. A block type of arrangement of the sulfate groups would be expected to lead to greater interaction with galactomannan than would regular or random arrangements. As yet, little work has been carried out on this type of fine structure in carrageenans.[194]

3. Interactions with the Agar Group of Polysaccharides

The fundamental, chemical-repeat structure (15) of the agar groups

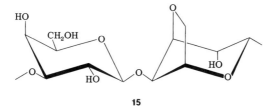

15

Idealized, Chemical Repeat of Agarose

of polysaccharides is very similar to that of the carrageenans.[182,195] A major difference in structure is the replacement of α-D-linked 3,6-anhydro-D-galactopyranosyl residues in the carrageenans by the L-enantiomer in the agars. Agarose is the non-substituted polysaccharide. As in the carrageenans, modification of the fundamental

(193) D. E. Koshland, *Pure Appl. Chem.*, **25**, 119–123, (1971).
(194) W. Yaphe, *Can. J. Bot.*, **37**, 751–757 (1959).
(195) M. Duckworth, K. C. Hong, and W. Yaphe, *Carbohyd. Res.*, **18**, 1–9 (1971).

structure occurs naturally; thus, depending on the botanical source, various proportions of 6-methyl ether, 4,6-pyruvic acetal, and sulfation occur. The degree of sulfation is generally much less than in the carrageenans, but *Gloiopeltis furcata* and *Gloiopeltis cervicornis* yield agars sulfated on O-6 of every D-galactose residue and on O-2 of ~15% of the 3,6-anhydro-L-galactose residues.[196-198] Agarose and its natural 6-O-methylated derivatives form gels at very low concentrations (0.1%) compared with those for the carrageenans. As the degree of substitution of the charged groups increases, the concentration necessary for gelation increases. *Gloiopeltis furcata* agar gels only at concentrations[178] greater than 2–3%.

Gelation and liquefaction of the agars may readily be monitored by the optical rotation. For agarose, a typical graph of optical rotation *versus* temperature is shown in Fig. 5. As with the carrageenans, both heating and cooling curves have a sigmoidal form indicative of cooperative, conformational changes; the form differs from that of κ-

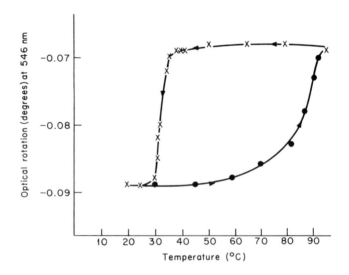

FIG. 5.—Optical Rotation Changes during the Setting and Liquefaction of an Agarose Gel (0.2%). [Heating and cooling curves are distinguished by the arrows. Measurements were made at 546 nm.]

(196) D. J. Stancioff and N. F. Stanley, *Proc. Intern. Seaweed Symp. 6th Madrid, 1968,* 595 (1969).

(197) S. Hirase and K. Watanabe, *Proc. Intern. Seaweed Symp. 7th Sapporo, 1971,* 451 (1973).

(198) A. Penman and D. A. Rees, *J. Chem. Soc. Perkin I,* 2182–2187 (1973).

carrageenan only in that the signs of the shifts are opposite, and the hysteresis for agarose is much greater.[40] Increase in the charged-group substitution leads to a decrease in the hysteresis, but it is still observable for *Gloiopeltis furcata* agar.[178]

Agarose and its *O*-methyl and *O*-sulfate derivatives have been shown by X-ray diffraction analysis to adopt a 3-fold, right-handed, parallel and exactly staggered, double helix (having a pitch of 19 Å) as the ordered, tertiary structure in the solid state. The changes in optical rotation calculated by assuming this tertiary structure are closely similar to the changes in optical rotation experimentally observed on gelation of these agaroses.[199] It is, therefore, envisaged that the chain associations necessary for agarose-gel formation occur by way of such double-helical junction-zones (see Fig. 2, p. 287). The extent of aggregation in gels of agarose and its methylated derivatives is very much greater, even, than for furcellaran; as the content of charged-group substituents increases, the aggregation decreases.[178]

Segmented agars can be made in the same way as segmented carrageenans. Optical rotation measurements for segmented *Gloiopeltis furcata* agars indicated that they still undergo helix–coil transitions, although they are completely non-gelling. Segmented agarose also forms the ordered conformation on cooling, but immediately aggregates and precipitates. Modification of the fundamental agarose structure can, therefore, be seen to alter the bulk properties markedly. The extent of interaction with galactomannan will be shown later (see p. 297) to be similarly affected by the degree of substitution of agarose.

Figure 5 shows the change in optical rotation on cooling and reheating a 0.2-% agarose gel; this is a very weak gel. Agarose at a concentration of 0.05% does not gel on cooling; however, the optical rotation behavior on cooling and reheating is identical.[40] This suggested that the same change in conformation occurs, but that there are insufficient agarose chains to form a complete network, and only gel islands are formed. Further support for this interpretation was obtained by the fact that the gel islands are precipitated on standing for 48 hours, possibly by the aggregation step shown in Fig. 2 (see p. 287).

On addition of certain galactomannans to non-gelling concentrations of agarose and to the non-gelling, segmented agarose, firm, rubbery gels are obtained.[40] This effect is qualitatively the same as that observed with κ-carrageenan and furcellaran, but the interaction is

(199) I. C. M. Dea, R. Moorhouse, and W. E. Scott, unpublished results.

very much greater for agarose. Thus, the minimum concentrations of locust-bean gum (D-mannose:D-galactose ratio, 3.35:1) and tara gum (ratio of 3:1) that are effective in forming gel structures with the potassium salt of κ-carrageenan, at the non-gelling concentration of 1%, are 1 and 3% , respectively. In contrast, the minimum concentrations of locust-bean gum, tara gum, guar gum (D-mannose:D-galactose ratio, 1.56:1), and fenugreek-seed mucilage (ratio of 1.08:1) that are effective in forming gel structures with 0.05% agarose are 0.05, 0.1, 0.7, and 1.0%, respectively.[40] Thus, it must be assumed that galactomannans cross-link agarose-gel islands, and form a complex network with the segmented agarose; and the most effective galactomannans in this gelling interaction are those containing the least D-galactose. As with the carrageenans, it was concluded that galactomannan chains are incorporated into the gel structure in a process mainly involving the β-D-(1 → 4)-linked, sparsely substituted regions in the backbone. It was also found that *Crotalaria mucronata* galactomannan (D-mannose:D-galactose ratio, 2.7:1), in which 22% of the backbone linkages[67] are not β-D-(1 → 4), cannot gel non-gelling concentrations of agarose,[178] even at 1% levels; this result emphasizes the importance of sparsely substituted regions of β-D-(1 → 4)-linked D-mannosyl residues in the backbone for the interaction.

Addition of galactomannan to agarose–water systems profoundly alters the optical rotation behavior. Thus, both in the absence and presence of locust-bean gum, there is, below 35°, a sharp shift in optical rotation. However, these shifts are opposite in sign, as shown in Fig. 6. The reheating curves also differ, in that, when locust-bean gum is present, a butterfly-shaped hysteresis curve is obtained.[40]

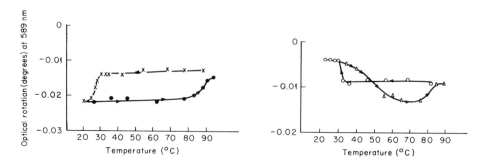

FIG. 6. — Comparison of the Variation of Optical Rotation with Temperature (left) for Agarose at a Non-gelling Concentration (0.05%) and (right) for a Gelling Mixture of Agarose (0.05%) and Locust-bean Gum (0.1%). [To facilitate comparison, the lower curve has been adjusted by subtracting the contribution expected from locust-bean gum alone. Measurements were made at 589 nm.]

These results have been interpreted as follows.[40] The binding of galactomannan is triggered by the change in conformation of agarose to its ordered, tertiary structure; this explains why the shifts in optical rotation shown in Fig. 6 occur in the same temperature range. The positive direction of the optical rotation shift arises because, superimposed on the negative, agarose transition, there is a conformational change of the galactomannan backbone that makes an overriding, positive contribution. On reheating, the galactomannan–agarose associations melt, with hysteresis, eventually leaving unbound agarose helices. The final part of the heating curve for the mixture corresponds to the behavior of agarose (alone) in the same temperature range; this confirms the presence of agarose helices in the mixed gel.

With a given concentration of agarose, the progressive addition of galactomannan causes the optical rotation shift to become increasingly positive until there is reached a limit that seems to represent a saturation effect (see Fig. 7). The shift depends similarly on the concentration of the agarose. This can be explained by an interaction between agarose and galactomannan, where the agarose and the galactomannan have a finite number of interaction sites. Calculations show that a right-handed, threefold conformation of the mannan backbone, four periods of which could match with three of agarose, is consistent with the optical rotation results.[40,200]

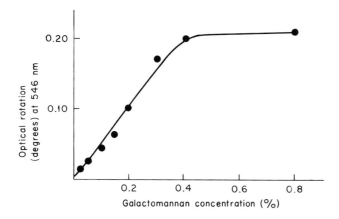

FIG. 7—Variation of Shift in Optical Rotation in Mixtures of Agarose (0.05%) and Locust-bean Gum. [Measurements were made at 589 nm.]

(200) D. A. Rees, *J. Chem. Soc. (B)*, 877–884 (1970).

It was, therefore, concluded that agarose interacts with galactomannans by the same type of junction zones as postulated for carrageenan–galactomannan mixed gels (see Fig. 4, p. 290). The ordered conformation of the mannan backbone in the agarose–galactomannan junction-zones inferred from optical rotation measurements can only be tentative, and serves only to give a general idea of the type of interaction that probably exists. Detailed knowledge of the molecular geometry must await examination of these polysaccharide associations in the solid state by X-ray diffraction. It is, however, apparent that the ordered conformation of the mannan backbone that interacts with agarose is probably different from that which interacts with the carrageenans, as the repeat periods of κ-carrageenan and furcellaran are significantly different from that of agarose.[184,185,199] This conclusion suggests a versatility in the way in which galactomannan associates with other polysaccharides. The mixed junction-zones of agars and carrageenans with galactomannans, zones that involve association between the tertiary structures of the agars and carrageenans with a secondary structure of the galactomannan, provide the first example of quaternary structure of a polysaccharide.

The variation in optical rotation contribution with content of D-galactose is shown by the behavior of various fractions of locust-bean gum, and of α-D-galactosidase-treated galactomannans.[178] Extraction of whole, locust-bean gum with cold water yields a galactomannan fraction having a D-mannose: D-galactose ratio of 3:1. Extraction of the residue in hot water yields a fraction having a ratio of 4:1. The optical rotation behavior of these fractions in the presence of agarose was compared with that of whole, locust-bean gum (D-mannose: D-galactose ratio, 3.35:1), and the results are shown in Fig. 8. Similarly, the optical rotation behavior, in the presence of agarose, of tara gum (D-mannose: D-galactose ratio, 3:1) and α-D-galactosidase-treated tara gum (ratio of 5:1) was compared, and the enzyme-modified material was found to give a positive, optical rotation contribution that was more than twice that of the untreated galactomannan.[178]

Although a number of other explanations are possible, a plausible reason why agarose interacts very much better than the carrageenans with galactomannans is because it is completely non-substituted. There is evidence from the carrageenan–galactomannan interactions that the increase in sulfate content from furcellaran to κ-carrageenan decreases the interaction, presumably owing to steric reasons. The same phenomenon has been observed in the agar series.[178] The most

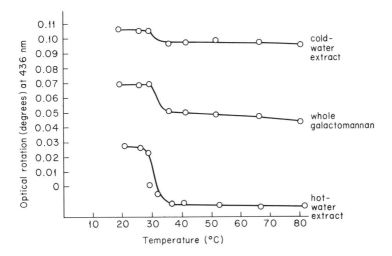

FIG. 8.—Comparison of the Variation of Optical Rotation with Temperature for (top) a Gelling Mixture of Agarose (0.05%) and a Cold-water-soluble, Locust-bean Gum Fraction (0.4%), (middle) a Gelling Mixture of Agarose (0.05%) and Locust-bean Gum (0.4%), and (bottom) a Gelling Mixture of Agarose (0.01%) and a Hot-water-soluble Locust-bean Gum Fraction (0.4%). [Measurements were made at 436 nm.]

common natural substituents in the agar series are methyl groups;[201] Guiseley reported properties of agars having methyl contents ranging[202] from 0.43 to 6.59%. On using Guiseley's samples, it was found that even low degrees of methylation significantly hinder the interaction with locust-bean gum.[178] A good estimate of the extent of interaction with locust-bean gum is obtained by calculating the positive contribution to the change in specific rotation for locust-bean gum. Fig. 9 shows how this quantity varies with increase in methyl content. It may be seen that even methyl contents as low as 0.43% cause a decrease in the interaction with locust-bean gum. With methyl contents greater than 6% (~30% of the sugar residues monomethylated), there is no perceptible, positive contribution to the optical rotation change from the locust-bean gum; bulk measurements indicated, however, that an interaction still occurs, but that it is very weak. Although the most common position of methylation in the agar series is believed to be O-6 of the D-galactose residues, methylation at O-2 of the 3,6-anhydro-L-galactose residues has been reported.[203]

(201) C. Araki, *Proc. Intern. Seaweed Symp. 5th Halifax, 1965,* 3 (1966).

(202) K. B. Guiseley, *Carbohyd. Res.,* **13,** 247–256 (1970).

(203) A. I. Usov, R. A. Lotov, and N. K. Kochetkov, *Zh. Obshch. Khim.,* **41,** 1154–1166 (1971).

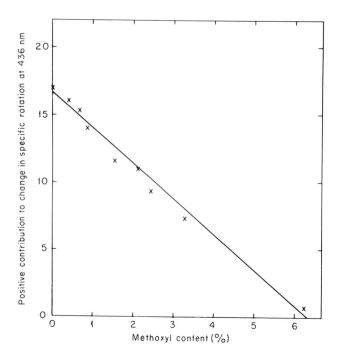

FIG. 9.—Comparison of the Positive Contribution to Change in Specific Rotation, Due to Locust-bean Gum, on Cooling Mixtures of Agarose (0.1%) and Locust-bean Gum (0.3%), with Variation in the O-Methyl Content of the Agarose. [Measurements were made at 436 nm.]

Further studies are, therefore, needed to permit ascertaining whether the position of methylation is important to the inhibition of the galactomannan interaction. Such knowledge may prove useful to a further understanding of the molecular geometry of the agar–galactomannan complex.

Extensive sulfation is uncommon in the agar series; the effect of sulfation on the interaction with galactomannans has been investigated[178] with the agar from *Gloiopeltis cervicornis*, which is sulfated on O-6 of every D-galactose residue and on O-2 of ~ 15% of the 3,6-anhydro-L-galactose residues. Non-gelling concentrations are gelled by addition of locust-bean gum, which appears to interact with galactomannans to the same extent as κ-carrageenan. As with κ-carrageenan, optical rotation measurements gave no evidence of a positive, optical rotation contribution from galactomannans in this interaction, but indicated that the melting of the associations is stabilized on heating. The fact that this degree of sulfation decreases the extent of interaction of the agar with galactomannan to that of κ-carrageenan is further, compelling evidence that substitution of the polysaccharide backbone of agar hinders association with galactomannans.

It has been shown that other polysaccharides interact with agarose in the same way as galactomannans. These all have backbones closely related structurally to the $(1 \rightarrow 4)$-β-D-mannan main-chain of galactomannans. The partially acetylated β-D-$(1 \rightarrow 4)$-linked glucomannan from *Amorphophallus konjac* tubers (konjac mannan) interacts particularly well with agarose.[178] In all respects, the extent of interaction of konjac mannan with agarose approximated closely to that of locust-bean gum. The acetate content of the material used corresponded to about one acetate group per six hexose residues, and so the degree of substitution of the glucomannan backbone is similar to that of the mannan backbone in locust-bean gum. Konjac mannan has been reported[204] to comprise glucose and mannose in the ratio of $1:1.6$. It is not known whether any region of the non-acetylated backbone of konjac mannan can associate with agarose, or whether there is favored association for $(1 \rightarrow 4)$-β-D-mannan regions, $(1 \rightarrow 4)$-β-D-glucan regions, or mixed regions.

Polysaccharides based solely on a $(1 \rightarrow 4)$-β-D-glucan backbone also interact with agar in the same way as galactomannans.[178] Thus, several commercial preparations of sodium O-(carboxymethyl)-cellulose cause non-gelling concentrations of agarose to gel, and increase the strength of agarose gels, as shown by rheological measurements. Those tested interacted with agarose to about the same degree as guar gum. Optical rotation studies demonstrated that the molecular association of agarose and sodium O-(carboxymethyl)-celluloses resulted in the same type of positive contribution to the change in optical rotation on cooling as was observed for the galactomannans. A number of seed amyloids also interact with agarose.[178] Optical rotation studies of agarose–amyloid mixtures indicated that this interaction is also accompanied by a positive contribution to the change in optical rotation attributable to the amyloid.

Attempts to demonstrate an interaction between agarose and barley β-glucan have failed[178]. Barley β-glucan is a linear D-glucan containing both β-D-$(1 \rightarrow 4)$- and β-D-$(1 \rightarrow 3)$-linkages,[205] and enzymic analysis indicated that a single β-D-$(1 \rightarrow 3)$-linkage alternates

(204) N. Sugiyama, H. Shimahara, and T. Andoh, *Bull. Chem. Soc. Jap.*, **45**, 561–563 (1972).

(205) G. O. Aspinall and R. G. J. Telfer, *J. Chem. Soc.*, 3519–3522 (1954).

with either two or three β-D-$(1 \rightarrow 4)$-linkages.[206] Its failure to interact with agarose therefore fits in well with the theory that blocks of β-D-$(1 \rightarrow 4)$-linked residues are essential for the molecular association with agarose.

Finally, some polysaccharides based on a $(1 \rightarrow 4)$-β-D-xylan backbone have been shown to interact with agarose.[178] Thus, by using gelation and optical rotation criteria, the xylan-based polysaccharides from *Watsonia pyrimidata* corm-sacs[138] and from sapote gum[207] were found to interact, albeit weakly, with agarose. It is relevant that solution studies with hemicellulose xylans indicated that the arabino-furanosyl side-chains have an unusual function, in that they do not cause the termination of binding sites involved in association with each other, which is the usual role of side chains;[208] instead, the associations remain, but in a modified form.[209] A similar role for side chains in *Watsonia pyrimidata* polysaccharide and sapote gum would explain why these heavily substituted D-xylans interact as well as they do with agarose.

An interesting interaction was noticed between agarose and water-soluble, esparto xylan.[178] The esparto xylan used was prepared by Hirst and coworkers;[210] it had an arabinose:xylose ratio of $1:11$ and a very low molecular weight. On cooling a non-gelling concentration of agarose (0.05%) with the xylan (0.75–1.5%), rapid precipitation occurred, instead of gelation. In this interaction, the xylan chains are envisaged to associate with the agarose tertiary structure in the normal way, but, owing to the very low molecular weight, the extra cross-linking necessary for gelation cannot occur. Instead, the binding of the xylan chains causes the agarose to be precipitated much faster than it would otherwise be.

These few examples show that molecular interaction with agarose is not a phenomenon peculiar to galactomannans, but is common to polysaccharides based on equatorial–equatorial β-D-$(1 \rightarrow 4)$-linked D-glycan backbones. None of these other systems have been studied in the same detail as the agarose–galactomannan interaction, and so

(206) F. W. Parrish, A. S. Perlin, and E. T. Reese, *Can. J. Chem.*, **38**, 2094–2104 (1960).

(207) R. D. Lambert, E. E. Dickey, and N. S. Thompson, *Carbohyd. Res.*, **6**, 43–51 (1968).

(208) D. A. Rees, *Biochem. J.*, **126**, 257–273 (1972).

(209) I. C. M. Dea, D. A. Rees, R. J. Beveridge, and G. N. Richards, *Carbohyd. Res.*, **29**, 363–372 (1973).

(210) S. K. Chanda, E. L. Hirst, J. K. N. Jones, and E. G. V. Percival, *J. Chem. Soc.*, 1289–1297 (1950).

the necessary data are not available to indicate whether these different types of backbone interact to the same or different extents with agarose. Elucidation of the effect of changes in the structure of the β-D-(1 → 4)-linked backbone on the interaction with agarose could be useful in the development of a more detailed model for the molecular geometry of the associations.

4. Interactions with Other Polysaccharides

It has been discovered that galactomannans interact with a number of bacterial polysaccharides.[168] The most studied has been the interaction with the exo-polysaccharide from *Xanthomonas campestris*. Mixtures of this non-gelling polysaccharide with locust-bean gum form firm, rubbery gels at total-polysaccharide concentrations[181,211] greater than 0.5%. The gels are firmest at a *Xanthomonas* polysaccharide:locust-bean gum ratio of ~1:3.

This interaction is even less well understood than that of galactomannans with carrageenans and agars, as the primary structure of *Xanthomonas* polysaccharide is not yet clear. *Xanthomonas* polysaccharide comprises D-glucose, D-mannose, D-glucuronic acid, acetate, and pyruvate in[212] the ratios 2.8:3.0:2.0:1.7:0.6. Structural investigations have been interpreted in terms of a 16-sugar repeat-unit[213,214] and a 13-sugar repeat-unit,[215] in which a backbone containing all three sugar constituents bears D-mannose and 4-6-O-(1-carboxyethylidene)-D-glucose side-stubs. The position of the O-acetyl groups is not fully known, although it has been suggested that they are on O-6 of the D-mannose side-chains. Oriented and crystalline X-ray diffraction data have been obtained for the polysaccharide from *Xanthomonas campestris*[199] that confirm the existence of a chemical-repeat structure, which may be masked, for the polysaccharide, although it may be simpler than the 16- and 13-sugar repeat-units proposed earlier.[213–215] The existence of such a chemical repeat for *Xanthomonas* polysaccharide is consistent with its strong interaction with galactomannans, which, like the interactions of galactomannans with carrageenans and agars, is best envisaged as a co-operative association of two polysaccharides in ordered conformations.

(211) E. R. Morris and D. A. Rees, forthcoming publication.

(212) J. H. Sloneker and A. (R.) Jeanes, *Can. J. Chem.*, **40**, 2066–2071 (1962).

(213) J. H. Sloneker, D. G. Orentas, and A. (R.) Jeanes, *Can. J. Chem.*, **42**, 1261–1269 (1964).

(214) J. H. Sloneker and D. G. Orentas, *Can. J. Chem.*, **40**, 2188–2189 (1962).

(215) I. R. Siddiqui, *Carbohyd. Res.*, **4**, 284–291 (1967).

The extent of interaction between galactomannans and *Xanthomonas* polysaccharide decreases with the increase in the D-galactose content of the galactomannan. Thus, mixtures of *Xanthomonas* polysaccharide and tara gum (D-mannose : D-galactose ratio, 3.0 : 1) form gels, but more tara gum than locust-bean gum is needed in order to effect gelation, and the resulting gels are less rubbery and resilient than *Xanthomonas* polysaccharide–locust-bean gum gels.[211] Guar gum–*Xanthomonas* polysaccharide mixtures do not form gels, but an interaction still occurs, as is apparent from the synergistic increase in viscosity of these mixtures.[179] Thus, in this respect, the galactomannan–*Xanthomonas* polysaccharide interaction is very similar to the interaction of galactomannans with carrageenans and agars, and is indicative of a likelihood that galactomannans associate with *Xanthomonas* polysaccharide through the D-galactose-poor regions.

As with agar and carrageenan, substitution in the *Xanthomonas* polysaccharide structure inhibits the interaction with galactomannan. Thus, removal of the acetyl groups by saponification yields a product that interacts much better with galactomannans.[211] Less locust-bean gum is required to gel deacetylated *Xanthomonas* polysaccharide, and mixtures of locust-bean gum and deacetylated *Xanthomonas* polysaccharide form a gel at lower, total-polysaccharide concentrations.

The *Xanthomonas* polysaccharide–galactomannan interaction has been further investigated by optical rotation.[211] On cooling a *Xanthomonas* polysaccharide solution, a negative optical-rotation transition of a type similar to, but less sharp than, that of agar was observed[211] (see Fig. 10); this result indicated that *Xanthomonas* polysaccharide may adopt an ordered conformation in solution on cooling, but that this does not lead to gelation. Such a conformational change could be the basis of the strange viscosity behavior of *Xanthomonas* polysaccharide, in which the viscosity first diminishes with increased temperature, then rises sharply, and finally falls again[208,216] (see Fig. 10). Deacetylated *Xanthomonas* polysaccharide forms a gel, albeit weakly, at concentrations greater than 3%, and shows the same type of optical rotation behavior.

Gelling mixtures of *Xanthomonas* polysaccharide with locust-bean gum or tara gum do not give reproducible optical rotation results, possibly because of gel strain, but examination of non-gelling

(216) A. (R.) Jeanes, J. E. Pittsley, and F. R. Senti, *J. Appl. Polym. Sci.*, **15**, 519–526 (1961).

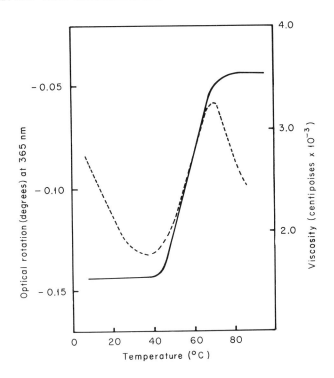

FIG. 10.—Comparison of the Changes in Viscosity (---) and Optical Rotation (—) when a Solution of *Xanthomonas* Polysaccharide (1%) is Heated and Cooled.

Xanthomonas polysaccharide–guar gum mixtures indicated subtle changes in optical rotation behavior. Thus, although this system has been studied less extensively, the evidence suggested that gels of mixed *Xanthomonas* polysaccharide–locust-bean gum are based on mixed junction-zones similar to that shown in Fig. 4 (see p. 290). However, as this system is still poorly characterized, both as regards the primary structure and the nature of the ordered conformation, detailed understanding of the association is lacking.

The interaction with polysaccharides closely related structurally to galactomannans has been studied less for *Xanthomonas* polysaccharide than for agar. No interaction is observed between *Xanthomonas* polysaccharide and sodium *O*-(carboxymethyl)-cellulose (degree of substitution 0.7), probably because of electrostatic repulsion between the two negatively charged polysaccharides.[178] *Xanthomonas* polysaccharide can, however, interact with polysaccharides based on β-D-(1 → 4)-linked D-glucan backbones, as is shown by the enhancement of gels of microcrystalline cellulose[182] by the presence of *Xanthomonas* polysaccharide.[178] A very strong interaction is observed in mixtures of *Xanthomonas* polysaccharide and konjac mannan, the partially acetylated β-D-(1 → 4)-linked glucomannan from *Amorphophallus konjac* tubers. The interaction is, in fact, stronger than that observed for *Xanthomonas* polysaccharide–locust-bean gum mixtures, as the resulting gels melts 20° higher than comparable *Xanthomonas* polysaccharide–locust-bean gum gels, and can form at much lower total-polysaccharide concentrations. Indeed, *Xanthomonas* polysaccharide–konjac mannan gels are formed at concentrations as low as 0.05%, a concentration at which agar no longer gels.[178] It is significant that konjac mannan and locust-bean gum interact with agar to similar extents, whereas konjac mannan interacts much better than locust-bean gum with *Xanthomonas* polysaccharide. Further study of this interaction might, therefore, lead to a better understanding of the molecular associations involved.

Jeanes and her coworkers searched for interactions of the same type between galactomannans and other bacterial polysaccharides by using viscometry.[217] Viscosities were measured for a series of ratios of bacterial polysaccharide–locust-bean gum (all at 1% of total polysaccharide), and synergistic increases in viscosity indicative of molecular interaction were observed for the extracellular polysaccharides from two strains of *Arthrobacter viscosus* (NRRL B-1797 and B-1973) and one strain of *Arthrobacter stabilis* (NRRL B-3225). These interactions are, however, not strong enough to result in gelation, and are probably similar in strength to the *Xanthomonas* polysaccharide–guar gum interaction. Nevertheless, the results are promising, and examination of the interactions by the optical rotation method may lead to a better understanding of the process.

An interaction has been reported[218] between the mucilage extracted from hulls of yellow-mustard seed (*Brassica hirta*) and each of two galactomannans (locust-bean gum and guar gum) that resulted in a synergistic increase in viscosity. Subsequently, white-mustard seed mucilage (*Brassica sinapsis alba*) was shown to interact with locust-bean gum and guar gum in a similar way.[219] Both of the mustard mucilages interact strongly with O-(carboxymethyl)cellulose,[218,219] suggesting that the interaction depends on

(217) A. (R.) Jeanes and J. E. Pittsley, personal communication.
(218) F. E. Weber, S. A. Taillie, and K. R. Stauffer, to be published.
(219) D. Thom, unpublished results.

equatorial–equatorial β-D-(1 → 4)-linked D-glycan backbones. The mustard mucilages dissolve in cold water to give cloudy solutions, and are known to comprise cellulose crystallites that are solubilized by association with pectic material.[220] Because polysaccharides based on β-D-(1 → 4)-D-glucan backbones have been reported to bind strongly to cellulose,[133] it seems likely that the basis of this synergistic increase in viscosity is ordered binding by β-D-(1 → 4)-linked chains of galactomannans and O-(carboxymethyl)cellulose to the cellulose crystallites of the mucilage.

In a number of food systems, mixtures of galactomannans and starch are found to have beneficial properties. For example, Cowley[221] found that addition of locust-bean gum and guar gum to gluten–starch loaves results in a desirable increase in loaf volume. Preliminary results on the solution and retrogradation properties of amylose indicated a strong interaction with galactomannans and with such (structurally similar) soluble cellulose derivatives as sodium O-(carboxymethyl)cellulose.[222] As with the interactions already described in this Section, the extent of interaction decreases with increase in content of D-galactose of the galactomannan. It is possibly significant that starch and certain β-D-(1 → 4)-linked polysaccharides often co-exist in food stores of plants.[153] For example, pentosans and starch co-exist in the wheat endosperm, and konjac mannan and starch co-exist in *Amorphophallus konjac* tubers (see p. 342 of Ref. 8). Although much more work on this system is needed, it seems possible that some of the desirable properties of starch–galactomannan mixtures may result from molecular association of the components.

5. Conclusions and Implications

Non-reserve polysaccharides seem to function in biological tissues through the part they play in cohesion, the retention of water and salts, the physical organization, and the elasticity and general texture. Polysaccharide conformation and association, as well as chemical structure, are obviously involved in the control of such properties. The polysaccharide–polysaccharide interactions considered in this Section can be regarded (in the nomenclature of protein biochemistry) as showing secondary, tertiary, and quaternary structure.[40,208] The

(220) G. T. Grant, C. McNab, D. A. Rees, and R. J. Skerrett, *Chem. Commun.*, 805–806 (1969).
(221) R. W. Cawley, *J. Sci. Food Agr.*, **15**, 834–838 (1964).
(222) I. C. M. Dea, G. Robinson, and A. Suggett, unpublished results.

secondary structure is the shape of the single polysaccharide D-mannan backbone (compare the polypeptide α-helix); the tertiary structure is the arrangement of two chains in the carrageenan double-helix (compare collagen, and the arrangement within one subunit of globular protein); and quaternary structure is the association of unlike polysaccharide chains in the cooperative binding of galactomannan with carrageenan double-helices (compare the binding of different polypeptide subunits in the hemoglobin structure). The systems discussed in this Section are the first examples of this type of organization for polysaccharides. A further analogy with protein behavior is that, in combination with galactomannans, the carrageenan double-helix can be caused to form, or persist, when it would otherwise be unfavorable. This influence can be considered to be an example of ligand induction of polysaccharide conformation (see Ref. 193).

The unsubstituted regions of galactomannans are similar in structure to such important, skeletal polysaccharides as hemicelluloses and cellulose, suggesting that the associations with the carrageenans and agars occur in imitation of natural associations between the latter polysaccharides and flexible chains bound in the microfibrillar structure with which they co-exist. Examples of structural polysaccharides whose interaction with agars and carrageenans could be important *in vivo* are the unsubstituted β-D-(1 → 4)-linked D-xylans[223] and D-mannans,[224] which frequently occur in red seaweeds. The structural variations in the carrageenan and agar series not only affect the adoption of their tertiary structures, and, therefore, the bulk gelling properties, but also materially alter the extent of association with skeletal polysaccharides. In addition, results from the model systems indicated that the degree of association is also dependent on variation in structure of the β-D-(1 → 4)-linked backbone. The occurrence of occasional β-D-(1 → 3)-linkages in certain linear, β-D-(1 → 4)-linked D-xylans of red seaweed[225] might be a natural method of modifying the extent of association with agars and carrageenans.

Xanthomonas campestris causes leaf blight in cabbage, and, generally, *Xanthomonas* species are plant pathogens. In this invasion of the plant, the viscous, extracellular polysaccharides may bind strongly to the cell-wall polysaccharides of the vascular system, and

(223) J. R. Turvey and E. L. Williams, *Phytochemistry*, **9**, 2383–2388 (1970).
(224) E. Frei and R. D. Preston, *Proc. Roy. Soc., Ser. B*, **160**, 293–313 (1964).
(225) A. S. Cerezo, A. Lezerovich, R. Labriola, and D. A. Rees, *Carbohyd. Res.*, **19**, 289–296 (1971).

act as a strong anchor for the bacterial cells. The gelling interaction of *Xanthomonas* polysaccharide with galactomannans is considered to mimic this natural association. Association of *Xanthomonas* polysaccharide with cellulose or hemicelluloses in Nature could explain why konjac mannan, which contains ~40% of D-glucose residues, interacts so much better than locust-bean gum with *Xanthomonas* polysaccharide.

The interaction of galactomannans and *O*-(carboxymethyl)cellulose with mustard mucilage suggests one method whereby polysaccharides may function in the physical organization of tissues of the higher plants. Galactomannans and *O*-(carboxymethyl)cellulose are similar in structure to important structural polysaccharides based on $(1 \rightarrow 4)$-β-D-glucan, -D-mannan, and -D-xylan main-chains; this suggests that their interaction with the mustard mucilages may mimic natural associations at the cell walls of higher plants.

For conformational transitions to occur in polysaccharide solutions and for the biological associations of polysaccharides to be readily and conveniently temperature-reversible[226] is unusual. Hence, it is fortunate that systems have been discovered in which both of these types of behavior are modelled and can be observed together. It is to be expected that such systems will prove useful in the further development of the theory of conformation in respect of biological function.

VI. COMMERCIAL EXPLOITATION OF GALACTOMANNANS

1. Introduction

In the early 1940's, when the war in Europe caused restriction of the supplies of locust-bean gum, it became a commercial necessity to find a satisfactory substitute therefor, and a large-scale assessment of seeds of leguminous plants for extractable gums was undertaken that led to the emergence of guar gum as a commercial rival to locust-bean gum. Most of the studies on the chemistry of the seed galactomannans resulted from the commercial interest in these gums, and it is, therefore, appropriate to conclude this article with a brief appraisal of the industrial uses of galactomannans.

(226) D. A. Rees, "The Shapes of Molecules: Carbohydrate Polymers," Oliver and Boyd, Edinburgh, 1967.

World usage of locust-bean gum is now of the order[227] of 12,500 tons per annum, and of guar gum, 37,000 tons per annum. Considerable though this tonnage is, it is minute relative to that of cellulose (800 million tons per annum). The usage is distributed amongst the various industries as follows: food, 40; paper, 30; textiles, 20; and oil, paint, explosives, and adhesives, 10%. Apart from guar and locust-bean gums, the galactomannans have been hardly exploited at all, although tara (*Caesalpinia spinosa*)[228] and fenugreek (*Trigonella foenum-graecum*)[228] gums show some promise in the food industries[10] and *Gleditsia triacanthos* in the cosmetics industries.[229]

The great advantage of the galactomannans (especially guar gum and locust-bean gum) is their ability, at relatively low concentrations, to form very viscous solutions that are only slightly affected by pH, added ions, and heating and cooling cycles. The rheological properties of locust-bean gum and guar gum have been discussed in Section IV (see p. 282). One of the problems with the industrial application of these polysaccharides is the variability of the quality of the gums, which range from the crude flour obtained by grinding the whole seed to highly purified polysaccharide containing less than 2% of impurities. This problem was emphasized in 1957 by Schlakman and Bartilucci[230] and it is still to a large extent in evidence. The tendency of some manufacturers to purify their products by alcohol precipitation has improved the situation, but there are still many brands on the market that are very impure by modern standards. The disadvantages of these gums include lessened stability of solutions due to endogenous enzymic activity, and, particularly in the food industry, poor clarity and unacceptable flavor.

2. Industrial Applications

The patent literature pertaining to the industrial exploitation of guar gum from 1948 to 1962 has been comprehensively collated by Saxena,[34] and will not be considered here. Saxena's compilation[34] covers more than 70 patents relating to 15 industries, and includes gum extraction and purification, and modification of gum properties. A selection, taken from the literature over the past decade or so, of applications of galactomannans is given in Table VI. It may be seen that guar gum is, indeed, a versatile product. Some of the more im-

(227) M. Stacey, *Chem. Ind.* (London), 222–226 (1973).
(228) M. Glicksman and E. H. Farkas, Fr. Pat. 2,119,365 (1972); *Chem. Abstr.*, **78**, 96,287 (1973).
(229) J. I. Gonzales, Fr. Pat. 2,067,649 (1971); *Chem. Abstr.*, **78**, 20,125 (1973).
(230) I. A. Schlakman and A. J. Bartilucci, *Drug Stand.*, **25**, 149 (1957).

portant, industrial derivatives thereof include the carboxymethyl ether[231] and hydroxyalkyl ethers.[248]

The chemical and physical properties of the galactomannans are important in their industrial applications. Thus, the ability to imbibe water, to form a thick, mucilaginous paste that is impervious to further water, is utilized by the mining and related industries, both as a method of waterproofing explosives[256-258] and as a plugging composi-

(231) General Mills, Inc., Ger. Pat. 2,104,743 (1971); *Chem. Abstr.*, **76**, 35,433 (1972).

(232) P. Kovacs, *Food Technol.*, **27**, 26–30 (1973).

(233) H. R. Schuppner, Can. Pat. 824,635 (1969); *Dairy Sci. Abstr.*, **32**, 1488 (1970).

(234) A. J. Leo and E. Bielskis, U. S. Pat. 3,396,039 (1968); *Chem. Abstr.*, **69**, 66,286 (1968).

(235) B. Weinstein, U. S. Pat. 2,856,289 (1958); *Chem. Abstr.*, **53**, 1,583e (1959).

(236) L. L. Little (Battelle Development Corp), U. S. Pat 3,370,955 (1968); *Chem. Abstr.*, **68**, 94,742 (1968).

(237) H. Burton, H. R. Chapman, and D. J. Jayne-Williams, *Proc. Intern. Dairy Congr. 16th, Copenhagen*, **3**, 82 (1962).

(238) E. Nuernberg, Ger, Pat. 1,290,661 (1969); *Chem. Abstr.*, **70**, 118,096 (1969).

(239) Laboratories Dausse S. A., Fr. Pharm. Pat. M. 7794 (1970); *Chem. Abstr.*, **76**, 131,509 (1972).

(240) E. Nuernberg, E. Rittig, and H. Mueller, Ger. Pat. 2,130,545 (1972); *Chem. Abstr.*, **78**, 62,171a (1973).

(241) E. Nuernberg, H. Mueller, H. Nowak, and P. Luecker, Ger. Pat. 2,017,495 (1971); *Chem. Abstr.*, **76**, 17,808 (1972).

(242) Synthelabo S. A., Fr. Pat. 2,073,254 (1971); *Chem. Abstr.*, **77**, 39,247 (1972).

(243) E. Merck A. G., Neth. Pat. 6,504,974 (1965); *Chem. Abstr.*, **64**, 14,042g (1966).

(244) R. Nordgren, U. S. Pat. 3,225,028 (1965); *Chem. Abstr.*, **64**, 6891c (1966).

(245) J. W. Opie and J. L. Keen, U. S. Pat. 3,228,928 (1966); *Chem. Abstr.*, **64**, 11,430a (1966).

(246) J. D. Chrisp, U. S. Pat. 3,301,723 (1967); *Chem. Abstr.*, **67**, 92,485u (1967).

(247) M. H. Yueh and E. D. Schilling, Fr. Pat. 2,080,462 (1971); *Chem. Abstr.*, **77**, 103,610 (1972).

(248) J. Fath and M. Rosen, U. S. Pat. 3,700,612 (1972); *Chem. Abstr.*, **78**, 59,856 (1973).

(249) D. J. Pettitt, U. S. Pat. 3,658,734 (1972); *Chem. Abstr.*, **77**, 35,846 (1972).

(250) G. Benz (for Chemische Fabrik Gruenau, G.m.b.H.), Ger. Pat. 1,206,777 (1965); *Chem. Abstr.*, **64**, 6290c (1966).

(251) R. E. Walker, U. S. Pat. 3,215,634 (1965); *Chem. Abstr.*, **64**, 3820e (1966).

(252) H. N. Black and L. L. Melton, U. S. Pat. 3,227,212 (1966); *Chem. Abstr.*, **64**, 7941b (1966).

(253) V. V. Horner and R. E. Walker, U. S. Pat. 3,208,524 (1965); *Chem. Abstr.*, **64**, 503b (1966).

(254) W. C. Browning, A. C. Perricone, and K. A. C. Elting, U. S. Pat. 3,677,961 (1972); *Chem. Abstr.*, **77**, 128,458 (1972).

(255) F. B. Knop, S. Afr. Pat. 69 06946 (1970); *Chem. Abstr.*, **73**, 111,681 (1970).

(256) J. J. Yancik, R. E. Schulze, and P. H. Rydlund, U. S. Pat. 3,640,784 (1972); *Chem. Abstr.*, **76**, 101,828 (1972).

(257) P. R. Goffart, Fr. Pat. 1,533,471 (1968); *Chem. Abstr.*, **71**, 23,404 (1969).

(258) E. I. du Pont de Nemours and Co., Inc., Fr. Pat. 1,537,625 (1968); *Chem. Abstr.*, **72**, 45,652 (1970).

TABLE VI

Industrial Uses of Galactomannans

Industry	Product	Gum	Function	References
Food		carboxymethylated guar gum	thickening and gelling agents	231
	desserts	carboxymethylated locust-bean gum	soluble and insoluble fibers	228
		Xanthonomas campestris gum–tara gum mixtures	synergistic effects	232,233
	jelled puddings	*Xanthonomas campestris* gum–locust-bean gum mixtures	synergistic effects	232,233
	ice cream	guar gum with carrageenan, and O-(carboxymethyl)-cellulose	stable thixotropic-stabilizer–emulsifier system	234,235
	acidified dairy product	guar gum–locust-bean gum		236
	ice cream	locust-bean gum with alginate and gelatin	allows high-temperature treatment	237
Pharmaceuticals	vitamin B$_{12}$ preparation	guar gum	stable, water-soluble preparation	238
	preparation for treatment of gastrointestinal ulcers and diarrhea	guar gum	synergistic activity with bismuth salt	239
	sustained-release drugs	guar gum	guar gum (M.W. 280,000) releases ascorbic acid after 2.75 hours; M.W. 50,000, after 2.00 hours. No gum–drug released after 30 minutes	240
	microencapsulation of drugs (sulfonamides)	guar gum	high resorptivity of drug	241

	composition for treatment of gastrointestinal disorders	guar gum		242
	tablet preparation		dry binder	243
Cosmetics	hair decrimper			229
Paper	hand towels, tissues	*Gleditsia triacanthos* gum galactomannan	treated with acrolein to give paper strengthener	244
		oxidized guar or locust-bean gum	improves wet strength of paper	245
Paper–textiles–explosives	sizing	guar–locust-bean gum	cross-linked with transition-metal ions; viscosity increases one hundred-fold	246
Paper		calcium salts of carboxymethylated galactomannans	degree of substitution 0.6–1.6. Viscosity doubled, and thixotropic surface increased from 0–$16,766$ dynes, cm^{-2} $sec.^{-1}$	247
Paint	paint	hydroxyalkyl ethers of galactomannans	viscosity stabilization of poly(vinyl acetate) latex paints	248,249
Building	plaster	guar gum–locust-bean gum	thickening agent	250
Well-drilling	oil bores	guar gum cross-linked with borate	stable, superelastic liquid with lessened temperature-sensitivity, for control of lost circulation in oil-field drilling operation	251
	oil, gas, water bores		for plugging leaks	252,253
	oil, gas, water bores	guar gum–*Xanthomonas campestris* mixtures	stabilization of cross-linkage	254
Coal mining		guar gum with boric acid and borax	used for shock impregnation of coal seams; makes possible a high build-up of pressure inside shock area	255
Explosives	explosive gel	guar gum	improved resistance to water and aging	256–258
Fire-fighting	air-drop, forest-fire control	guar gum with decyl sulfate and ammonium phosphate solution	provides viscosity stability	259

tion for leaking wells.[251-254] The same property is responsible for the use of guar gum to increase the wet strength of paper.[246,247]

The slow hydration of guar gum makes it a suitable base for delayed-release drugs,[240] although it also has direct medicinal uses, for example, as a laxative. The very high viscosity attainable with locust-bean gum and guar gum solutions has been put to good use in the food industry, where both gums are in great demand as thickeners for soups, desserts, pie-fillings, sauces, and mayonnaise.[10] The viscosity of the gum solutions can be greatly enhanced by the addition of borate ions[251] or transition-metal ions,[246] which provide cross-linkages between the polysaccharide chains, but such chemically modified gums are not permitted in foodstuffs.

Galactomannans have been used as flocculents (for the purification of ores, for example) and a study has been made of the gum structure by electron microscopy. The smallest fiber-diameter for an evaporated film of guar gum was found[260] to be 4 nm, but, when a kaolinite–guar gum flocculate was examined, the diameter of the smallest fiber that was attached to the kaolin was found to be ~10 nm. It is suggested that flocculation is caused by polymer bridging of the clay particles.

Perhaps the most interesting applications of these gums arise from an understanding of (a) detailed molecular structure as discussed in Section V (see p. 284) and, in particular, (b) the effect, on the properties, of the distribution of D-galactose along the mannan chain. Thus, novel dessert textures can be obtained by exploiting the synergistic interactions of locust-bean gum[232,233] or tara gum[228] with *Xanthomonas* polysaccharide.

Dugal and Swanson[171] examined the effect of an unsubstituted mannan backbone on the effectiveness of modified guar gums as beater adhesives. A single guar gum was treated with enzymes isolated from sprouted guar seeds, the effect of which was to hydrolyze off D-galactosyl groups, giving a series of guar gums having identical chain-lengths but variable D-mannose:D-galactose ratios. Burst factor, breaking length, tear factor, opacity, and scattering coefficient were all examined in terms of the D-mannose:D-galactose ratio. In most cases, there was a dramatic effect on these properties as the D-mannose:D-galactose ratio increased from about 1.5:1 to 2:1, after which, the effect reached a maximum at a D-mannose:D-galactose ratio of 3:1, and then decreased again.

(259) W. W. Morgenthaler, U. S. Pat. 3,634,234 (1972); *Chem. Abstr.*, **76**, 143,009 (1972).
(260) A. Audsley and A. Fursey, *Nature*, **208**, 753–754 (1965).

THE INTERACTION OF HOMOGENEOUS, MURINE MYELOMA IMMUNOGLOBULINS WITH POLYSACCHARIDE ANTIGENS*

By Cornelis P. J. Glaudemans**

Laboratory of Chemistry, National Institute for Arthritis, Metabolism, and Digestive Diseases, National Institutes of Health, Bethesda, Maryland 20014

I. Introduction

When certain foreign chemicals (antigens) or bacteria, covered with antigenic surface components, enter the animal body, an immune response is initiated, leading to the production of antibodies or immunoglobulins. These proteins can combine non-covalently with their antigen molecules in a specific way. The antigenicity and im-

* Dedicated to Professor Michael Heidelberger.

** I am grateful to Dr. M. Potter for the many discussions I have had with him in the course of writing this article, and to Dr. S. Rudikoff for checking the manuscript.

munogenicity (that is, the ability to elicit an immune response) of polysaccharides has been known for many years, beginning with the classical studies of Avery, Heidelberger, and Kendall.[1,2] Antibodies to polysaccharides of known or partially known structures have been useful in identifying unknown components and intersaccharidic linkages in carbohydrate polymers.[3] Most of the work has involved heterogeneous antibodies, that is to say, sera containing antibody populations that, although specific for one antigen, are made up of differing molecular species of immunoglobulins. Different immunoglobulins in the specific-antibody population within one serum may bind the same antigenic grouping (hapten) in different ways, or different immunoglobulins may bind small or large sections of the repetitive pattern on the macromolecular antigen (polysaccharide) chain.

A new potential in polysaccharide immunochemistry is provided by homogeneous immunoglobulins that bind carbohydrate polymers. Here, the specific immunoglobulin–hapten interaction can be studied, and characterized in detail. Thus, the discipline of carbohydrate chemistry can make a real contribution to the elucidation of the structure of immunoglobulins. In return, carbohydrate chemistry may find a tool that can increasingly be applied to the unraveling of its own unsolved problems in the structural analysis of polysaccharides.

II. General Features Concerning Immunoglobulins

1. Structure

Extensive reviews have been written on the structure of immunoglobulins (Ig's).[4,5] For understanding of the present article, a brief description of immunoglobulin structure is essential.

There are five classes of immunoglobulins that can be distinguished by antigenic determinants on the heavy chains (see later), namely, IgM, IgG, IgA, IgD, and IgE. Antibody activity has been demonstrated in each of the classes. The molecular organization of immunoglobulins is based on a four-chain, unitary structure con-

(1) M. Heidelberger and O. T. Avery, *J. Exp. Med.*, **38**, 73–80 (1923).

(2) M. Heidelberger and F. E. Kendall, *J. Exp. Med.*, **50**, 809–823 (1929).

(3) M. J. How, J. S. Brimacombe, and M. Stacey, *Advan. Carbohyd. Chem.*, **19**, 303–358 (1964); O. Lüderitz, A. M. Staub, and O. Westphal, *Bacteriol. Rev.*, **30**, 192–255 (1966).

(4) J. A. Gally, in "The Antigens," M. Sela, ed., Academic Press, New York, 1973, pp. 162–285.

(5) *Ann. N. Y. Acad. Sci.*, **190** (1971).

sisting of two, identical, heavy (H) chains, each having a molecular weight of ~50,000, and two, identical, light (L) chains, each having a molecular weight of ~23,000 (see Fig. 1A).[4,6]

Some immunoglobulin molecules are functionally completed in four-chain, monomeric units (the IgG, IgD, and IgE classes). Others are formed into various sized polymers of this four-chain unit. The mammalian IgM molecules are pentamers having five units joined together by disulfide linkages. Mammalian IgA molecules are com-

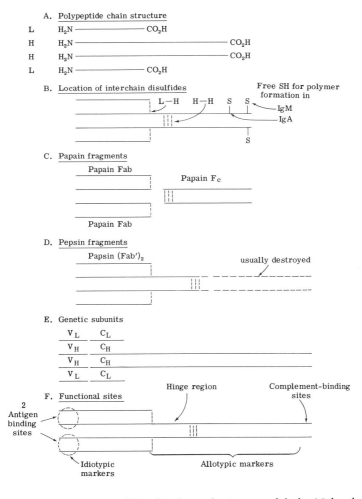

FIG. 1.—Various Structural Landmarks on the Immunoglobulin Molecule.

(6) G. M. Edelman, *Sci. Amer.*, **223**, 34–42 (1970).

pleted either as monomers, dimers, or trimers, which also are held together by disulfide bonds (see Fig. 1B). All light chains, regardless of the class of heavy chain with which they are associated, belong either to the κ or the λ type (see Fig. 2), which are differentiated by their constant regions (see later). The class specificity of an immunoglobulin is conventionally based on heavy-chain, antigenic determinants specific for each class.[4] In addition to the four-chain, unitary structure, other types of polypeptide chains are known to be involved in polymer formation. "J-chain" (Ref. 7) is present in both IgM and IgA polymers. The IgA secreted across epithelial linings into the lumen of the gastrointestinal and respiratory tracts has two four-chain units joined in a special arrangement by a "T" or transport piece;[8] this is called "secretory IgA."

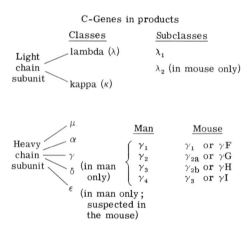

FIG. 2.—Classes of Immunoglobulins.

The immunoglobulin monomer of molecular weight 150,000 is organized in three, large, molecular regions that can artificially be dissociated by digestion with proteolytic enzymes into fragments (F).[9,10] In the presence of cysteine, papain (see Fig. 1C) splits the Ig molecules into a total of three fragments: an Fc fragment, so named because it crystallizes readily (molecular weight ~50,000), and two, identical, Fab fragments (molecular weight also 50,000), so named

(7) M. S. Halpern and M. E. Koshland, *Nature*, **228**, 1276–1278 (1970).
(8) T. B. Tomasi and J. Bienenstock, *Advan. Immunol.*, **9**, 1–87 (1968).
(9) R. R. Porter, *Biochem. J.*, **46**, 479–484 (1950).
(10) A. Nisonoff and J. G. Thorbecke, *Ann. Rev. Biochem.*, **33**, 355–402 (1964).

because they each retain one combining site (antibody = ab). Each four-chain monomer therefore contains two antigen-combining sites. Pepsin splits the H-chain more towards the carboxyl end,[11] and subsequently decomposes the Fc portion, leaving a divalent $(Fab)_2'$ fragment (see Fig. 1D). The $(Fab)_2'$ fragment may be reduced by 2-mercaptoethanol or 1,4-dithiothreitol, thus cleaving the H–H disulfide linkage and yielding two Fab' fragments, each possessing one combining site.

Genetic control of the biosynthesis of the Ig molecule differs from that of most other proteins so far studied, in that each of the chain types L and H is controlled by two different structural genes, the V (variable) and C (constant, see later) genes.[4,12] (Generally, for other proteins, one gene codes for one polypeptide chain, that is, one continuous stretch of polypeptide synthesis.) Thus, four different genes control a single Ig monomer; these are V_L, C_L, V_H, and C_H. (V_L means the gene controlling the variable part of the light chain, and so on; the corresponding, polypeptide segments of the immunoglobulin molecule controlled by the respective genes are shown in Fig. 1E.) The V_L and V_H segments involve up to residues 110 and 120, respectively, from the amino-terminal end of the chains. The V polypeptide segments were so named because they are highly variable within a class. The C (constant) polypeptides were so named because they are fewer in number, and, with minor exceptions, are identical within a given class; that is, the C regions of all IgM molecules in a given species are identical. A single, C-polypeptide segment may be associated with a vast number of different V-polypeptides. As the C-polypeptides differ strikingly from species to species, this circumstance raises the problem of how such a complex system could have evolved. It is unlikely that the multiple Ig gene system for each species is based upon a large number of copies of the same, species-specific gene. A reasonable approach[13] is that one gene controls the C-polypeptide, and an additional gene controls each of the V-polypeptides. To try to explain, then, the synthesis of the L and H chains, having continuous peptide bonds, it was postulated that the corresponding C and V genes are joined by a translocation mechanism on the chromosome, and that a template 2'-deoxyribonucleic acid (DNA) is made, from which a covalently linked m-RNA is transcribed.

(11) A. Nisonoff, F. C. Wissler, and L. N. Lipman, *Science*, **132**, 1770–1771 (1960).
(12) *Bull. World Health Organ.*, **41**, 975–978 (1969).
(13) W. J. Dreyer and J. C. Bennett, *Proc. Nat. Acad. Sci. U. S.*, **54**, 864–869 (1965).

2. Location of the Antigen-combining Sites

The combining sites of immunoglobulins are situated at the end of each of the two arms of the molecule,[14] that is, on the Fab fragments. The two sites per four-chain unit are identical, and there is evidence that they may measure up to some $34 \times 12 \times 7$ Å ($3.4 \times 1.2 \times 0.7$ nm) in size.[15] The combining sites are formed by the interaction of the V_L and V_H polypeptides (see Fig. 1F). Current crystallographic studies on immunoglobulins have shown that the Fab' fragment contains four domains, made up by the V_L, C_L, V_H, and C_H regions. The combining site seems to be located between the V_L and V_H domains.[16]

3. Antigenic Markers

Antigenic determinants (markers) found on the C-polypeptide segments are usually called allotypic markers. In general, antigenic markers found on the V-regions are usually called idiotypic markers, although, in the rabbit, allotypic markers have been demonstrated in the V region. For a given myeloma protein, idiotypic determinants refer to those antigenic determinants on the variable region that are unique to that protein; that is, if an idiotypic antiserum is made to protein A, and the serum is then treated with proteins A, B, C, D, and so on, only A will give a positive reaction.

III. POLYCLONAL, RESTRICTED, AND MONOCLONAL TYPES OF IMMUNOGLOBULIN-PRODUCING, CELLULAR RESPONSES

The cellular component of the immune system that synthesizes and secretes immunoglobulins[17] differentiates in a highly specialized way. It is considered that, essentially, lymphoid stem-cells divide and differentiate into lymphocytes. It was originally supposed that an antigen could order a non-committed lymphocyte to synthesize antibody against the antigen, and that any antigen could so order any lymphocyte. Following the theory of clonal selection (see later for the definition of a clone),[18,19] it is now generally supposed that each

(14) R. C. Valentine and W. M. Green, *J. Mol. Biol.*, **27**, 615–617 (1967).

(15) E. A. Kabat, *J. Amer. Chem. Soc.*, **76**, 3709–3713 (1954); *J. Immunol.*, **77**, 377–385 (1956); **84**, 82–85 (1960).

(16) E. A. Padlan, D. M. Segal, T. F. Spande, D. R. Davies, S. Rudikoff, and M. Potter, *Nature*, **245**, 165–166 (1973).

(17) H. N. Claman, *Bioscience*, **23**, 576–581 (1973).

(18) D. W. Talmage, *Ann. Rev. Med.*, **8**, 239–256 (1957).

(19) F. M. Burnet, "The Clonal Selection Theory of Acquired Immunity," Vanderbilt Univ. Press, 1959.

lymphocyte can synthesize (probably) only one immunoglobulin. Thus, each lymphocyte is specialized to produce an immunoglobulin having only a single variety of V_L and V_H polypeptide. Operationally, there are so many V_L, V_H pairing possibilities that it is fair to state that each cell that is specialized to produce only a single V_L, V_H pair will probably bind only a single variety of antigenic determinants.

The clonal-selection theory is now widely accepted, but certain aspects, particularly those concerning the generation of diversity, are still very much in doubt. For instance, a lymphocyte could carry all of the genetic information necessary for the synthesis of a multitude of immunoglobulins. Alternatively, it is possible that stem cells carry the genetic information for only a relatively few V_L and V_H pairs, and that mutation occurring during somatic differentiation generates the diversity that accounts for the ability of an animal to respond to a wide range of antigens. These possibilities have not yet been resolved.[20-27]

An immune response is initiated when an antigen is introduced into an organism either naturally (by the normal, microbial flora, or by infection), or artifically (by injection). Cells elaborating immunoglobulins having specificity for the antigen are selected for by the antigen from the great potential pool of differentiated lymphocytes. By a process that is not yet clear, this results in cell proliferation and further differentiation to an immunoglobulin-secreting cell-type (the plasma cell, or, sometimes, the lymphocytoid-plasma cell). All of the cells that have originated from a specialized lymphocyte are said to belong to one clone. In the usual immune response, many different lymphocytes, each specialized towards different aspects of the molecular structure of the antigen, are so stimulated; in this case, the response is *polyclonal*. Many species of immunoglobulins are produced by such a polyclonal response, and, usually, no one molecular variety predominates.

In certain animals, if the immunization process is hyper-stimulated by prolonged injection of antigens, the response may narrow down to

(20) G. P. Smith, L. Hood, and W. M. Fitch, *Ann. Rev. Biochem.*, **40**, 969–1012 (1971).
(21) L. Hood and D. W. Talmage, *Science*, **168**, 325–334 (1970).
(22) N. Hilschman, H. Ponstingl, M. Hess, L. Suter, H. U. Barnikol, K. Bazcko, S. Watanabe, and D. Braun, *Angew. Chem. Int. Ed. Engl.*, **9**, 534 (1970).
(23) M. Cohn, in "Nucleic Acids in Immunology," O. J. Plescia and W. Braun, eds., Springer Verlag, New York, 1967 pp. 671–717.
(24) N. K. Jerne, *Eur. J. Immunol.*, **1**, 1–9 (1971).
(25) S. Brenner and C. Milstein, *Nature*, **211**, 242–243 (1966).
(26) M. Cohn, *Cell. Immunol.*, **5**, 1–20 (1972).
(27) M. Cohn, *Cell. Immunol.*, **1**, 468–475 (1971).

a relatively few clones that do predominate, and this behavior is called a *restricted response*.[28] In this restricted response, immunoglobulins from specific clones can be identified serologically by anti-idiotypic sera, or, physicochemically, by iso-electric focusing in cellulose acetate electrophoresis. By special methods, it is also possible to isolate homogeneous antibodies by specifically adsorbing them from the restricted population.

In the most-restricted, cellular response, only one clone is stimulated: this is a *homogeneous response*. This response is rarely seen during immunization, but it does occur in proliferative disorders of plasma cells (that is, plasma-cell tumors) that result in the expansion of a single clone. These tumors can be readily transplanted to normal hosts, where they continue to produce large quantities of homogeneous immunoglobulin. In the inbred, BALB/C strain of mice, it is possible to *induce* plasma-cell tumors in high frequency by the intraperitoneal injection of mineral oil, or by X-irradiation.[29] Hundreds, or thousands, of different, homogeneous immunoglobulins have thus been made available. If these proteins are screened against a battery of test antigens, it is possible to find some (less than 5%) that demonstrate specific-binding activity; this low incidence is probably attributable to the limited number of suitable test-antigens available.

IV. General Features of Antigen–Antibody Interactions

1. The Nature of the Precipitin Reaction

The precipitin theory of Heidelberger and Kendall[2,30] predicted that the multiple determinants of a polysaccharide antigen[30a] play a decisive role in the formation of the immune precipitate. The theory stated that, if a soluble antigen is added to serum containing homologous (that is, specific for the antigen in question) antibodies, a precipitate will form that will increase in quantity upon the addition of more antigen, until the amount of precipitated antibody–antigen complex reaches a maximum. Addition of more antigen causes the amount of complex precipitated to level off, and addition of yet more antigen causes dissolution of some of the complex (see Fig. 3). These three areas of the curve (Fig. 3) have been referred to as the area of

(28) R. M. Krause, *Fed. Proc.*, **29**, 59–65 (1970).

(29) M. Potter, *Physiol. Rev.*, **52**, 631–719 (1972).

(30) M. Heidelberger, "Lectures in Immunochemistry," Academic Press, New York, 1956.

(30a) Historically, the theory was developed by using polysaccharide antigens, not protein antigens.

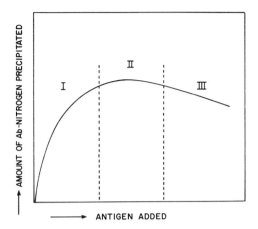

FIG. 3.—Typical Precipitin Curve for an Antigen–Antibody Combination.

antibody excess (I), equivalence (II), and antigen excess (III). In inhibition studies, small ligands (haptens) are tested for their ability to inhibit, competitively, the precipitation of antigen with antibody in the equivalence zone.

Antigens and their corresponding antibodies precipitate by crosslinking to form an insoluble network. Polysaccharides have multiple, repetitive immunodeterminants and virtually none have demonstrable tertiary structure in solution (except, perhaps, under viscous stress). The number of these immunodeterminant groupings on each macromolecule is large. In the case of dextran, for instance, there are several thousand of them (if the dextran has a molecular weight of several million), even if the determinant involves the heptasaccharide. There is, thus, ample opportunity to form a precipitating, crosslinked complex with divalent (or polyvalent) antibody molecules.

If some of the antibodies in the antibody population of a serum against antigen A also precipitate with another antigen B, the process is called a cross-reaction between A and B. Cross-reactions are caused by antigenic determinants that A and B have in common.

2. The Interaction of Homogeneous Immunoglobulins with Antigens

In contrast to polysaccharides, protein antigens *do* have tertiary structure in solution; this leaves only a limited number of amino acid sequences exposed at the surface of the molecule. Of these determinants, few, if any, will be repetitive. It would therefore be expected that antiprotein antibodies will only precipitate with their homologous antigen if the antibodies are of such a sufficiently heterogeneous population that the antigen molecules can be crosslinked by a series of several molecular species of immunoglobulins. There is good evidence that this process does, in fact, occur.[31] Therefore, a *homogeneous* immunoglobulin would not be expected to precipitate with its *protein* antigens (although it will bind).[32]

As polysaccharides are an abundant source of naturally occurring antigens having multiple determinants, it is clear that the role of anti-carbohydrate, homogeneous immunoglobulins will be very valuable in providing an insight into the problems of antigen–antibody interaction.

As already noted, most immunoglobulins (apart from those belonging to the IgG class) do not occur in the monomeric, 4-chain unit. The potential binding-valence of an immunoglobulin is equal to the total number of combining sites per molecule, namely, 10 for an IgM, 4 or more for a dimeric or polymeric IgA, and so on. However, in a number of instances, it has been observed that homogeneous, bivalent, myeloma immunoglobulins will not precipitate with a polyvalent antigen. A case in point is IgA MOPC 315, which—although it has a K_a for ϵ-(2,4-dinitrophenyl)-L-lysine of 10^7 liter. mol^{-1}—does not precipitate with polyvalent DNP-antigen.[33] It has been suggested,[29] that owing to restricted rotation, one of the binding

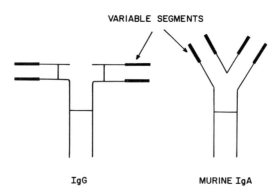

VARIABLE SEGMENTS

IgG   MURINE IgA

FIG. 4.—The General Appearance of IgG and IgA Molecules, Showing the Less Flexible Hinge-region of the IgA Molecule.

(31) M. Heidelberger and F. E. Kendall, *J. Exp. Med.*, 62, 697–720 (1932).
(32) M. Potter, E. B. Mushinski, and C. P. J. Glaudemans, *J. Immunol.*, 108, 295–300 (1972).
(33) M. Potter, in "Multiple Myeloma and Related Disorders," H. A. Azar and M. Potter, eds., Harper and Row, Hagerstown, Md., 1973, p. 197.

sites of the IgA monomer may be sterically hindered from binding a second antigenic determinant (see Fig. 4). Crosslinking would, however, be possible in IgA dimers.

V. Purification of Homogeneous Immunoglobulins from Sera or Ascites

Immunoglobulins, be they myeloma proteins, antibodies, or cold hemagglutinins, are optimally purified from ascites or serum by affinity chromatography, that is, immunoadsorption on an inert support, followed by elution with soluble hapten. In 1953, Lerman first used affinity chromatography[34] for the purification of mushroom tyrosinase, and he subsequently applied the method to the isolation of antibodies.[35] For both enzymes[36] and antibodies,[37,38] it has been observed that the substrate or ligand must be extended at some distance from the column support in order to assure adequate binding. Thus, "spacer" groups are usually attached to the support resin, and the ligands are attached to the "spacer" groups. The valence of the immunoglobulin may be another important factor in successful adsorption. Univalent, bivalent 4-chain monomer, and polyvalent polymer forms of the same immunoglobulin may be adsorbed to a different extent for stereochemical reasons. It appears that derivatized agarose gels are not so efficient for immunoadsorption as are derivatized Sepharose gels.[37,39-41] The former frequently bind albumin.[37]

Immunoadsorbents can be made with whole antigens or haptens. For example, Haber[42] reported the preparation of a *Diplococcus pneumoniae* Type III immunoadsorbent. The capsular polysaccharide (S III) was *p*-nitrobenzylated to a low degree of substitution by the method of Avery and Goebel.[43] The resultant ether was re-

(34) L. S. Lerman, *Proc. Nat. Acad. Sci. U. S.*, **39**, 232–236 (1953).

(35) L. S. Lerman, *Nature*, **172**, 635–636 (1953).

(36) P. Cuatrecasas, M. Wilchek, and C. B. Anfinsen, *Proc. Nat. Acad. Sci. U. S.*, **61**, 636–643 (1968).

(37) M. Potter and C. P. J. Glaudemans, *Methods Enzymol.*, **28**, 388–395 (1972).

(38) A. Sher and H. Tarikas, *J. Immunol.*, **106**, 1227–1234 (1971).

(39) M. E. Jolley, S. Rudikoff, M. Potter, and C. P. J. Glaudemans, *Biochemistry*, **12**, 3039–3044 (1973).

(40) M. E. Jolley and C. P. J. Glaudemans, *Carbohyd. Res.*, **33**, 377–382 (1974).

(40a) See also, Ref. 80.

(41) S. Rudikoff, E. B. Mushinski, M. Potter, C. P. J. Glaudemans, and M. E. Jolley, *J. Exp. Med.*, **138**, 1095–1106 (1973).

(42) E. Haber, *Fed. Proc.*, **29**, 66–71 (1970).

(43) O. T. Avery and W. F. Goebel, *J. Exp. Med.*, **54**, 437–447 (1931).

duced to the *p*-amino derivative, which was then coupled to bromo-acetylated cellulose (the insoluble support). Potter and Glaudemans

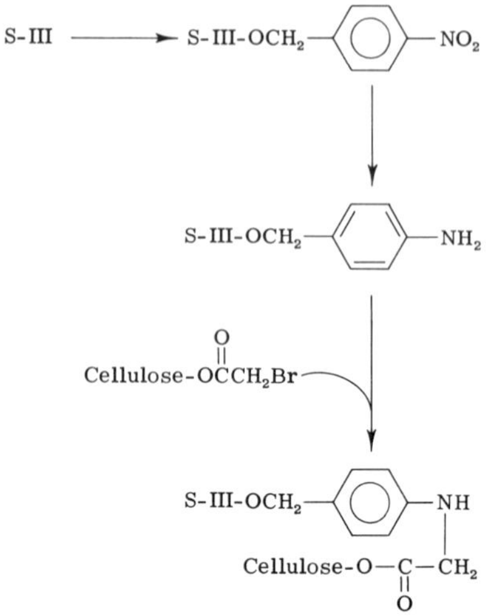

reported the preparation of an immunoadsorbent having *p*-aminophenyl 1-thio-β-D-galactopyranoside coupled by means of its amino group to agarose bearing carboxyl groups as substituents.[37]

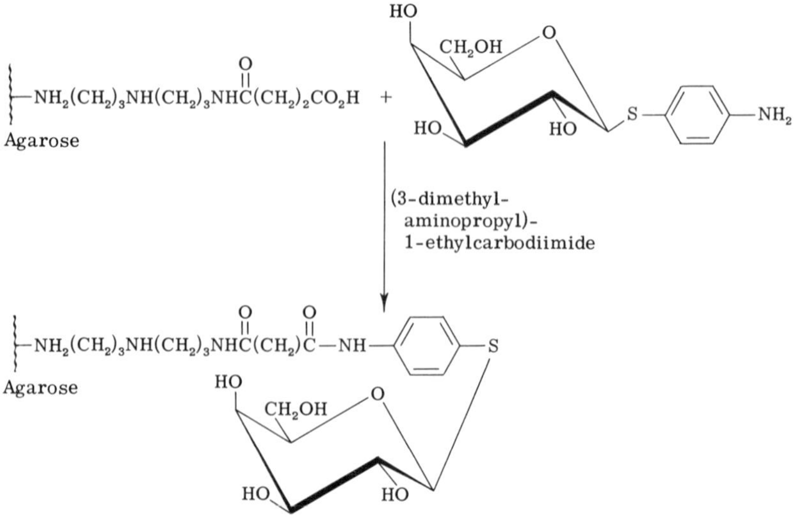

An immunoadsorbent that avoided the problem of albumin binding was obtained when the *p*-aminophenyl 1-thio-β-D-galactopyranoside was first diazotized and then coupled to BSA (bovine serum albumin), followed by coupling of the thioglycoside–BSA conjugate to cyanogen bromide-activated Sepharose[44] to yield an insoluble Sepharose–BSA–1-thio-D-galactoside.[37] Young and coworkers[45]

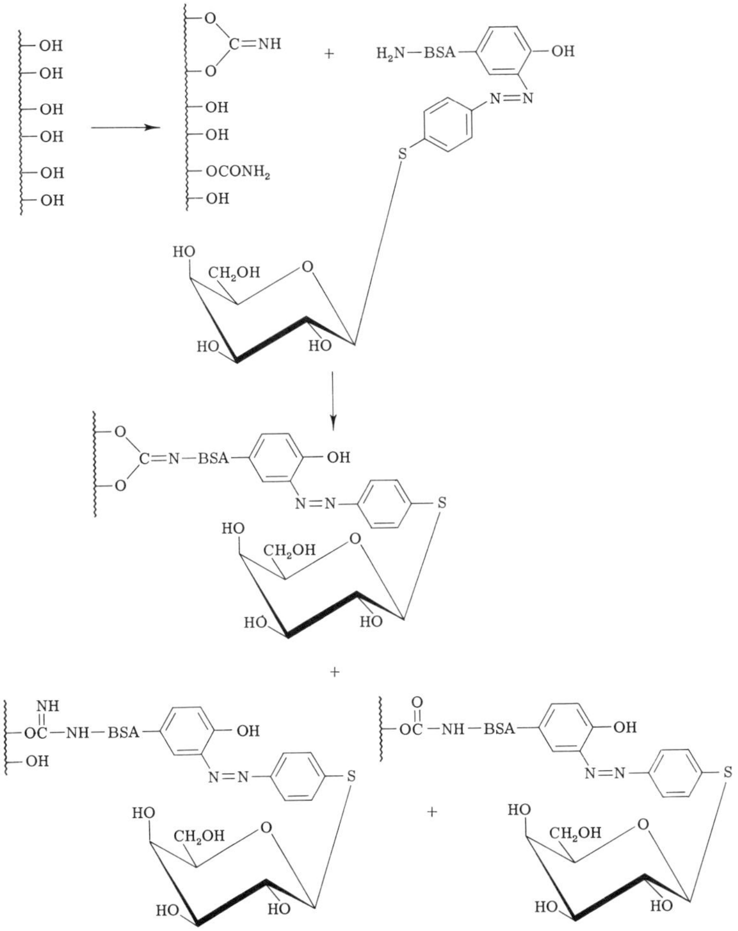

used dextran to purify by immunoprecipitation a homogeneous, myeloma immunoglobulin having anti-dextran specificity. A review [46] on affinity chromatography may be helpful to the reader.

(44) R. Axen, J. Porath, and S. Ernback, *Nature*, **214**, 1302–1304 (1967).
(45) N. M. Young, I. B. Jocius, and M. A. Leon, *Biochemistry*, **10**, 3457–3460 (1971).
(46) H. Guilford, *Quart. Rev.* (London), **2**, 249–270 (1973).

VI. The Known Anti-carbohydrate, Myeloma Immunoglobulins

1. Anti-(2 → 6)- and (2 → 1)-β-D-fructans

It may be seen from Table I that one of the largest groups of anti-carbohydrate, myeloma immunoglobulins is that of antifructans having specificity for β-D-(2 → 1)-linkages. It is convenient to combine the discussion of this group with that of the anti-(2 → 6)-β-D-fructan immunoglobulins.

In the first sub-group, the proteins having anti-(2 → 6)-β-D-fructosyl specificity, namely, proteins Y5476 and UPC10, have been found to react with a levan from perennial rye-grass, a polysaccharide having mainly β-D-(2 → 6)-linked D-fructofuranosyl residues. Inulin, consisting of β-D-(2 → 1)-linked D-fructofuranosyl residues, does not react with these proteins.[47,48] Immunoglobulin J606 displays[48] a rather dual specificity. It is precipitated well by Hestrin levans [mainly containing β-D-(2 → 6)-linked D-fructofuranosyl residues], whereas inulin [having mostly β-D-(2 → 1)-linkages] is a poor precipitating substance.[49] However, when inhibition of the J606–levan precipitin system was attempted, oligosaccharides containing (2 → 1)-β-D-fructosyl residues terminated by D-glucose residues proved to be the best inhibitors.[49] As no inhibition studies were made with the β-D-(2 → 6)-linked D-fructofuranosyl oligosaccharides, the exact specificity of myeloma immunoglobulin J606 remains to be established. Both immunoglobulin J606 (Ref. 48) and UPC10 were found to belong to the IgG class. All other homogeneous anti-fructans appear to belong to the IgA class.

The subgroup of homogeneous, anti-(2 → 1)-β-D-fructans contains six immunoglobulins in addition to J606. Proteins UPC61 and W3082 both precipitate with levans and with inulin. The specificity seems directed against β-D-(2 → 1)-linkages: perennial rye-grass levan, whose structure is mostly[50,51] β-D-(2 → 6), hardly reacts. In inhibition studies on the precipitin systems UPC61–levan and W3082–levan, these systems were best inhibited by β-D-(2 → 1)-linked D-fructofuranosyl oligosaccharides. It is interesting that both UPC61 and W3082 precipitated the same amount, or more, of anti-

(47) J. Cisar, E. A. Kabat, J. Liao, and M. Potter, *J. Exp. Med.*, **139**, 159–179 (1974).

(48) H. M. Grey, J. W. Hirst, and M. Cohn, *J. Exp. Med.*, **133**, 289–304 (1971).

(49) A. Lundblad, R. Stellar, E. A. Kabat, J. W. Hirst, M. Weigert, and M. Cohn, *Immunochemistry*, **9**, 535–544 (1972).

(50) R. A. Laidlaw and S. G. Reid, *J. Chem. Soc.*, 1830–1834 (1951).

(51) B. Lindberg, J. Lönngren, and J. L. Thompson, *Acta Chem. Scand.*, in press.

TABLE I: **Murine, Myeloma Immunoglobulins of Known Specificity**

Tumor, yielding the myeloma Ig	Specificity	Ig chain class H	Ig chain class L	Precipitable antigen	Inhibitors	References
S176						
J558		α	λ			
MOPC104E	$(1 \rightarrow 3)$-α-D-	μ	λ			49,59–
UPC102	glucopyranan	α	λ	dextran	nigerose	63
QUPC52		α	κ			
W3434	$(1 \rightarrow 6)$-α-D-	α	κ			
W3129	glucopyranan	α	κ	dextran	isomaltose	47
J606		γ_3	κ			
UPC10		γ	κ			
Y5476	$(2 \rightarrow 6)$-β-D-	α	κ			
AB-PC48	fructofuranan	α	κ	levan		47,48
AB-PC4		α	κ			
AB-PC45		α	κ			
AB-PC47		α	κ			
AM-PC1		α	κ			
UPC61						
W3082	$(2 \rightarrow 1)$-β-D-	α	κ			
J606	fructofuranan	γ_3	κ	levan, inulin	D-fructofuranose	49,52
J1		α	κ			
SAPC10		α	κ			
XRPC24		α	κ			
XRPC44		α	κ			
TEPC191		α	κ			
J539		α	κ			37,39–
TEPC601	$(1 \rightarrow 6)$-β-D-	α	κ	arabinogalactan,	β-D-galactopy-	41,53–
CBPC4	galactopyranan	α	κ	lung galactan	ranosides	54a
S63		α	κ			
S107		α	κ			
TEPC15		α	κ			
HOPC8		α	κ			
MCPC603		α	κ			
MOPC167		α	κ	pneumococcal		65–69,
MOPC511	phosphoryl-	α	κ	C-polysac-		73–75,
ALPC43	choline	γ	κ	charide	phosphorylcholine	77–81
S117	β-D-GlcNAc-p	α	κ	Streptococcus A polysaccha-ride	2-acetamido-2-deoxy-β-D-glucosides	
MOPC406	β-D-ManNAc-p	α	κ	Salmonella weslaco poly-saccharide	2-acetamido-2-deoxy-β-D-mannosides	73,82,84
MOPC384	α-D-Gal$p$	α	κ	Salmonella telaviv, Proteus mirabilis	Me α-D-Gal$p$	73,85

body nitrogen with levan as with inulin, although much more of the levan was required than of inulin. It is likely that proteins W3082 and UPC61 have specificity towards β-D-(2 → 1)-linked D-fructofuranosyl linkages, but that they precipitate with levans because these levans contain a small percentage of β-D-(2 → 1)-linkages.[47,52]

2. Anti-(1 → 6)-β-D-galactans

Another large group of anti-polysaccharide, myeloma proteins consists of those having anti-(1 → 6)-β-D-galactan activity (see Table I). This group of proteins that precipitate with galactans from wood,[32] as well as mammalian-lung galactan,[53] has been fairly extensively investigated.[32,37,39–41,53–54a]

The binding of some of the IgA proteins with β-D-(1 → 6)-linked oligosaccharides of D-galactose has been studied by tryptophanyl fluorescence titration.[39–40a] The method, which is applicable to homogeneous proteins only, is far more sensitive than equilibrium dialysis, the foremost alternative for the determination of association constants (K_a) between ligands and proteins.[55] The fluorescence titration method permits[40] rapid and accurate determination of association constants as low as 10^2. The magnitude of the K_a values found for two anti-galactan, mouse immunoglobulins (X24 and J539), as well as for their Fab' fragments, with the ligands (1 → 6)-β-D-galactotriose (Gal$_3$) and the corresponding (1 → 6)-β-D-galactotetraose (Gal$_4$) (see

TABLE II

Association Constants of Two Anti-galactan Immunoglobulins and Their Fab' Fragments as Determined by Fluorescence Titration[a]

Oligo-saccharide	Immunoglobulin			
	X24 whole ($\times 10^{-5}$)	X24 Fab' ($\times 10^{-5}$)	J539 whole ($\times 10^{-5}$)	J539 Fab' ($\times 10^{-5}$)
Gal$_3$	not determined	1.75 (± 0.03)	not determined	1.50 (± 0.02)
Gal$_4$	2.66 (± 0.1)	2.93 (± 0.05)	3.49 (± 0.04)	3.44 (± 0.07)

[a] Reprinted, with permission, from Ref. 39, copyright by the American Chemical Society.

(52) R. Z. Allen and E. A. Kabat, *J. Exp. Med.*, **105**, 383–394 (1957).

(53) C. P. J. Glaudemans, M. E. Jolley, and M. Potter, *Carbohyd. Res.*, **30**, 409–413 (1973).

(54) M. E. Jolley, C. P. J. Glaudemans, S. Rudikoff, and M. Potter, *Biochemistry*, **13**, 3179–3184 (1974).

(54a) B. N. Manjula, C. P. J. Glaudemans, E. Mushinski, and M. Potter, *Carbohyd. Res.*, **40**, 137–142 (1975).

(55) A. H. Sehon, *Methods Immunol. Immunochem.*, **3**, 375–394 (1971).

Table II) lies[39] between 1.5 and 3.5 × 10^5. This K_a value is very similar to that (2.5 × 10^5) found[56] for a rabbit IgG antipneumococcal polysaccharide with an octasaccharide fragment from *Diplococcus pneumoniae* SVIII. In a subsequent investigation, Jolley and co-workers[54] studied the binding affinities (K_a values) of the Fab' fragments from the same two proteins (X24 and J539) with a variety of derivatives of D-galactose (see Tables III and IV). Both proteins were clearly shown to be specific for terminal, nonreducing, β-D-linked galactopyranosyl groups. As methyl β-D-glucopyranoside, methyl β-D-gulopyranoside, and 2-deoxy-D-*lyxo*-hexose did not bind, the correct stereochemistry[56a] of C-2, C-3, and C-4 seems to be a prerequisite for binding with either antibody. However, the hydroxymethyl group on C-5 is not essential for binding, as methyl α-L-arabinopyranoside interacts with both proteins (see Table III). Changing the linkage in methyl β-D-galactopyranoside from β to α resulted in a total loss of binding. The presence of a β-D-linked glycosidic oxygen atom is not essential, as binding occurred with 1,5-anhydro-D-galactitol (see Table III).

From Table III, it may also be seen that, at the monosaccharide level, protein J539 is a significantly better binder of haptens than protein X24. It is only at the trisaccharide level that protein X24 attains the binding ability of protein J539. Table IV shows the contributions to the total binding-energy made by each D-galactose moiety in Gal$_2$, Gal$_3$, and Gal$_4$. These data were obtained from the K_a values by calculation of the free energies of binding ($-\Delta F = RT \ln K_a$). A number of observations were made. First, in all cases, the terminal, nonreducing D-galactosyl group (residue 1) contributed most to the binding of each oligosaccharide; this is consistent with the concept of immunodominance for terminal, nonreducing glycosyl groups.[15] Second, the terminal, nonreducing D-galactosyl group in these oligosaccharides contributed more to the binding energies for immunoglobulin J539 than for immunoglobulin X24. Third, the contributions of residues 2 and 3 to the binding of Gal$_3$ and Gal$_4$ with immunoglobulin X24 were significantly greater than with immunoglobulin J539. In both proteins, residue 2 contributed more than residue 3. Fourth, for both proteins X24 and J539, residue 4 contributes little to the binding energy of Gal$_4$. It would, therefore, appear that the limiting size for the combining sites corresponds to that of the tetraose.[54]

In addition, by using various D-galactose derivatives, deductions could be made with regard to the contributions of portions of the Gal$_2$ molecule to its total binding-energy with these two im-

(56) J. Spyer and A. Pappenheimer, cited in E. Haber, *Fed. Proc.*, **29**, 66–71 (1970).
(56a) The carbon atoms in the D-galactose units of Fig. 6 are numbered.

TABLE III

The Binding Constants of a Number of Derivatives of D-Galactose with Two IgA Myeloma Proteins Having Anti-galactan Specificity[a]

Ligand	Formula	Ka (Fab')$_{X-24}$	Ka (Fab')$_{J-539}$
Gal$_4$	β-D-Gal$p$-[1($\rightarrow$ 6-D-Gal$p$-1)$_2$ $\rightarrow$ 6]-D-Gal$p$	2.93 ($\pm$0.05) $\times$ 10^5	3.44 ($\pm$0.07) $\times$ 10^5
Gal$_3$	β-D-Gal$p$-(1 $\rightarrow$ 6)-D-Gal$p$-(1 $\rightarrow$ 6)-D-Gal$p$	1.75 ($\pm$0.03) $\times$ 10^5	1.50 ($\pm$0.02) $\times$ 10^5
Gal$_2$	β-D-Gal$p$-(1 $\rightarrow$ 6)-D-Gal$p$	5.75 ($\pm$0.07) $\times$ 10^3	1.14 ($\pm$0.01) $\times$ 10^4
Methyl β-D-galactopyranoside	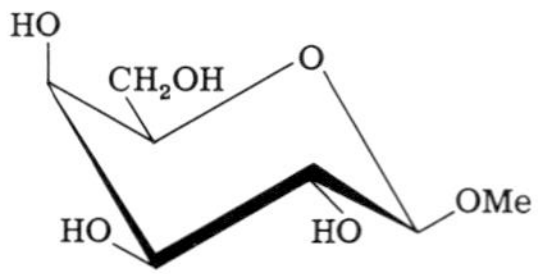	3.77 ($\pm$0.07) $\times$ 10^2	10.01 ($\pm$0.21) $\times$ 10^2
1,5-Anhydro-D-galactopyranose	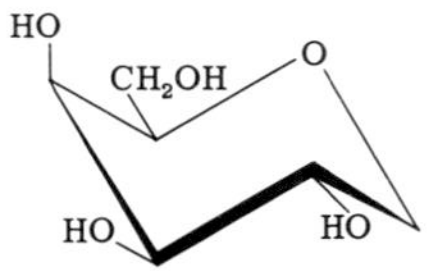	1.07 ($\pm$0.04) $\times$ 10^2	4.09 ($\pm$0.05) $\times$ 10^2
Isopropyl 1-thio-β-D-galactopyranoside	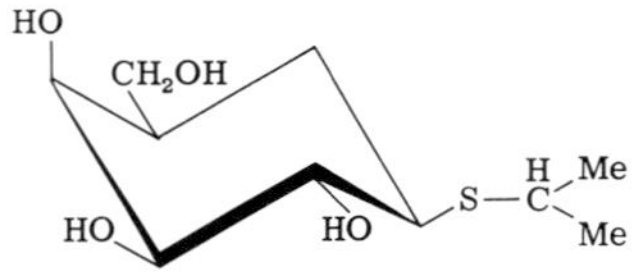	4.71 ($\pm$0.06) $\times$ 10^2	1.06 ($\pm$0.02) $\times$ 10^3

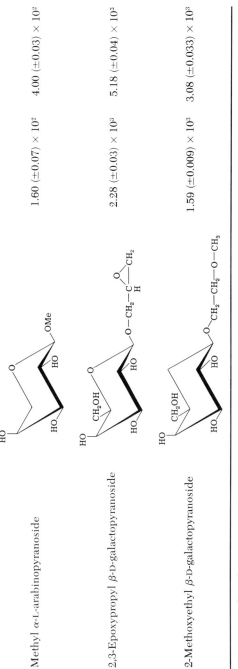

Methyl α-L-arabinopyranoside

2,3-Epoxypropyl β-D-galactopyranoside

2-Methoxyethyl β-D-galactopyranoside

1.60 (±0.07) × 10²   4.00 (±0.03) × 10²

2.28 (±0.03) × 10³   5.18 (±0.04) × 10³

1.59 (±0.009) × 10³   3.08 (±0.033) × 10³

TABLE IV

The Percentage Contribution of Each D-Galactose Moiety[a] to the Binding Energies
of Gal$_2$, Gal$_3$, and Gal$_4$ with Myeloma IgA Proteins X24 and J539[b]

Oligosaccharide	Residue no.			
	1	2	3	4
X24				
Gal$_2$	54.0	46.0		
Gal$_3$	38.7	33.0	28.3	
Gal$_4$	37.2	31.6	27.1	4.1
J539				
Gal$_2$	64.4	35.6		
Gal$_3$	50.5	27.9	21.6	
Gal$_4$	47.2	26.1	20.2	6.5

[a] Residue 1 is the terminal, nonreducing moiety.

[b] Reprinted, with permission, from Ref. 54, copyright by the American Chemical Society.

munoglobulins. Tables V and VI show the apparent, percentage contribution made by portions of Gal$_2$ to the binding of this hapten to proteins X24 and J539, as determined by their relative binding energies.[56b] Fig. 5 presents the data from the structural viewpoint.

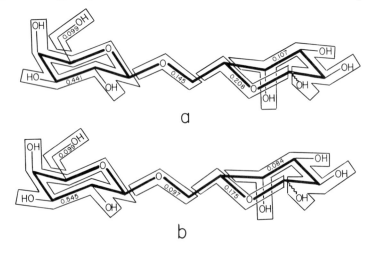

FIG. 5.—Fractional Contributions Made by Various Areas in Gal$_2$ to the Overall Binding of This Disaccharide to (a) Fab' X-24 and (b) Fab' J-539. Reprinted, with permission, from Ref. 54, copyright by the American Chemical Society.

(56b) Determined by comparison of structures and free-energy changes. The latter were determined from the expression $\Delta F = -RT \ln K_a$.

TABLE V

**Relative Contributions by Parts of the Disaccharide Gal₂ to Its
Overall Binding Energy with Protein X24[a]**

Ligand	Percentage contribution to binding energy of Gal₂	Portion of hapten	Percentage contribution to binding energy of Gal₂ by the indicated portion of hapten
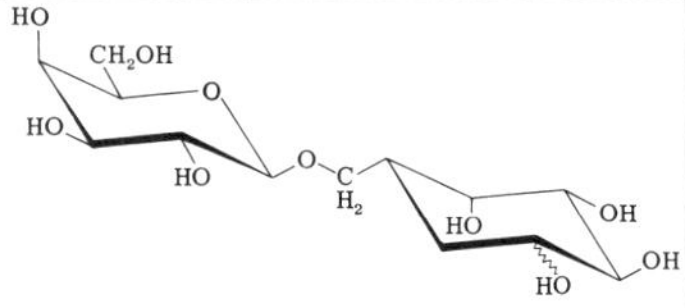	100		
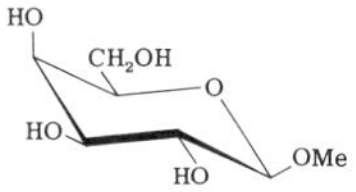	68.5		
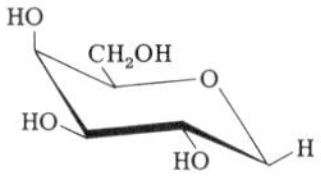	54.0	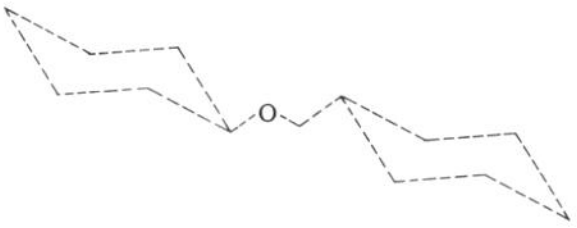	14.5
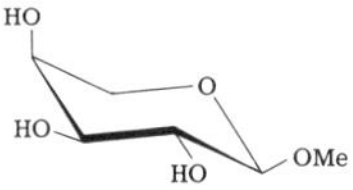	58.6	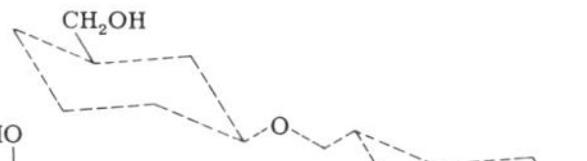	9.9
		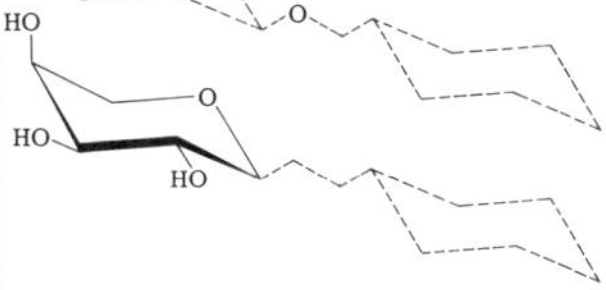	44.1
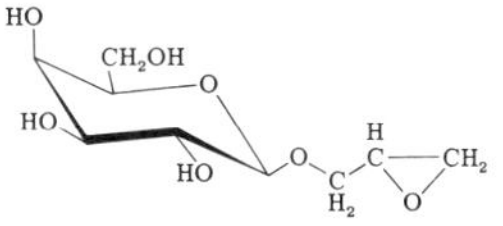	89.3	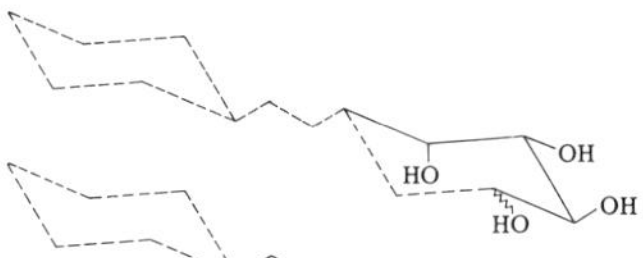	10.7
		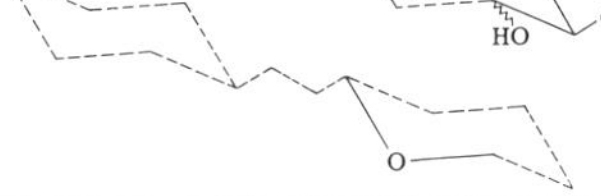	20.8

[a] Reprinted, with permission, from Ref. 54, copyright by the American Chemical Society.

The superior binding-energies of 2,3-epoxypropyl β-D-galactopyranoside and 2-methoxyethyl β-D-galactopyranoside (see Table III for formulas) compared to that of methyl β-D-galactopyranoside are, presumably, attributable to the presence of the epoxy and methoxyl oxygen atoms. By model building, it was determined that these oxygen atoms could take up the position of the ring oxygen atom in residue 2 of Gal₂. The increased binding-energies of these two compounds over that of methyl β-D-galactopyranoside is, presumably, not due to the increased bulk of the substituent at C-1, as isopropyl 1-

TABLE VI

Relative Contributions by Parts of the Disaccharide Gal₂ to Its Overall Binding Energy with Protein J539[a]

Ligand	Percentage contribution to binding energy of Gal₂	Portion of hapten	Percentage contribution to binding energy of Gal₂ by the indicated portion of hapten
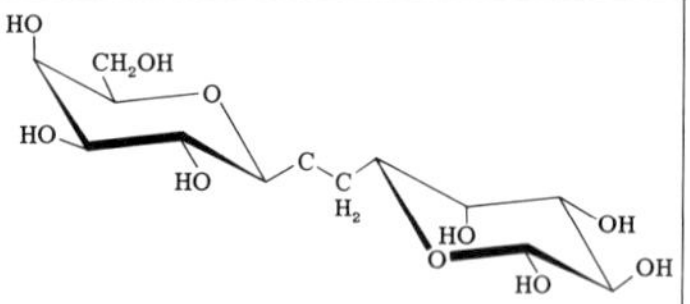 *(structure)*	100		
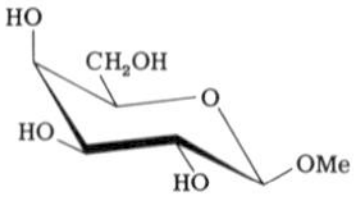 *(structure)*	74.1		
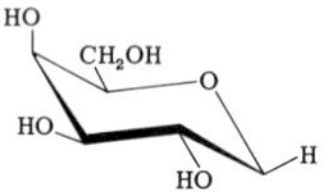 *(structure)*	64.4	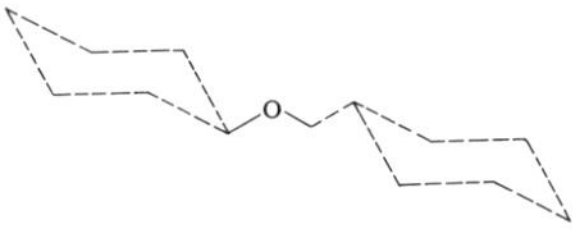 *(structure)*	9.7
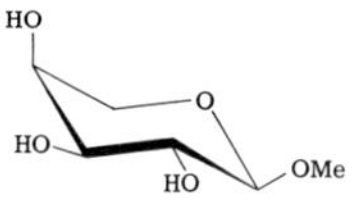 *(structure)*	64.2	 *(structure)*	9.9
		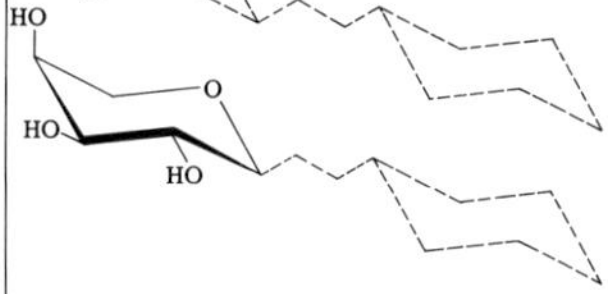 *(structure)*	54.5
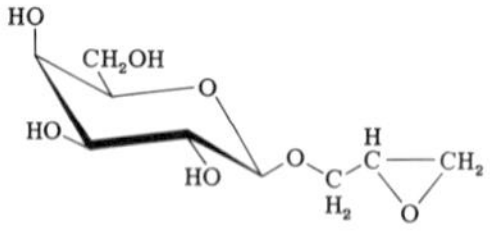 *(structure)*	91.6	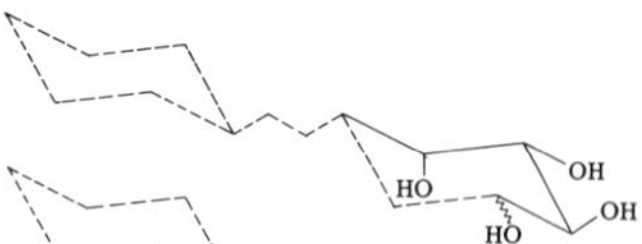 *(structure)*	8.4
		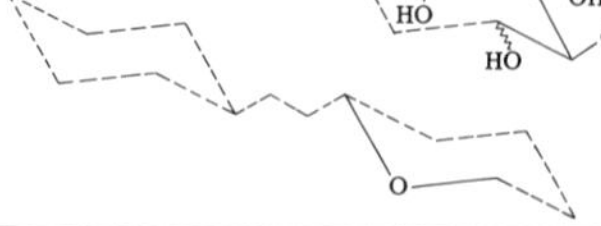 *(structure)*	17.5

[a] Reprinted, with permission, from Ref. 54, copyright by the American Chemical Society.

thio-β-D-galactopyranoside has a binding energy only slightly greater than that of methyl β-D-galactopyranoside. The higher binding of the 2,3-epoxypropyl glycoside compared to the 2-methoxyethyl could be caused by the lower degree of freedom of rotation of the former compound, resulting in a greater probability that the epoxy oxygen atom will be in a position similar to that occupied by the ring-oxygen atom of the "aglycon" moiety of Gal₂ at any instant in time. There is some freedom of rotation in the 2,3-epoxypropyl derivative, which, if les-

sened, might increase its association constants with the two immunoglobulins. Were this indeed so, it would indicate that the rest
of residue 2 contributes even less to the binding energy of Gal_2 than
the results suggest.

A number of points were noted with regard to Fig. 5. (1) For
both proteins X24 and J539, the series of atoms from the exocyclic,
C-1 oxygen atom to C-1 of residue 2 contributes significantly to the
binding energy of the whole Gal_2 molecule (27.2% for protein J539,
and 35.3% for protein X24). (2) The glycosidically bound 6-hydroxyl
group and its associated carbon atom contribute only approximately
10% in both cases to the total binding energy of Gal_2. (3) The second
residue (excepting the ring-oxygen atom and atoms C-1, C-5, and
C-6) contribute little to the binding energy of Gal_2 to both immunoglobulins (8.4% for protein J539, and 10.7% for protein X24).
(4) The ring-oxygen atom and hydroxyl groups of residue 1, excluding the C-6 hydroxyl group, provide a major contribution to the
binding energy of Gal_2 (54.5% for protein J539, and 44.1% for protein X24). (5) A conformation for Gal_2 that would avoid orbital
overlap between the intersaccharidic O, the O-ring (aglycon), and
O-4 (aglycon) is depicted in Figs. 5 and 6. If this conformation is correct, the region of greater binding is localized to one side of the Gal_2
molecule (see Fig. 6), indicating that the contact between amino
acids in the antibody combining-site and the carbohydrate occurs
along an extended surface-area complementary to the "underside" of
Gal_2.

It has been found[57] that the introduction of an O-acetyl group on
O-6 in methyl β-D-galactopyranoside in no way hinders the binding

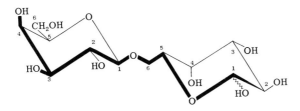

FIG. 6.—The Postulated Conformation of Gal_2. [The dark outline shows the bonds
most involved in binding to the IgA's (Fab' X-24 and Fab' J-539).] Reprinted, with
permission, from Ref. 54, copyright by the American Chemical Society.

(57) C. P. J. Glaudemans, E. Zissis, and M. E. Jolley, Carbohyd. Res., **40**, 129–135
(1975).

of this hapten to J539 Fab'. Moreover, when the Gal_2 conformation depicted in Figs. 5 and 6 was extended to Gal_4 (see Fig. 7), it was observed that residues 1, 2, and 3 all curved in the direction of that side of the tetrasaccharide which is the side accounting for most of the binding energy in Gal_2. The fourth D-galactose residue in Gal_4 projects away rather sharply from the surface of binding in this conformation. The proposed positioning of Gal_4 is consistent with the observation that the fourth saccharide moiety does not significantly improve the binding of the oligosaccharide to either immunoglobulin.[57a]

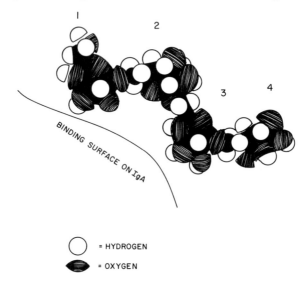

○ = HYDROGEN
● = OXYGEN

FIG. 7.—Proposed Conformation of Gal_4. [The suggested binding-surface of the immunoglobulin is also shown.] Reprinted, with permission, from Ref. 54, copyright by the American Chemical Society.

Polysaccharides containing immunodeterminant β-D-$(1 \rightarrow 6)$-linked D-galactosyl residues do occur in mammalian tissue such as has been isolated from beef.[53,58]

The group of proteins having anti-$(1 \rightarrow 6)$-β-D-galactan activity is particularly interesting, because the members exhibit a great deal of structural similarity, although each member clearly differs from the others. For the homogeneous, myeloma proteins obtained from tumors JPCl, SAPC 10, XRPC 24, XRPC 44, TEPC 191, and J 539,

(57a) Partially quoted, with permission, from Refs. 39 and 54, copyright by the American Chemical Society.
(58) M. L. Wolfrom, G. Sutherland, and M. Schlamowitz, *J. Amer. Chem. Soc.*, **74**, 4883–4886 (1952).

amino acid sequencing studies have shown that the light chains are identical to residue 23. The heavy chains from proteins S10, X44, T191, and J539 are identical through residue 30. Proteins J1 and X24 differ from this group at only a single position (a leucine-isoleucine interchange at position 5 in protein J1, and an unidentifiable residue at position 19 in X24).[41] However, when antisera were raised to each protein in the appropriate strain of mice, it was found that all of these immunoglobulins had unique idiotypic determinants.[41,54a] Thus, each myeloma immunoglobulin, despite having very similar amino terminal sequences, possessed an antigenic determinant(s) unique to the respective protein, indicating structural differences at points beyond the amino terminus.

3. Anti-dextrans

The anti-$(1 \rightarrow 3)$- and anti-$(1 \rightarrow 6)$-α-D-glucopyranans are combined in this group.

In 1965, McIntire and coworkers[59] reported an IgM mouse globulin (MOPC 104, now called MOPC 104E) that was subsequently found to bind dextrans.[60] (The protein has λ light chains, and not κ as reported earlier.) This protein is of historical interest, as it was the first homogeneous, murine immunoglobulin shown to bind a neutral, hydrophilic polysaccharide. Its specificity is not sharply defined; that is, the IgM myeloma protein binds dextran B 1254L, which has *no* $(1 \rightarrow 3)$-linkages [only $(1 \rightarrow 6)$ and $(1 \rightarrow 4) + (1 \rightarrow 2)$], as well as dextran B 1355 S1,3, which *does* have $(1 \rightarrow 3)$-linkages [also, $(1 \rightarrow 6)$ and $(1 \rightarrow 4) + (1 \rightarrow 2)$]. From these results, it would be expected that the binding between MOPC 104E and these dextrans would be mediated by $(1 \rightarrow 6)$, or $(1 \rightarrow 4)$, or $(1 \rightarrow 2)$ linkages. However, the binding is strongest for α-D-$(1 \rightarrow 3)$-linked D-glucopyranosyl residues,[60] as is evidenced by the fact that nigerose is a better inhibitor than kojibiose, maltose, or isomaltose in the inhibition of the agglutination of red cells, coated by either dextran B1254L or B1355 S1,3, by MOPC 104E. Extremely low concentrations of nigerose can inhibit the agglutination of B 1254L-coated red cells by the IgM, again showing the preference of the immunoglobulin site for $(1 \rightarrow 3)$-linked D-glucopyranosyl residues. Inhibition studies on the precipitin system MOPC 104E–dextran B1254L showed that nigerosyl-α-$(1 \rightarrow 4)$-nigerose or nigerosyl-α-$(1 \rightarrow 3)$-nigerose are the

(59) K. R. McIntire, R. M. Asofsky, M. Potter, and E. L. Kuff, *Science*, **150**, 361–363 (1965).

(60) M. A. Leon, N. M. Young, and K. R. McIntire, *Biochemistry*, **9**, 1023–1030 (1970).

best inhibitors.[61] In the work on hemagglutination,[60] it had been
shown that the MOPC 104E–dextran 1355 S1,3 system was much
harder to inhibit by nigerose (needing 30 times the concentration of inhibitor) than the system MOPC 104E–dextran B1254L. The
latter dextran lacks (1 → 3)-linkages, and this indicates that the protein preferably binds α-D-(1 → 3)-linked D-glucose residues. Thus,
nigerose readily displaces dextran B1254L. With dextran 1355 S1,3,
however, so many (1 → 3)-linkages occur in the polysaccharide
that nigerose cannot so effectively compete for the immunoglobulin,
such effective competition requiring a high concentration of inhibitor. These data again support the specificity of MOPC 104E as being
for α-D-(1 → 3)-linked D-glucose residues. In this work,[60] the authors
also found evidence that the immunoglobulin-combining region had
maximal complementarity for nigerotetraose. Equilibrium dialysis
employing [³H]nigerose-(1 → 3)-α-nigeritol gave[61] an association
constant of the MOPC 104E IgM for this ligand of 2.7×10^4, and a
valence of 10 combining sites (see Fig. 8).

A second protein having antidextran activity, J 558 (α,λ), precipitates well with dextrans containing high proportions of α-D-(1 → 3)-
linked D-glucopyranosyl residues.[49] Confirmation of this specificity
comes from the observation that oligosaccharides in the nigerose
series are excellent inhibitors. Kojibiose and isomaltose are very

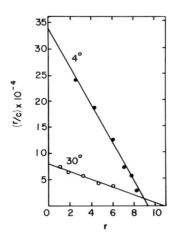

FIG. 8.—Equilibrium Dialysis of MOPC 104E IgM at 4° (●) and 30° (○) with
[³H]Nigerosyl-α-(1 → 3)-nigeritol. [Reprinted, with permission, from N. M. Young,
I. B. Jocius, and M. A. Leon, *Biochemistry*, **10**, 3457–3460 (1971), copyright by the
American Chemical Society.]

(61) N. M. Young, J. B. Jocius, and M. A. Leon, *Biochemistry*, **10**, 3457–3460 (1971).

poor inhibitors, but maltose is reasonably effective. From these results, it appears that the specificity of this anti-dextran is not sharply
defined, and that the protein is capable of binding with both α-D-
$(1 \rightarrow 3)$- and -$(1 \rightarrow 4)$-linked D-glucopyranosyl-containing polysaccharides. This interpretation is supported by the observation that J558
does precipitate with dextran B 1299 S3 [having no α-D-$(1 \rightarrow 3)$-link-
ages, only $(1 \rightarrow 6)$ and $(1 \rightarrow 4)$], but not with dextran B 512 [which has
no $(1 \rightarrow 4)$-linkages, but 95% of α-D-$(1 \rightarrow 6)$-linkages (see Fig. 9)].

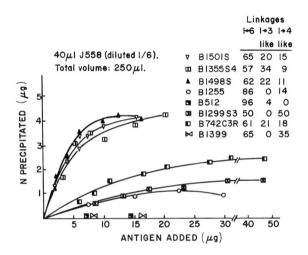

FIG. 9.—Quantitative Precipitin Curves of Mouse Myeloma Serum J558 with
Various Dextrans. [Reprinted, with permission, from A. Lundblad, R. Steller, E. A.
Kabat, J. W. Hirst, M. C. Weigort, and M. Cohn, *Immunochemistry*, 9, 535–543 (1972),
Pergamon Press.]

It may be seen[49] from Fig. 9 that all dextrans do not precipitate the
same amount of J 558 protein, and it may seem puzzling that a
homogeneous protein is only partly precipitated by the addition of
antigen. There are two possible explanations. One is that the J 558–B
1299 S3 precipitin system could eventually be able to precipitate
the same amount of antibody nitrogen as the J 558–B 1498S system, were high enough antigen concentrations used. The binding in
the former system may, however, be sufficiently weak to cause the
equilibrium in the following system to be displaced to the right. Addi-

$$(\text{J } 558\text{-B } 1299 \text{ S } 3) \rightleftharpoons (\text{J } 588) \text{ solution} \times (\text{B } 1299 \text{ S } 3) \text{ solution}$$
Precipitated gel

tion of more dextran should then, eventually, precipitate all of the IgA.

A more likely explanation is that dextran B 1299 S 3 does not precipitate all of the J 558, because together they form soluble complexes. The structure of the dextran may not be unique; that is to say,
there are populations of highly branched glucans among the total
average population. These highly branched molecules may have
some of their $(1 \to 3)$- or $(1 \to 4)$-linkages shielded in such a way that,
after binding one site of the IgA, steric hindrance makes it impossible for another dextran molecule to come sufficiently close to bind
to another site of that same molecule. Thus, there would be dextrans
in solution bound to IgA, and, hence, the precipitate would not account for all of the immunoglobulin; this could be experimentally
verified, as the supernatant liquor should then also contain soluble
antigen.

The idiotype of protein J 558 has been extensively studied,[62] and it
has been found to depend on the interaction between light and heavy
chains.[62a] The presence of ligands in the combining site inhibits the
interaction of J 558 and its anti-idiotypic antibody; this has also been
observed for other myeloma proteins. For example, the binding of
W 3129 [an IgA, κ, having anti-$(1 \to 6)$-α-dextran specificity] with its
anti-idiotypic antibody can be partially inhibited by isomaltose. Iso-
malto-triose, -tetraose, and -pentaose were significantly more effective than the disaccharide,[63] and no inhibition could be achieved by
maltose[63] or nigerose.[62] Thus, in these instances, the combining site
is clearly an idiotypic determinant.[63a]

Weigert and coworkers compared the idiotypes of three im-
munoglobulins having anti-$(1 \to 3)$-α-dextran activity,[63] namely,
IgA's UPC 102, J 558, and IgM MOPC 104E. Both MOPC 104E and
UPC 102 could completely inhibit the ^{125}I-labeled J 558–anti-J 558

(62) D. Carson and M. Weigert, *Proc. Nat. Acad. Sci. U. S.*, **70**, 235–239 (1973).
(62a) The anti-$(1 \to 3)$-α-dextrans J 558, UPC 102, and MPOC 104E are unusual in
that they possess λ light chains. In the mouse, only some 3% of the immunoglobulins have λ L-chains, and diversity is very limited. Hence, all λ L-
chains from IgA's are very similar (or, sometimes, identical), even if the specifi-
cities of the IgA's are unrelated. It is not surprising, then, that the idiotype may
depend on the interaction of V_L and V_H regions, but that the H-chains may contribute
most to the idiotype. See, for instance, M. Weigert, I. M. Cesari, S. J. Yonkovich,
and M. Cohn, *Nature*, **228**, 1045–1047 (1970); I. M. Cesari and M. Weigert, *Proc.
Nat. Acad. Sci. U. S.*, **70**, 2112–2116 (1973).
(63) M. Weigert, W. C. Raschke, D. Carson, and M. Cohn, *J. Exp. Med.*, **139**, 137–147
(1974).
(63a) For further discussion, see the Addendum on p. 346.

idiotype reaction, but less effectively than J 558 itself. It is interesting that both MOPC 104E and UPC 102 competed equally well for the anti-J 558 antibody. This cross-reaction between J 558, on the one hand, and MOPC 104E and UPC 102, on the other, for anti-J 558 antibody indicates that 104E and 102 are structurally very similar, and that all three proteins share some idiotypic determinants in common and thus probably have a considerable degree of structural similarity.

For three other myeloma proteins having anti-$(1 \rightarrow 6)$-α-D-glucopyranan specificity (W 3129, W 3434, and QUPC 52), the authors found a somewhat different pattern of cross-reactivity.[63] The reaction of W 3129 with its anti-idiotypic antibody was not totally inhibitable by even high concentrations of W 3434 protein, and not at all by protein QUPC 52. Thus, proteins W 3129 and W 3434 share some determinants, but are clearly different from each other and from QUPC 52.

Cisar and coworkers[47] studied dextran–antidextran immunoglobulin systems, employing inhibition by oligosaccharides as a means of exploring specificity. For protein W 3129 [anti-$(1 \rightarrow 6)$-α-D-glucan specificity], isomaltopentaose was found to be the best inhibitor of its precipitation by dextran B 512, no improvement being noted on employing either the corresponding hexaose or heptaose. This was confirmed by determination of the interactions of isomaltose oligosaccharides with protein W 3129 by measuring the quenching of its native tryptophan fluorescence upon binding ligand.[63a] Thus, the combining site may be maximally occupied by the pentaose. Branched derivatives of the hapten caused what appeared to be partial expulsion of the ligand from the immunoglobulin site, starting at the nonreducing end. Thus 3^4-α-D-glucosyl-isomaltohexaose (**1**) had the same inhibitory power as isomaltotriose (or tetraose). It has been suggested that the nonreducing end of an

$$\alpha\text{-D-Glc}p\text{-}(1\rightarrow 6)\text{-}\alpha\text{-D-Glc}p\text{-}(1\rightarrow 6)\text{-}\alpha\text{-D-Glc}p\text{-}(1\rightarrow 6)\text{-}\alpha\text{-D-Glc}p\text{-}(1\rightarrow 6)\text{-}\alpha\text{-D-Glc}p\text{-}(1\rightarrow 6)\text{-D-Glc}$$
$$3$$
$$\uparrow$$
$$1$$
$$\alpha\text{-D-Glc}p$$

1

oligosaccharide contributes more to the overall binding-energy than the other residues.[64] For oligosaccharide **1** to show binding equal to

(63a) E. A. Kabat, personal communication.
(64) E. A. Kabat, "Structural Concepts in Immunology and Immunochemistry," Holt, Rhinehart and Winston, New York, 1968, p. 87.

that of isomaltotriose, it seems reasonable to assume that this binding arises from the first three residues at the reducing end. The alternative would have to be that the bulky branch-point would not hinder the binding of the first three residues at the nonreducing end, and this would appear unlikely.

The inhibition of the precipitation of protein UPC 102 by dextran B 1355S4 was more effectively achieved by nigerose than by maltose, kojibiose, or isomaltose,[47] showing, again, the anti-$(1 \rightarrow 3)$-α-D-glucan specificity of this protein.[60] An interesting observation made by Cisar and coworkers[47] was that the monomeric form of the anti-dextran IgA is more readily inhibited from precipitating with its antigen than the polymeric form. It is unlikely that this reflects a difference in binding strength between the forms; more probably, it is a result of the polymeric form's being a superior crosslinking agent in the precipitation reaction.

4. Anti-pneumococcal C Polysaccharides

Eleven IgA myeloma proteins that precipitate with Pneumococcus C polysaccharide (PnC) have been described.[65-69] This antigen has a structure whose general features are known,[70,71] and it is a somatic, species-specific polysaccharide.

$$
\begin{array}{c}
\text{O} \\
\| \\
\text{H}_2\text{C}-\text{O}-\text{P}-\text{O}- \\
| \qquad\qquad | \\
\text{HOCH} \qquad \text{ONa} \\
| \\
\text{HOCH} \\
|
\end{array}
$$

$\rightarrow (3,4)$-D-GalNAc-$(1 \rightarrow 6)$-D-Glc-$(1 \rightarrow 3)$-trideoxy-NAc-hexosamine-O$-$CH

$$
\begin{array}{c}
\text{O} \qquad \text{CH}_2 \\
\| \qquad / \\
\overset{+}{\text{Me}_3\text{NCH}_2\text{CH}_2\text{O}}-\text{P}-\text{O} \\
| \\
\text{O}^-
\end{array}
$$

Leon and Young found that the specificity of anti-pneumococcal C myeloma immunoglobulins is for the phosphorylcholine or choline determinant of this antigen.[72] Many other antigens, such as polysaccharides from groups O and H streptococcus,[66] Ascaris,[73] Lac-

(65) M. Cohn, Cold Spring Harbor Symp. Quant. Biol., 32, 211–221 (1967).
(66) M. Potter and M. Leon, Science, 162, 369–371 (1968).
(67) M. Cohn, G. Notani, and S. A. Rice, Immunochemistry, 6, 111–123 (1969).
(68) M. Potter and R. Lieberman, J. Exp. Med., 132, 737–751 (1970).
(69) A. Sher, E. Lord, and M. Cohn, J. Immunol., 107, 1226–1234 (1971).
(70) D. E. Brundish and J. Baddiley, Biochem. J., 105, 30C–31C (1967).
(71) D. E. Brundish and J. Baddiley, Biochem. J., 110, 574–582 (1968).
(72) M. Leon and N. M. Young, Biochemistry, 10, 1424–1429 (1971).
(73) M. Potter, Fed. Proc., 29, 85–91 (1971).

tobacillus acidophilus antigen,[68] and some fungi[5] (all of which contain choline or phosphorylcholine) have been shown to precipitate with these proteins. An antigen obtained from *Proteus morganii* precipitates only[5] with McPC 603. It is possible that the latter antigen contains a phosphorylcholine-like group, as it was found that the McPC 603–PnC precipitin system is most effectively inhibited by phosphorylcholine.[69] Strictly speaking, the haptenic determinant, then, is not a carbohydrate. The structural diversity of some of these homogeneous immunoglobulins has been demonstrated by a study of their idiotypy.[68] Three globulins (MOPC 167, MOPC 511, and McPC 603) have unique idiotypes. The other proteins (HOPC 8, TEPC 15, MOPC 299, S 63, and S 107) all showed idiotypic identity; that is, an antiserum made to any one of the proteins in this group reacted equally well with all members, but not with MOPC 167, MOPC 511, or McPC 603. The observation of shared idotypy suggests that these proteins are structurally very similar, if not identical.

McPC 603 is a key myeloma protein, as it is the first homogeneous immunoglobulin with known specificity to be crystallized.[74] Crystallographic data on the structure of the Fab′ fragment from this protein at a resolution of 4.5 Å have been reported.[16] The molecule consists of four, clearly defined regions of electron density corresponding to the four domains (C_L, C_H, V_L, and V_H) in the Fab′ fragment. The hapten-binding site is located in a cleft that extends through most of the variable region. Subsequently, the same authors[75] showed (data from 3.1-Å resolution) that the cleft in which the hapten is situated is formed by loops comprised of the three hypervariable regions of the heavy chain and two of the hypervariable regions of the light chain. (Hypervariable regions are those regions of the L and H chain variable-segments having an unusually high incidence of variability in their amino acid composition. It has been suggested that these hypervariable regions contribute particularly to the formation of the combining site.[76]) The predominant, spe-

(74) S. Rudikoff, M. Potter, D. M. Segal, E. A. Padlan, and D. R. Davies, *Proc. Nat. Acad. Sci. U. S.*, **69**, 3689–3693 (1972).

(75) (a) E. A. Padlan, D. M. Segal, G. H. Cohen, D. R. Davies, S. Rudikoff, and M. Potter, in "The Immune System," E. E. Sercarz, A. R. Williamson, and C. F. Cox, eds., Academic Press, New York, 1974, pp. 7–14. (b) D. M. Segal, E. A. Padlan, G. H. Cohen, S. Rudikoff, M. Potter, and D. R. Davies, *Proc. Nat. Acad. Sci. U.S.*, **71**, 4298–4302 (1974).

(76) T. T. Wu and E. A. Kabat, *J. Exp. Med.*, **132**, 211–250 (1970).

cific interactions appear to occur between the hapten and the heavy chain. The phosphate group is in close proximity to both the tyrosine residue at position 33 and the arginine residue at position 52 to which two of the oxygen atoms are apparently hydrogen-bonded. The choline group is involved in Van der Waals interactions with residues in the third hypervariable region of the heavy chain. From the light chain, only the third hypervariable region appears close enough to the ligand to provide points of interaction.

It is clear that phosphorylcholine fills only a small portion of the cleft, and that there would be ample room for a polysaccharide containing side chains of phosphorylcholine (such as PnC) to approach the combining region (see Fig. 10).

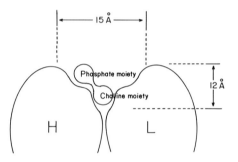

FIG. 10.–Schematized Combining Region of McPC 603, Showing the General Location of Phosphorylcholine. [From E. A. Padlan, D. M. Segal, G. H. Cohen, D. R. Davies, S. Rudikoff, and M. Potter.[75(a)]]

That the H-chain seems to account for a major part of the binding with the phosphorylcholine ligand in McPC 603 is consistent with the sequence data obtained for several anti-phosphorylcholine proteins,[77] including HOPC 8, TEPC 15, S 107, McPC 603, and MOPC 167. Heavy chains from all five proteins have identical amino acid sequences through the first hypervariable region to residue 36, except for position 4 of MOPC 167. For the proteins that share idiotypic determinants (TEPC 15, HOPC 8, and S 107), it was found that the L-chains, also, are identical through the first hypervariable region. Proteins having unique, idiotypic determinants (McPC 603, and MOPC 167) have differing L-chain sequences. This conservation of the heavy chain is particularly interesting in that, as already

(77) P. Barstad, S. Rudikoff, M. Potter, M. Cohn, W. Konigsberg, and L. Hood, Science, 183, 962–964 (1970).

pointed out, the major specific points of contact are between the hapten and the heavy chain. In contrast to this work, Chesebro and Metzger[78] found, by affinity labeling, that reaction with *p*-diazoniumphenylphosphorylcholine labeled the L-chain (tyrosine) of TEPC 15 almost exclusively (99%). Actually, the reactive group of this affinity label is at the end of the phosphorylcholine molecule, and it could project outward[75] from the H-chain of the immunoglobulin in McPC 603 to interact with parts of the L-chain of the globulin that are quite remote from the combining region.

Association constants between a number of these anti-PnC myeloma proteins and phosphorylcholine have been determined[79] by equilibrium dialysis and found to range from 10^4 to 10^5 liter.mol^{-1}. Values of K_a obtained by fluorescence titration have been found to be in close agreement.[80,81]

5. Anti-2-acetamido-2-deoxy-D-hexoses

Two myeloma proteins have been reported that have activity to 2-acetamido-2-deoxy-D-hexoses. Protein S 117 has specificity for 2-acetamido-2-deoxy-β-D-glucopyranosyl residues,[82] whereas MOPC 406 has specificity for 2-acetamido-2-deoxy-β-D-mannopyranosides.[73] Protein S 117 precipitates with streptococcal group A polysaccharide, with β-teichoic acid, and with partially degraded blood-group H substance. The disaccharide 2-acetamido-2-deoxy-β-D-glucopyranosyl-(1 → 3)-D-galactose is a good inhibitor of this reaction, and the protein also precipitates with conjugates containing 2-acetamido-2-deoxy-β-D-glucopyranosyl residues. After periodate oxidation and Smith degradation, blood-group H substance was found to be the best precipitant for S 117, but only 50% of the immunoglobulin was precipitated at equivalence. When the same polysaccharide was used as an insoluble immunoadsorbent, 81% of the immunoglobulin was adsorbed, showing that a substantial proportion of the protein was not precipitated by the polysaccharide, although it does bind.[82] It was suggested that IgA monomers would bind, but not cause precipi-

(78) B. Chesebro and H. Metzger, *Biochemistry*, **11**, 766–771 (1972).
(79) H. Metzger, B. Chesebro, N. M. Hadler, J. Lee, and N. Otchin, *Prog. Immunol.*, *Proc. Intern. Congr. Immunol. 1st.*, 253–267 (1971).
(80) R. Pollet and H. Edelhoch, *J. Biol. Chem.*, **248**, 5443–5447 (1973).
(81) R. Pollet, H. Edelhoch, S. Rudikoff, and M. Potter, *J. Biol. Chem.*, **249**, 5188–5194 (1974).
(82) G. Vicari, A. Sher, M. Cohn, and E. A. Kabat, *Immunochemistry*, **7**, 829–838 (1970).

tation, and thus, in effect, act as competitive inhibitors for the precipitation of S 117 polymers and polysaccharide.

An anti-2-acetamido-2-deoxy-D-mannose immunoglobulin, MOPC 406, is an IgA that binds[73] polysaccharides from *Salmonella weslaco* and *Escherichia coli* 031. These polysaccharides are known to contain 2-acetamido-2-deoxy-D-mannose,[83] and it has been shown that the specificity of this protein is directed towards β-D-linked 2-acetamido-2-deoxy-D-mannopyranosyl residues.[84] Inhibition studies revealed that the apparent affinity of the methyl β-D-glycoside for the combining region of MOPC 406 is high.

6. Anti-α-D-galactopyranosides

As may be seen from Table I (see p. 327), MOPC 384 precipitates with the lipopolysaccharides from *Proteus mirabilis* sp2, *Salmonella tranaroa*, *Escherichia coli* 070, and *Salmonella telaviv*.[73] Precipitation with the latter polysaccharide can be inhibited with methyl α-D-galactopyranoside. No inhibition could be achieved by using *p*-aminophenyl 1-thio-α-D-galactopyranoside,[85] but the specificity of this protein nevertheless appears to be for α-D-linked D-galactopyranosyl residues.

VII. ADDENDUM[86]

Recombinants of J 558 H-chains and various L-chains have been studied.[62] $H^{J\ 558}\ L^{S\ 104}$ was indistinguishable from protein J 558 by inhibition studies. This is not surprising, as the S 104 L-chain has the same amino acid sequence as the J 558 L-chain.[62a] Subtle differences in the L-chain amino acid sequence were found to have no bearing on the idiotype; for example, recombinant $H^{J\ 558}\ L^{S\ 176}$ was indistinguishable from protein J 558, although L S 176 has an asparagine instead of a serine residue at position 25. Even two amino acids could be changed in the L-chain without effect. However, a λ L-chain having three amino acids (positions 25, 52, and 97) different from $L^{J\ 558}$ formed a recombinant with $H^{J\ 558}$ that had a significantly altered idiotype when compared to protein J 558.

(83) O. Lüderitz, J. Gmeinev, B. Kickhoffen, H. Mager, O. Westphal, and R. W. Wheat, *J. Bacteriol.*, **95**, 490–494 (1968).

(84) L. Rovis, E. A. Kabat, and M. Potter, *Carbohyd. Res.*, **23**, 223–227 (1972).

(85) M. E. Jolley, C. P. J. Glaudemans, and M. Potter, unpublished results.

(86) See p. 340.

BIBLIOGRAPHY OF CRYSTAL STRUCTURES OF CARBOHYDRATES, NUCLEOSIDES, AND NUCLEOTIDES 1973

By George A. Jeffrey* and Muttaiya Sundaralingam**

Department of Crystallography, University of Pittsburgh, Pittsburgh, Pennsylvania 15260; and Department of Biochemistry, College of Agricultural and Life Sciences, University of Wisconsin, Madison, Wisconsin 53706

I. Introduction

The format is the same as for the bibliography[1] for 1970–1972. For tetrahydrofuran and 1,4-dioxolane rings, the conformational descriptor, E or T, is followed (in parentheses) by the value of the pseudo-torsional phase-angle, P, defined with P = 0° for ${}^{3}_{2}T$ for ribofuranoses (as in Fig. 3 of Ref. 2), ${}^{4}_{3}T$ for fructofuranoses, and ${}^{2}_{1}T$ for 1,4-dioxolane rings.

For the preliminary communications (Section IV,2), the conformation of the furanoid sugar is denoted by the atom most displaced

* Work supported by NIH Grant GM-11293. Present address: Department of Chemistry, Brookhaven National Laboratory, Upton, Long Island, New York 11973.

** Work supported by NIH Grant GM-17378.

(1) G. A. Jeffrey and M. Sundaralingam, *Advan. Carbohyd. Chem. Biochem.*, **30**, 445–466 (1974).

(2) C. Altona and M. Sundaralingam, *J. Amer. Chem. Soc.*, **94**, 8205–8212 (1972).

from the average ring-plane; the glycosyl disposition, by *anti* or *syn;* and the exocyclic, C-4′–C-5′ torsion-angle, by g^-, g^+, or *t*. The latter angles are with reference to the ring-oxygen atom, O-4′–C-4′–C-5′–O-5′, and not with reference to the ring (backbone) carbon atom C-3′–C-4′–C-5′–C-5′–O-5′. Exact values of these conformational parameters will be supplied when the atomic coordinates are published.

II. Data for Carbohydrates

$C_5H_{10}O_5 \cdot CaCl_2 \cdot 3H_2O$ α-D-Xylopyranose · calcium chloride, trihydrate[3]

$P2_1$; Z = 4; D_x = 1.51; R = 0.079 for 1,240 intensities (film measurements at −193°). The α-D-xylopyranose molecule is in the $^1C_4$ conformation, as in the crystal structure of α-D-xylopyranose. The calcium atom is seven-coordinated by a distorted, pentagonal bipyramid of oxygen atoms consisting of three from water molecules and two from a hydroxyl group of each of two sugar molecules. The ring-oxygen atom is excluded from this first coordination shell.

$C_6H_8O_6$ D-Isoascorbic acid; D-*arabino*-ascorbic acid[4]

Correction of typographical errors in an atomic parameter and in structural data in a previous paper.[5] In Table I thereof, the *x* parameter of O-6 should be 0.1292(6), not 0.2192(6). In Table III, the hydrogen-bonding distance $d(jk)$ of O-2–H → O-6a should be 1.76 Å (176 pm) not 176 Å, and the torsion angle O-6–C-6–C-5–O-5 should be 67.2°, not 70.7°.

$C_6H_{10}O_5$ 2,6-Anhydro-β-D-fructofuranose[6]

$P2_12_12_1$; Z = 8; D_x = 1.58, R = 0.028 for 1,481 intensities. Both the furanose and the anhydride ring have an envelope conformation, ^0E (96°) and $_{0-2}E$ (54°), respectively, with the same oxygen atom, O-2, displaced. The conformation of the pyranoid moiety is $^3H_4$. The hydrogen bonding involves both hydroxyl groups and is intermolecular.

$C_6H_{12}O_4S$ Methyl 1-thio-α-D-ribopyranoside,[7] m.p. 155–156°C

(3) G. F. Richards, *Carbohyd. Res.*, **26**, 448–449 (1973).
(4) N. Azarnia, H. M. Berman, and R. D. Rosenstein, *Acta Crystallogr.*, **B29**, 1170 (1973).
(5) N. Azarnia, H. M. Berman, and R. D. Rosenstein, *Acta Crystallogr.*, **B28**, 2157–2161 (1972).
(6) W. Dreissig and P. Luger, *Acta Crystallogr.*, **B29**, 1409–1416 (1973).
(7) R. L. Girling and G. A. Jeffrey, *Acta Crystallogr.*, **B29**, 1006–1011 (1973).

$P2_12_12_1$; $Z = 4$; $D_x = 1.472$; $R = 0.028$ for 830 intensities. The pyranoside has the $^1C_4$ conformation, with an intramolecular, *syn*-axial hydrogen-bond between O-2–H and O-4. The absolute configuration was confirmed.

$C_6H_{12}O_4S$ Methyl 5-thio-α-D-ribopyranoside,[8] m.p. 74–76°C

C2; $Z = 4$; $D_x = 1.484$; $R = 0.03$ for 681 intensities. The conformation of the pyranoside is $^4C_1$. There is no intramolecular hydrogen-bond, and the ring is slightly strained by the *syn*-axial repulsion between O-1 and O-3.

$C_6H_{12}O_4S$ Methyl 5-thio-β-D-ribopyranoside,[8] m.p. 127–129°C

C2; $Z = 4$; $D_x = 1.474$; $R = 0.04$ for 693 intensities. The conformation of the pyranoside is $^4C_1$. All hydrogen-bonding is intermolecular.

$C_6H_{12}O_6 \cdot H_2O$ α-D-Glucose monohydrate[9]

$P2_1$; $Z = 2$; $D_x = 1.512$; $R = 0.030$ for 906 intensities. Refinement of a previous determination.[10]

$C_6H_{12}O_6 \cdot CaBr_2 \cdot 3H_2O$ α-D-Galactopyranose · calcium bromide, trihydrate[11]

$P2_12_12_1$; $Z = 4$; $D_x = 1.961$; $R = 0.048$ (number of intensities not reported). The conformation of the pyranose is $^4C_1$, although the most symmetrical chair-form is the $^3C_0$ conformation. The calcium ions are eight-coordinated by a distorted anti-prism of five hydroxyl groups belonging to three D-galactopyranose molecules and three water molecules. Hydrogen bonding is from D-galactopyranose to bromide ions or to water molecules. The absolute configuration was confirmed.

$C_6H_{12}O_6 \cdot CaBr_2 \cdot 5H_2O$ *myo*-Inositol · calcium bromide, pentahydrate[12]

$P\bar{1}$; $Z = 2$; $D_x = 1.910$; $R = 0.039$ for 2,724 intensities. The Ca^{2+} is eight-coordinated by a distorted, square anti-prism of oxygen atoms from two inositol and four water molecules. The *myo*-inositol mole-

(8) R. L. Girling and G. A. Jeffrey, *Acta Crystallogr.*, **B29**, 1102–1111 (1973).
(9) E. Hough, S. Neidle, D. Rogers, and P. G. H. Troughton, *Acta Crystallogr.*, **B29**, 365–366 (1973).
(10) R. C. G. Killean, W. G. Ferrier, and D. W. Young, *Acta Crystallogr.*, **15**, 911–912 (1962).
(11) W. J. Cook and C. E. Bugg, *J. Amer. Chem. Soc.*, **95**, 6442–6446 (1973).
(12) W. J. Cook and C. E. Bugg, *Acta Crystallogr.*, **B29**, 2404–2411 (1973).

cule is slightly distorted, with ring torsion-angles ranging from 51 to 60°.

$C_6H_{12}O_6 \cdot MgCl_2 \cdot 4H_2O$ *myo*-Inositol · magnesium chloride, tetrahydrate[13]

$P2_1/c$; $Z = 4$; $D_x = 1.56$; $R = 0.073$ for 2,433 intensities. The Mg^{2+} ion is octahedrally coordinated by four water molecules and two vicinal hydroxyl groups, leading to small distortions of the torsion angles of the cyclohexane ring, 50–60°. The Cl^- ions are four- and five-coordinated by water and hydroxyl oxygen atoms. There is also extensive, intermolecular hydrogen-bonding.

$C_9H_{16}O_6$ 2,3,6-Tri-*O*-methyl-D-galactono-1,4-lactone[14]

$P2_12_12_1$; $Z = 4$; $D_x = 1.272$; $R = 0.10$ for 1,457 intensities (film measurements at 120 K). The lactone ring is unusual, as it has the E_3 (196°) conformation. The acyclic chain has an extended conformation.

$C_{10}H_{10}Cl_2O_4$ Methyl 2,6-dichloro-2,6-dideoxy-3,4-*O*-isopropylidene-α-D-altropyranoside[15]

$P2_12_12_1$; $Z = 4$; $D_x = 1.365$; $R = 0.047$ for 736 intensities. The pyranoside has a distorted $^2S_0$ or $B_{2,5}$ conformation; the dioxolane ring is $_{0-4}E$ (52°), such that all substituents are equatorially or quasi-equatorially attached.

$C_{11}H_{16}O_8 \cdot H_2O$ Ranuncoside monohydrate;[16] 1,2-*O*-[2-(*S*)-2-(2-oxotetrahydro-5-furylidene)]ethylene-α-D-glucopyranose monohydrate

$P2_1$; $Z = 2$; $D_x = 1.493$; $R = 0.051$ for 1,161 intensities. The configuration assigned to the natural product was confirmed. The conformation consists of two fused chairs, $^4C_1$, and $^1C_4$, with a nonplanar lactone ring having the ^3_2T (0°) conformation.

$C_{11}H_{19}NO_9 \cdot 2H_2O$ Sialic acid dihydrate; *N*-acetylneuraminic acid dihydrate; 5-acetamido-3,5-dideoxy-β-D-*glyc-ero*-D-*galacto*-nonulopyranosonic acid dihydrate[17]

(13) G. Blank, *Acta Crystallogr.*, **B29**, 1677–1683 (1973).
(14) B. Sheldrick, *Acta Crystallogr.*, **B29**, 2631–2632 (1973).
(15) G. H. Lin, M. Sundaralingam, and J. Jackobs, *Carbohyd. Res.*, **29**, 439–449 (1973).
(16) R. A. Mariezcurrena and S. E. Rasmussen, *Acta Crystallogr.*, **B29**, 1030–1035 (1973).
(17) J. L. Flippen, *Acta Crystallogr.*, **B29**, 1881–1885 (1973).

$P2_1$; $Z = 2$; $D_x = 1.42$; $R = 0.049$ for 1,400 intensities. The β-D configuration was confirmed. The pyranose has the $^2C_5$ conformation, with all substituent groups equatorially attached, except the 2-hydroxyl group. All hydrogen-bonding is intermolecular.

$C_{12}H_{20}O_6$ 1,2:4,5-Di-O-isopropylidene-β-D-fructopyranose,[18] m.p. 119°C

$P2_1$; $Z = 2$; $D_x = 1.288$; $R = 0.047$ for 1,281 intensities. That the compound is β-D was confirmed. The pyranose has the $^5C_2$ conformation partially flattened at C-5. Both isopropylidene rings are envelopes, ^5E (25°) and $_{0-1}E$ (118°). The intermolecular hydrogen-bonding, involving the hydroxyl group and O-4, forms a polar helix.

$C_{12}H_{21}NO_9 \cdot H_2O$ Sialic acid, methyl ester, monohydrate; N-acetylneuraminic acid, methyl ester, monohydrate; methyl 5-acetamido-3,5-dideoxy-β-D-*glycero*-D-*galacto*-nonulopyranosonate monohydrate[19]

$P2_12_12_1$; $Z = 4$; $D_x = 1.428$; $R = 0.042$ for 2,081 intensities. The crystal contained 86 percent of the ester and 14 percent of the free acid. The pyranose has the $^2C_5$ conformation, with major substituents equatorially attached and the anomeric hydroxyl group axially attached. The methyl ester and acetamido groups and the carbon chain of the *glycero* group are almost planar. There is an intramolecular hydrogen-bond from a *glycero* hydroxyl group to the acetoxyl oxygen atom.

$C_{12}H_{21}O_{12} \cdot CaBr_2 \cdot 4H_2O$ Calcium bromide lactobionate tetrahydrate; calcium bromide 4-O-β-D-galactopyranosyl-D-gluconate tetrahydrate[20]

$P2_12_12_1$; $Z = 4$; $D_x = 1.75$; $R = 0.058$ for 1,981 intensities. The conformation of the D-galactopyranosyl group is $^4C_1$; the D-gluconate residue has a sickle, chain conformation. The linkage angles* are −70°, +131° (Gal → Glc). The calcium ion is eight-coordinated to three water molecules, one carboxylate oxygen atom, and four hydroxyl oxygen atoms of the gluconate in a distorted-square anti-prism.

(18) S. Takagi, R. Shiono, and R. D. Rosenstein, *Acta Crystallogr.*, **B29**, 1177–1186 (1973).
(19) A. M. O'Connell, *Acta Crystallogr.*, **B29**, 2320–2328 (1973).
(20) W. J. Cook and C. E. Bugg, *Acta Crystallogr.*, **B29**, 215–222 (1973).
   * Linkage bonds are defined with reference to the atoms having the highest Klyne–Prelog priority,[21] *i.e.*, O–C–O–C–C(CHOH)$_2$CO$_2^-$.
(21) W. Klyne and D. Prelog, *Experientia*, **16**, 521–523 (1960).

$C_{12}H_{22}O_{11}$ Sucrose; β-D-fructofuranosyl α-D-glucopyranoside; α-D-glucopyranosyl β-D-fructofuranoside[22,23]

$P2_1$; $Z = 2$; $D_x = 1.590$; $R = 0.029$ for 3,280 intensities (3 sets of data from 2 different crystals)[22]; $R = 0.033$ for 2,813 intensities by neutron diffraction (3 sets of data from 3 different crystals).[23] Refinement of a previous refinement.[24] This is the most accurate carbohydrate structure-determination performed to date; standard errors C–C and C–O, 0.0012–0.0019 Å (120–190 fm); C–H, 0.0022–0.0033 Å (220–300 fm); O—H, 0.0028–0.0042 Å (280–420 fm). Linkage bonds (Glc → Fru) + 107.8°, −44.7°. Intramolecular H-bonds with H⋯O, 1.851 Å (185.1 pm), 1.895 Å (189.5 pm); O–H⋯O, 158°, 167°.

$C_{12}H_{22}O_{11} \cdot H_2O$ Isomaltulose monohydrate; palatinose monohydrate; 6-O-α-D-glucopyranosyl-α-D-fructofuranose monohydrate[25]

$P2_12_12_1$; $Z = 4$; $D_x = 1.53$; $R = 0.039$ for 1,706 intensities. The conformation of the D-glucopyranosyl group is $^4C_1$, and that of the D-fructofuranose residue is $^4T_3$ (12°). There is an intramolecular hydrogen-bond between O-2 of Glc and O-2 of Fru. The linkage bonds are (Glc → Fru) +77°, −65°.

$C_{12}H_{22}O_{11} \cdot CaBr_2 \cdot 7H_2O$ Lactose · calcium bromide, heptahydrate; 4-O-β-D-galactopyranosyl-D-glucopyranose · calcium bromide, heptahydrate[26]

$P2_12_12_1$; $Z = 4$; $D_x = 1.678$; $R = 0.043$ for 2,905 intensities. The crystal structure is a mixture of the α and β anomers in the ratio of 22:3. The linkage bonds are −77°, +106° (Gal → Glc). There is an intramolecular hydrogen-bond O-3′–H⋯O-5. The calcium ions are eight-coordinated to four hydroxyl groups of two lactose molecules and four water molecules.

$C_{12}H_{22}O_{11} \cdot CaCl_2 \cdot 7H_2O$ Lactose · calcium chloride, heptahydrate[27]

(22) J. C. Hanson, L. C. Sieker, and L. H. Jensen, *Acta Crystallogr.*, **B29**, 797–808 (1973).
(23) G. M. Brown and H. A. Levy, *Acta Crystallogr.*, **B29**, 790–797 (1973).
(24) G. M. Brown and H. A. Levy, *Science*, 141, 921–923 (1963).
(25) W. Dreissig and P. Luger, *Acta Crystallogr.*, **B29**, 514–521 (1973).
(26) C. E. Bugg, *J. Amer. Chem. Soc.*, **95**, 908–913 (1973).
(27) W. J. Cook and C. E. Bugg, *Acta Crystallogr.*, **B29**, 907–909 (1973).

$P2_12_12_1$; $Z = 4$; $D_x = 1.509$; $R = 0.046$ for 2,386 intensities. The structure studied contained 95 percent of the α and 5 percent of the β anomer. The linkage bonds are $-76°$, $+108°$ (Gal $\rightarrow$ Glc). There is an intramolecular hydrogen-bond between O-3-H and O-5.

$C_{12}H_{26}O_4S_3$ 2-S-Ethyl-2-thio-D-mannose diethyl dithioacetal[28]

$P2_1$; $Z = 4$; $D_x = 1.25$; $R = 0.065$ for 2,585 intensities. The molecules have the extended carbon-chain conformation. The S-2 atom is gauche to both S-1 atoms.

$C_{13}H_{18}O_9$ 1,2,3,5-Tetra-O-acetyl-β-D-ribofuranose[29]

$P2_12_12_1$; $Z = 4$; $D_x = 1.354$; $R = 0.043$ (number of intensities not stated). The conformation of the molecule is $^3T_2$ ($2°$). The acetyl–ring torsion-angles are $-78°$ at C-5, $-96°$ at O-2, $-156°$ at O-3, and $-70°$ at O-5.

$C_{14}H_{21}ClO_7$ 5-O-(Chloroacetyl)-1,2:3,4-di-O-isopropylidene-α-D-glucoseptanose,[30] m.p. 117–118°C

$P2_12_12_1$; $Z = 4$; $D_x = 1.25$; $R = 0.066$ for 1,234 intensities. The conformation of the septanose is a $^{5,6}C_2$ (D) twist-chair. Both dioxolane rings are envelopes, ^2E ($17°$) and ^4E ($16°$). The chloroacetate group is almost planar, with the largest deviation (of $11°$) in the C-7–O-5 torsion-angle. There is no hydrogen-bonding, which accounts for the relatively low density.

$C_{14}H_{22}O_8$ 4,5-Di-O-acetyl-1,2-O-isopropylidene-3-O-methyl-α-D-guloseptanose,[31] m.p. 111–112°C

$P2_1$; $Z = 2$; $D_x = 1.239$; $R = 0.038$ for 1,774 intensities. The conformation of the septanose lies between $^5C_{1,2}$ and $^{3,4}TC_{5,6}$ in which the symmetry axis is through C-1. The conformation of the dioxolane ring is $E^{0.2}$.

$C_{16}H_{24}O_{10}$ 3,4,6-Tri-O-acetyl-1,2-O-(1-exo-ethoxyethylidene)-α-D-glucopyranose[32]

$P2_1$; $Z = 2$; $D_x = 1.357$; $R = 0.068$ for 1,680 intensities (film measure-

(28) A. Ducruix, C. Pascard-Billy, D. Horton, and J. D. Wander, *Carbohyd. Res.*, **29**, 276–279 (1973).
(29) V. J. James and J. D. Stevens, *Cryst. Struct. Commun.*, **2**, 609–612 (1973).
(30) J. Jackobs, M. A. Reno, and M. Sundaralingam, *Carbohyd. Res.*, **28**, 75–85 (1973).
(31) J. F. McConnell and J. D. Stevens, *Cryst. Struct. Commun.*, **2**, 619–623 (1973).
(32) J. A. Hertman and G. F. Richards, *Carbohyd. Res.*, **28**, 180–182 (1973).

ments at $-193°$). The conformation of the pyranose is a distorted $^3S_5$. The dioxolane ring is $_{0-2}E$ ($42°$). The bonds to the 3- and 4-acetoxyl groups are axial and quasi-axial, respectively.

$C_{17}H_{24}O_{12}$ ($\pm$)-2-(Acetoxymethyl)-1,3,4,6-tetra-O-acetyl-epi-ino-
sitol[33]

$P2_1/c$; $Z = 4$; $D_x = 1.392$; $R = 0.22$ (all crystals were twinned) for 3,298 intensities. The structure determination was undertaken in order to establish the configuration. The cyclohexane ring is slightly distorted, owing to syn-axial O-2–O-4 interaction. The acetate groups are planar within $\pm6°$.

$C_{17}H_{28}O_{15} \cdot 3H_2O$ O-(4-O-Methyl-α-D-glucopyranosyluronic acid)-
$(1 \to 2)$-O-β-D-xylopyranosyl-$(1 \to 4)$-α-D-xylo-
pyranose trihydrate,[34] m.p. 179°C

$P2_1$; $Z = 2$; $D_x = 1.525$; $R = 0.066$ for 2,232 intensities (film measurements at $-193°$). The conformation of all three pyranoid moieties is $^4C_1$. The linkages are $79°$, $96°$ (Glc $\to$ Xyl); $-98°$, $+162°$ (Xyl $\to$ Xyl). The anomeric C–O bonds are short, 1.395 Å (139.5 pm), in both D-xylose moieties. There is no intramolecular hydrogen-bonding.

$C_{22}H_{30}O_{15}$ β-D-$(1 \to 4)$-Xylobiose hexaacetate[35]

$P2_12_12_1$; $Z = 4$; $D_x = 1.29$; $R = 0.070$ for 2,366 intensities. The conformations of the pyranose moieties are $^4C_1$. The linkage bonds are $-104°$, $+89°$, $(1 \to 4)$. The C–O ring bond–lengths differ by 0.044 Å (4.4 pm) in both rings, O-5–C-1 being the shorter. There is no hydrogen-bonding.

$C_{36}H_{60}O_{30} \cdot I_2 \cdot 4H_2O$ Cyclohexaamylose–iodine tetrahydrate[35a]

$P2_12_12_1$; $Z = 4$; $D_x = 1.704$; $R = 0.148$ for 2,872 intensities. The six D-glucose residues are all $^4C_1$ with ring torsion angles varying from 45 to 66°. The linkages are 97 to 112°, $-82°$ to $-123°$ $(1 \to 4)$. Four intramolecular hydrogen-bonds are formed between O-2 and O-3 of adjacent D-glucose units. The I_2 molecules are co-axial with the cyclohexaamylose molecules. The four water molecules fill space and are involved in an intricate hydrogen-bonding network.

(33) H. Sternglanz and C. E. Bugg, *Acta Crystallogr.*, **B29**, 1536–1538 (1973).
(34) R. A. Moran and G. F. Richards, *Acta Crystallogr.*, **B29**, 2770–2782 (1973).
(35) F. Leung and R. H. Marchessault, *Can. J. Chem.*, **51**, 1215–1222 (1973).
(35a) R. K. McMullan, W. Saenger, J. Fayos, and D. Mootz, *Carbohyd. Res.*, **31**, 211–227 (1973).

III. Data for Nucleosides and Nucleotides

$C_8H_{11}N_3O_6$ 6-Azauridine[36]; 2-β-D-ribofuranosyl-*as*-triazine-3,5-(2H,4H)-dione

$P2_12_12_1$; Z = 8; D_x = 1.624; R = 0.04 for 1,998 intensities. There are two symmetry-independent molecules that are conformationally similar. The conformation of the D-ribosyl groups is $^3T_4$, the glycosyl dispositions are *anti* (81°, 76°), and the exocyclic C-4'–C-5' bond torsion-angles are 74° and 69°. The two independent molecules are approximately related to each other by a pseudo-diad that is almost parallel to the crystal's two-fold screw axis. The bases are stacked. Both molecules show a hydrogen bond between N-3–H of the base and the ring-oxygen atom (O-4') of the D-ribosyl group.

$C_8H_{12}N_2O_4$ N^1-β-D-Ribofuranosylimidazole[37]

$P2_1$; Z = 2; D_x = 1.473; R = 0.03 for 1,280 intensities. The conformation of the D-ribosyl group is $^3T_4$, the glycosyl disposition is *syn* (−98°), and the exocyclic C-4'–C-5' bond torsion-angle is 64°.

$C_9H_{10}N_2O_5$ 2,2'-Anhydro-1-β-D-arabinofuranosyluracil   ("cyclo-Ara-U")[38,39]

$P2_12_12_1$; Z = 8; D_x = 1.618; two independent investigations, R = 0.045 for 1,762 intensities[38]; R = 0.033 for 2,593 intensities.[39] There are two symmetry-independent molecules; the conformations of the arabinosyl groups are $^4T_3$ and $_3T^4$. The glycosyl dispositions are *anti* (65°, 70°) and the exocyclic, C-4'–C-5' bond torsion-angles are 56° and 55°. The two molecules form stacked dimers that possess a pseudo-twofold axis, and are linked to each other by two hydrogen bonds involving the O-5'–H groups and the carbonyl O-4 atoms. The carbonyl C-4–O-4 bonds of one molecule overlap the ring system of the other.

$C_9H_{11}ClN_2O_5$ 5-Chloro-2'-deoxyuridine[40]

$P2_1$; Z = 2; D_x = 1.494; R = 0.038 for 1,228 intensities. This structure is isomorphous with that of 5-bromo-2'-deoxyuridine.[41] The confor-

(36) C. H. Schwalbe and W. Saenger, *J. Mol. Biol.*, **75**, 129–143 (1973).
(37) M. N. G. James and M. Matsushima, *Acta Crystallogr.*, **B29**, 838–846 (1973).
(38) D. Suck and W. Saenger, *Acta Crystallogr.*, **B29**, 1323–1330 (1973).
(39) L. T. J. Delbaere and M. N. G. James, *Acta Crystallogr.*, **B29**, 2905–2912 (1973).
(40) D. W. Young and E. M. Morris, *Acta Crystallogr.*, **B29**, 1259–1264 (1973).
(41) J. Iball, C. H. Morgan, and H. R. Wilson, *Proc. Roy. Soc.* (London), **A302**, 225–236 (1968).

mation of the 2-deoxy-D-*erythro*-pentosyl group is $^2T_1$, the glycosyl disposition is *anti* (41°), and the exocyclic C-4′–C-5′ bond torsion-angle is 50°.

$C_9H_{12}N_2O_4S \cdot H_2O$[*]   1-β-D-Arabinofuranosyl-4-thiouracil mono-hydrate[41a]

$P2_1$, Z = 2, D_x = 1.57, R = 0.046 for 1,444 intensities. The conformation of the D-arabinosyl group is $^3T_2$, the glycosyl disposition is *anti* (36°), and the exocyclic C-4′–C-5′ bond torsion-angle is −56°. The sulfur atom is hydrogen-bonded to the hydroxyl groups O-2′–H and O-3–H of different molecules.

$C_9H_{12}N_2O_6$ 1-β-D-Arabinofuranosyluracil[42]

$P2_12_12_1$; Z = 4; D_x = 1.652; R = 0.058 for 1,164 intensities. The conformation of the D-arabinosyl group is $^2T_1$, the glycosyl disposition is *anti* (34°), and the exocyclic C-4′–C-5′ bond torsion-angle is −63°. There is an intramolecular hydrogen-bond in the arabinosyl group that involves O-2′–H and O-5′. The C-2–O-2 carbonyl groups and the bases form alternate stacks.

$C_9H_{12}N_2O_7$ 5-Hydroxyuridine[43]

$P2_1$; Z = 2; D_x = 1.687; R = 0.026 for 915 intensities. The conformation of the D-ribosyl group is $^2T_1$, the glycosyl disposition is *anti* (42°), and the exocyclic C-4′–C-5′ bond torsion-angle is −66°. The only inter-base hydrogen-bond is between N-3–H of the ring and the hydroxyl oxygen atom of the base.

$C_9H_{13}N_3O_5 \cdot HCl$ 1-β-D-Arabinofuranosylcytosine hydrochloride[44]

$P2_1$; Z = 2; D_x = 1.551; R = 0.022 for 2,490 intensities. The conformation of the D-arabinosyl group is $_1T^2$, the glycosyl disposition is *anti* (27°), and the exocyclic, C-4′–C-5′ bond torsion-angle is 69°.

$C_{10}H_{13}NO_5$ 3-Deazauridine, 4-hydroxy-1-β-D-ribofuranosyl-2-pyridinone[45]

[*] Omitted from the 1970–1972 bibliography.

(41a) W. Saenger, *J. Amer. Chem. Soc.*, **94**, 621–626 (1972).

(42) P. Tollin, H. R. Wilson, and D. W. Young, *Acta Crystallogr.*, **B29**, 1641–1647 (1973).

(43) U. Thewalt and C. E. Bugg, *Acta Crystallogr.*, **B29**, 1393–1398 (1973).

(44) J. S. Sherfinski and R. E. Marsh, *Acta Crystallogr.*, **B29**, 192–198 (1973).

(45) C. H. Schwalbe and W. Saenger, *Acta Crystallogr.*, **B29**, 61–69 (1973).

$P2_12_12_1$; $Z = 4$; $D_x = 1.501$; $R = 0.04$ for 998 intensities. The conformation of the D-ribosyl group is ^2E, the glycosyl disposition is *anti* (52°), and the exocyclic, C-4'–C-5' torsion-angle is −64°. The compound exists in the enol form, which is also the preponderant form in solution. Adjacent bases are linked by O-4–H···O-2 hydrogen bonds of length 2.549 Å (254.9 pm). The O-3'–H group forms bifurcated hydrogen-bonds; one of them is an intramolecular hydrogen-bond to an adjacent, hydroxyl oxygen atom (O-2').

$C_{10}H_{11}N_5O_3 \cdot H_2O$ Formycin monohydrate[46]; 7-amino-3-β-D-ribofuranosylpyrazolo[4,3-d]pyrimidine monohydrate

$P2_12_12_1$; $Z = 4$; $D_x = 1.523$; $R = 0.035$ for 937 intensities. The conformation of the D-ribosyl group is $^2T_1$, the C-glycosyl disposition is *syn* (110°), and the exocyclic C-4'–C-5' torsion-angle is 58°. Atom N-8 of the base is not involved in hydrogen bonding. There is an intramolecular hydrogen-bond between the ring-oxygen atom O-4' and the amino nitrogen atom (N-6) of the base. The bases are stacked in columns, with considerable overlap of the rings, but the interplanar separation of 3.755 Å (375.5 pm) is beyond the normal range of Van der Waals contact.

$C_{10}H_{13}N_5O_4 \cdot HCl$ Adenosine hydrochloride[47]

$P2_1$; $Z = 2$; $D_x = 1.562$; $R = 0.037$ for 1,895 intensities. The conformation of the D-ribosyl group is $^2T_3$, the glycosyl disposition is *anti* (43°), and the exocyclic C-4'–C-5' bond torsion-angle is −61°. The N-1 atom of the adenine residue is protonated.

$C_{10}H_{13}N_5NaO_6P \cdot 6H_2O$ 2'-Deoxyadenosine 5'-(sodium phosphate), hexahydrate[48]

$P2_12_12_1$; $Z = 4$; $D_x = 1.489$; $R = 0.060$ for 1,640 intensities. The conformation of the 2-deoxy-D-*erythro*-pentofuranosyl group is $^2T_1$, the glycosyl disposition is *anti* (63°), and the exocyclic C-4'–C-5' bond torsion-angle is −72°. The sodium ion is coordinated to six water

(46) P. Prusiner, T. Brennan, and M. Sundaralingam, *Biochemistry*, **12**, 1196–1203 (1973).
(47) K. Shikata, T. Ueki, and T. Mitsui, *Acta Crystallogr.*, **B29**, 31–38 (1973).
(48) B. Swaminatha Reddy and M. A. Viswamitra, *Cryst. Struct. Commun.*, **2**, 9–13 (1973).

molecules, and the anionic phosphate oxygen atoms are in the second coordination sphere.

$C_{10}H_{14}N_2O_6$ 1-β-D-Arabinofuranosylthymine (and the isomorphous 1-β-D-arabinofuranosyl-5-bromouracil)[49]

$P2_1$; Z = 2; D_x = 1.590; R = 0.075 for 982 intensities (film measurements). The conformation of the D-arabinosyl group is $_1T^0$, the glycosyl disposition is *anti* (24°), and the exocyclic C-4′–C-5′ bond torsion-angle is −60°.

$C_{11}H_{14}N_4O_4$ Tubercidin[50,51]; 4-amino-7-β-D-ribofuranosylpyrrolo-[2,3-d]pyrimidine

$P2_1$; Z = 2; D_x = 1.464. Two independent investigations, R = 0.11 for 1,117 intensities,[50] and R = 0.027 for 1,050 intensities.[51] The conformation of the D-ribosyl group is $^2T_1$, the glycosyl disposition is *anti* (73°), and the exocyclic C-4′–C-5′ bond torsion-angle is 62°. The bases are partially stacked.

$C_{12}H_{13}ClN_2O_4$ 2-Chloro-1-β-D-ribofuranosylbenzimidazole[52]

$P2_1$; Z = 2; D_x = 1.521; R = 0.054 for 1,058 intensities. The conformation of the D-ribosyl group is $^2T_1$, the glycosyl disposition is *syn* (−117°), and the exocyclic C-4′–C-5′ bond torsion-angle is −72°. The O-3′–H group forms an intermolecular hydrogen-bond to the O-4′ atom. The chlorine atom is not hydrogen-bonded. The molecular packing consists of well defined domains of hydrophilic and hydrophobic regions.

$C_{12}H_{14}N_2O_4S \cdot H_2O$ 1-β D-Ribofuranosyl-2-benzimidazolethiol monohydrate[53]

$P2_12_12_1$; Z = 4; D_x = 1.422; R = 0.04 for 1,194 intensities. The conformation of the D-ribosyl group is $^2T_1$, the glycosyl disposition is *syn* (−110°), and the exocyclic C-4′–C-5′ bond torsion-angle is 60°. The molecular packing consists of zones of stacked bases interleaved by D-ribofuranosyl groups. The sulfur atom is hydrogen-bonded to neighboring molecules.

(49) P. Tougard, *Acta Crystallogr.*, **B29**, 2227–2232 (1973).
(50) R. M. Stroud, *Acta Crystallogr.*, **B29**, 690–696 (1973).
(51) J. Abola and M. Sundaralingam, *Acta Crystallogr.*, **B29**, 697–703 (1973).
(52) S. Sprang and M. Sundaralingam, *Acta Crystallogr.*, **B29**, 1910–1916 (1973).
(53) P. Prusiner and M. Sundaralingam, *Acta Crystallogr.*, **B29**, 2328–2334 (1973).

IV. PRELIMINARY COMMUNICATIONS

1. Carbohydrates

$C_6H_{12}O_6$ DL-Glyceraldehyde dimer[54]

$C_7H_{12}O_5$ Methyl 3,6-anhydro-α-D-galactopyranoside[55]

$C_{12}H_{22}O_{11} \cdot CaBr_2 \cdot 3H_2O$ α,α-Trehalose · calcium bromide, trihydrate[56]

$C_{16}H_{28}CuN_2O_{14}$*   Di-O-β-D-xylopyranosyl-L-serinatocopper(II)[57]

2. Nucleosides and Nucleotides

$C_8H_{12}N_4O_5$ Virazole; 1-β-D-ribofuranosyl-1,2,4-triazole-3-carboxamide[58]

There are two crystalline, polymorphic forms of virazole, $P2_12_12_1$; Z = 4; D_x = 1.652 (form 1); R = 0.056; D_x = 1.585 (form 2); R = 0.036. In form 1, the conformation of the D-ribosyl group is 3'-*endo*, the glycosyl disposition is *anti*, and the exocyclic, C-4'–C-5' bond torsion-angle is g^-. In form 2, the conformation of the D-ribosyl group is $_2T^1$, the glycosyl disposition is *syn*, and the exocyclic C-4'–C-5' bond torsion-angle is g^+. Form 1 displays no inter-base hydrogen-bonding, whereas form 2 shows an intermolecular hydrogen-bond between the carboxamide amino group and the ring nitrogen atom N-4. In both structures, one of the carboxamide amino hydrogen atoms is "free."

$C_8H_{12}N_3O_9P \cdot H_2O$ 6-Azauridine 5'-phosphate, monohydrate[59]

$P2_12_12_1$; Z = 4; D_x = 1.486; R = 0.07. The conformation of the D-ribosyl group is 3'-*endo*, the glycosyl disposition is *anti*, and the exocyclic C-4'–C-5' bond torsion-angle is g^+.

(54) M. Senna, Z. Taira, K. Osaki, and T. Taga, *Chem. Commun.*, 880–881 (1973).
 * Omitted from the 1970–1972 bibliography.
(55) B. Lindberg, B. Lindberg, and S. Svensson, *Acta Chem. Scand.*, **27**, 373–374 (1973).
(56) W. J. Cook and C. E. Bugg, *Amer. Crystallogr. Assoc. Meeting*, Gainesville, Fla., Jan., 1973, Abstr. G-2.
(57) L. T. J. Belbaere, M. Higham, B. Kamenar, P. W. Kent, and C. K. Prout, *Biochim. Biophys. Acta*, **286**, 441–444 (1972).
(58) P. Prusiner and M. Sundaralingam, *Nature New Biol.*, **244**, 116–118 (1973).
(59) W. Saenger and D. Suck, *Nature New Biol.*, **242**, 810–812 (1973).

$C_9H_{11}N_3O_4 \cdot HCl$ 2,2'-Anhydro-1-β-D-arabinofuranosylcytosine hydrochloride ("cyclo-Ara-C")[60]

$P2_12_12_1$; $Z = 4$; $D_x = 1.535$; $R = 0.042$. The conformation of the D-arabinosyl group is 4'-*endo*, the glycosyl disposition is *anti*, and the exocyclic C-4'–C-5' torsion-angle is g^+. The hydroxyl oxygen atom O-5' exhibits short contacts to C-2, O-2, and N-1 (of the base) of 2.72, 2.92, and 2.94 Å, respectively.

$C_9H_{13}N_3O_5$ 1-β-D-Arabinofuranosylcytosine ("Ara-C")[61,62]

$P2_12_12_1$; $Z = 4$; $D_x = 1.608$; two independent investigations; $R = 0.032$ for 978 intensities[61]; $R = 0.10$ for 1,200 intensities.[62] The conformation of the D-arabinosyl group is 2'-*endo*, the glycosyl disposition is *anti*, and the exocyclic C-4'–C-5' bond torsion-angle is g^-. There is an intramolecular hydrogen-bond in the D-arabinosyl group involving O-2'–H and O-5'.

$C_9H_{14}N_3O_8P \cdot 3H_2O$ Cytidine 2'-phosphate, trihydrate[63]

$P1$; $Z = 2$; $D_x = 1.591$; $R = 0.13$ for 3,472 intensities. There are two independent molecules; the conformations of the D-ribosyl groups are 2'-*endo*, the glycosyl dispositions are *anti*, and the exocyclic, C-4'–C-5' bond torsion-angles are g^-. The cytosine bases are protonated at N-3, and the phosphate groups are singly ionized. The two molecules are linked, through the phosphate oxygen atoms, by a strong hydrogen-bond of length 2.48 Å (248 pm).

$C_{10}H_{11}N_5NaO_7P \cdot 4H_2O$ Guanosine 3',5'-(sodium monophosphate), tetrahydrate[64]

$P2_12_12_1$; $Z = 4$; $D_x = 1.665$; $R = 0.034$. The conformation of the D-ribosyl group is 4'-*exo*, the glycosyl disposition is *syn*, and the exocyclic C-4'–C-5' bond torsion-angle is t.

$2 (C_{10}H_{12}N_4O_5)$ Inosine[65]

$P2_12_12_1$; $Z = 8$; $D_x = 1.559$; $R = 0.065$ for 1,884 intensities. There are

(60) T. Brennan and M. Sundaralingam, *Biochem. Biophys. Res. Commun.*, **52**, 1348–1353 (1973).
(61) A. Ku Chwang and M. Sundaralingam, *Nature New Biol.*, **243**, 78–80 (1973).
(62) O. Lefebvre-Soubeyran and P. Tougard, *Compt. Rend.*, **276**, 403–406 (1973).
(63) G. Kartha, G. Ambady, and M. A. Viswamitra, *Science*, **179**, 495–496 (1973).
(64) A. Ku Chwang and M. Sundaralingam, *Nature New Biol.*, **244**, 136–137 (1973).
(65) E. Subramanian, J. J. Maden, and C. E. Bugg, *Biochem. Biophys. Res. Commun.*, **50**, 691–695 (1973).

two symmetry-independent molecules; the conformations of the D-ribosyl groups are 2'-*endo*, the glycosyl dispositions are *syn*, and the exocyclic, C-4'–C-5' bond torsion-angles are g^-. Both molecules display an intramolecular hydrogen-bond between O-5'–H and N-3.

$C_{10}H_{14}N_2O_5$ 2,5'-Anhydro-2',3'-*O*-isopropylideneuridine[66]

$C_{18}H_{25}N_4Na_2O_{15}P_2 \cdot 12H_2O$ Disodium (pT^1pT^2); disodium thymidylyl-(5' → 3')-5'-thymidylate, dodecahydrate[67]

$P2_12_12_1$; $Z = 4$; $D_x = 1.505$; $R = 0.12$ for 966 intensities. The two 5'-nucleotide halves of the molecule have very similar conformations, the bases are *anti*, the D-ribosyl residues have the 2'-*endo* conformation, and the exocyclic C-4'–C-5' bond torsion-angles are g^-. There is no intramolecular base-stacking; however, thymine 1 exhibits some intermolecular stacking with a twofold-symmetry-related base, while the carbonyl group C-2–O-2 of thymine 2 is stacked to a symmetry-related base. The phosphoric diester has the extended tg^- conformation. The structure does not show any interbase hydrogen-bonding. The sodium ions are coordinated to the carbonyl oxygen atoms (O-2) and to the oxygen atoms of the phosphoric diester. The presence of only two sodium atoms per molecule indicates that the 5'-phosphate is singly ionized. The second sodium atom is linked to the 5'-phosphate group by way of a water molecule.

$C_{19}H_{22}N_7NaO_{13}P \cdot 8H_2O$ Guanylyl-(3' → 5')-cytidine, sodium salt, octahydrate[68]

C2; $Z = 4$; $D_x = 1.478$; $R = 0.11$ for 2,276 intensities. Two molecules of the dinucleoside monophosphate, related by a crystallographic diad, form an anti-parallel, right-handed, double-helical segment with Watson–Crick complementary hydrogen-bonding between the guanine base atoms O-6, N-1–H, and N-2–H, and the cytosine base atoms N-4–H, N-3, and O-2, respectively. The sodium ion bridges adjacent phosphate groups; the water molecules complete the octahedral coordinates around the sodium ion. The guanosine and cytidine residues possess similar conformations: the D-ribosyl groups are 3'-*endo*; the glycosyl dispositions are *anti*; and the exocyclic,

(66) L. T. J. Delbaere, M. N. G. James, and R. U. Lemieux, *J. Amer. Chem. Soc.*, **95**, 9866–9868 (1973).
(67) N. Camerman, J. K. Fawcett, and A. Camerman, *Science*, **182**, 1142–1143 (1973).
(68) R. O. Day, N. C. Seeman, J. M. Rosenberg, and A. Rich, *Proc. Nat. Acad. Sci. U. S.*, **70**, 849–853 (1973).

C-4'–C-5' bond torsion-angles are g^-. The conformation of the phosphoric diester group is g^-g^-.

$(C_{19}H_{23}N_7NaO_{12}P)_2 \cdot 12H_2O$ Adenosyl-(3' → 5')-uridine phosphate, disodium salt, dodecahydrate[69]

$P2_1$; $Z = 4$; $D = 1.552$; $R = 0.091$. There are two independent molecules of the dinucleoside monophosphate, related by a noncrystallographic, two-fold axis. The adenine and uracil bases of one molecule are respectively base-paired to the uracil and adenine bases of the other, in the Watson–Crick complementary pairing scheme, N-6 of adenine to O-4 of uracil, and N-1 of adenine to N-3 of uracil. The base-paired complex forms a segment of a right-handed, antiparallel, double helix, and a sodium ion is trapped in the "minor groove" of the helix by complexation to the carbonyl oxygen atoms (O-2) of the uracil residues. The second sodium ion is coordinated to the phosphate oxygen atoms. The water molecules are concentrated around the sodium ions. The overall conformations of the two dinucleoside monophosphate molecules are very similar. The conformations of the two adenosine and uridine moieties are also very similar: the D-ribosyl groups are 3'-endo, the glycosyl dispositions are anti, and the exocyclic C-4'–C-5' bond torsion-angles are g^-. The conformation of the phosphoric diester group is g^-g^- in both molecules.

$C_{30}H_{37}N_{15}O_{16}P_2 \cdot 48H_2O$ Adenylyl-(3' → 5')-adenylyl-(3' → 5')-adenosine hydrate $(A^1pA^2pA^3)$[70]

$P4_12_12$; $Z = 8$; $D_x = 1.559$; $R = 0.07$. The three nucleotide moieties have similar conformations: the D-ribosyl groups are 3'-endo, the dispositions of the bases are anti, and the backbone C-4'–C-5' bonds are g^-. The conformation of the phosphoric diester in the A^1pA^2 portion is g^-g^-, and, in the A^2pA^3 portion, is g^+g^+. These conformations for the adjacent phosphate groups bring the terminal 3'-hydroxyl group of A^3 and the penultimate phosphate between A^1 and A^2 into juxtaposition for an intramolecular hydrogen-bond of 2.68 Å (268 pm) which folds the sugar–phosphate backbone into a loop. Only the A^2 and A^3 bases are protonated at N-1 by the ionized phosphate hydrogen atoms. These adenine residues form a right-handed, anti-parallel, base-paired structure with a crystallographic diad-related molecule. The base pairing involves N-6 and N-7 of both bases. A

(69) J. M. Rosenberg, N. C. Seeman, J. J. P. Kim, F. L. Suddath, H. B. Nicholas, and A. Rich, Nature, 243, 150–154 (1973).
(70) D. Suck, P. C. Manor, G. Germain, C. H. Schwalbe, G. Weimann, and W. Saenger, Nature New Biol., 246, 161–165 (1973).

further hydrogen-bond between N-6 and the phosphate anionic oxygen atom adds stabilization to the anti-parallel "duplex." The paired adenine bases are twisted at 31°. The neutral adenine A^1 is not involved in base pairing, but it stacks with the protonated adenine A^2. The protonated adenines A^2 and A^3 do not show intramolecular stacking, but display intermolecular stacking within the duplex.

V. List of Crystal Structures Reported for Carbohydrates, Nucleosides, and Nucleotides, 1935–1969

This list contains no structural or conformational information. Omitted are structure determinations that have been superseded by fuller publication or by more accurate published determinations.

1. Carbohydrates

$C_4H_{10}O_4$	Erythritol[71,72]
$C_5H_8N_4O_{12}$	Pentaerythritol tetranitrate[73]
$2C_5H_9O_6^- \cdot Ca^{2+} \cdot 5H_2O$	Calcium L-arabinonate pentahydrate[74]
$2C_5H_9O_6^- \cdot Sr^{2+} \cdot 5H_2O$	Strontium L-arabinonate pentahydrate[74]
$C_5H_9O_8P^{2-} \cdot Ba^{2+} \cdot 5H_2O$	Barium D-ribose 5-phosphate pentahydrate[75]
$C_5H_{10}O_4$	2-Deoxy-D-*erythro*-pentose[76]
$C_5H_{10}O_5$	β-DL-Arabinose[77]
	β-L-Arabinose[78]
	β-L-Lyxose[79]
$C_5H_{12}O_4$	Pentaerythritol[80–82]

(71) A. Shimada, *Bull. Chem. Soc. Jap.*, **32**, 325–333 (1959).
(72) A. Bekoe and H. M. Powell, *Proc. Roy. Soc.* (London), **A250**, 301–315 (1959).
(73) J. Trotter, *Acta Crystallogr.*, **16**, 698–699 (1963).
(74) S. Furberg and S. Helland, *Acta Chem. Scand.*, **16**, 2373–2383 (1962).
(75) S. Furberg and A. Mostad, *Acta Chem. Scand.*, **16**, 1627–1636 (1962).
(76) S. Furberg, *Acta Chem. Scand.*, **14**, 1357–1363 (1960).
(77) S.-H. Kim and G. A. Jeffrey, *Acta Crystallogr.*, **22**, 537–545 (1967).
(78) A. Hordvik, *Acta Chem. Scand.*, **15**, 16–30 (1961).
(79) A. Hordvik, *Acta Chem. Scand.*, **20**, 1943–1954 (1966).
(80) F. J. Llewellyn, E. G. Cox, and T. H. Goodwin, *J. Chem. Soc.*, 883–894 (1937).
(81) J. Hvoslef, *Acta Crystallogr.*, **11**, 383–388 (1958).
(82) R. Shiono, D. W. J. Cruickshank, and E. G. Cox, *Acta Crystallogr.*, **11**, 389–391 (1958).

$C_5H_{12}O_5$ DL-Arabinitol[83]
                                       Ribitol[84]
                                       Xylitol[85]
$C_6H_7O_6{}^-\cdot Na^+$ Sodium L-ascorbate[86]
$C_6H_8O_6$ β-D-Glucurono-6,3-lactone[87]
                                       L-Ascorbic acid[88,89]
$C_6H_9O_7{}^-\cdot K^+\cdot 2H_2O$ Potassium β-D-glucuronate,
                                       dihydrate[90]
$C_6H_9O_7{}^-\cdot Rb^+\cdot 2H_2O$ Rubidium β-D-glucuronate,
                                       dihydrate[90]
$2C_6H_9O_7{}^-\cdot Ca^{2+}\cdot 2H_2O$ Calcium D-xylo-hexos-
                                       ulosuronate dihydrate[91]
$C_6H_{10}O_6$ D-Galactono-1,4-lactone[92]
$2C_6H_{11}O_6{}^-\cdot Ca^{2+}$ Calcium "α"-D-isosac-
                                       charinate;[93] calcium
                                       3-deoxy-2-C-(hydroxy-
                                       methyl)-D-erythro-
                                       pentonate
$2C_6H_{11}O_6{}^-\cdot Sr^{2+}$ Strontium 3-deoxy-2-C-
                                       (hydroxymethyl)-D-
                                       erythro-pentonate[94]
$C_6H_{11}O_7{}^-\cdot K^+$ Potassium D-gluconate[95]
$C_6H_{11}O_9P^{2-}\cdot 2K^+\cdot 2H_2O$ Dipotassium α-D-glucosyl
                                       phosphate, dihydrate[96,97]

(83) F. D. Hunter and R. D. Rosenstein, Acta Crystallogr., B24, 1652–1660 (1968).
(84) H. S. Kim, G. A. Jeffrey, and R. D. Rosenstein, Acta Crystallogr., B25, 2223–2230 (1969).
(85) H. S. Kim and G. A. Jeffrey, Acta Crystallogr., B25, 2607–2613 (1969).
(86) J. Hvoslef, Acta Crystallogr., B25, 2214–2223 (1969).
(87) S.-H. Kim, G. A. Jeffrey, R. D. Rosenstein, and P. W. R. Corfield, Acta Crystallogr., 22, 733–743 (1967).
(88) J. Hvoslef, Acta Crystallogr., B24, 1431–1440 (1968).
(89) J. Hvoslef, Acta Crystallogr., B24, 23–35 (1968).
(90) G. E. Gurr, Acta Crystallogr., 16, 690–698 (1963).
(91) A. A. Balchin and C. H. Carlisle, Acta Crystallogr., 19, 103–111 (1965).
(92) G. A. Jeffrey, R. D. Rosenstein, and M. Vlasse, Acta Crystallogr., 22, 725–733 (1967).
(93) R. Norrestam, P. E. Werner, and M. von Glehn, Acta Chem. Scand., 22, 1395–1403 (1968).
(94) P. E. Werner, R. Norrestam, and O. Ronnquist, Acta Crystallogr., B25, 714–719 (1969).
(95) C. D. Littleton, Acta Crystallogr., 6, 775–781 (1953).
(96) C. A. Beevers and G. H. Maconochie, Acta Crystallogr., 18, 232–236 (1965).
(97) R. D. Rosenstein, Amer. Crystallogr. Assoc. Meeting, Buffalo, N. Y., August, 1968, Abstr. 92.

$C_6H_{12}O_4S$	Methyl 1-thio-β-D-xylopyranoside[98]
$C_6H_{12}O_5$	Methyl α-D-lyxofuranoside[99]
	Methyl β-D-xyloside[100]
$C_6H_{12}O_6$	β-D-Fructopyranose,[97] β-D-*arabino*-hexulose
	α-D-Glucose[101]
	β-D-Glucose[102,103]
	α-DL-Mannose[104]
	myo-Inositol[105]
	α-L-Sorbose,[106] α-L-*xylo*-hexulose
	α-D-Tagatose,[107] α-D-*lyxo*-hexulose
$C_6H_{12}O_6 \cdot 2H_2O$	*myo*-Inositol dihydrate[108]
$C_6H_{14}NO_5^+ \cdot Br^-$	2-Amino-2-deoxy-α-D-glucose hydrobromide[109,110]
$C_6H_{14}NO_5^+ \cdot Cl^-$	2-Amino-2-deoxy-α-D-glucose hydrochloride[109–111]
$C_6H_{14}O_6$	D-Mannitol (B-form)[112]
	D-Mannitol (K-form)[113]
	Galactitol (dulcitol)[114]

(98) A. M. Mathieson and B. J. Poppleton, *Acta Crystallogr.*, **21**, 72–79 (1966).

(99) P. Groth and H. Hammer, *Acta Chem. Scand.*, **22**, 2059–2070 (1968).

(100) C. J. Brown, E. G. Cox, and F. J. Llewellyn, *J. Chem. Soc., A*, 922–927 (1966).

(101) G. M. Brown and H. A. Levy, *Science*, **147**, 1038–1039 (1965).

(102) W. G. Ferrier, *Acta Crystallogr.*, **16**, 1023–1031 (1963).

(103) S. S. C. Chu and G. A. Jeffrey, *Acta Crystallogr.*, **B24**, 830–838 (1968).

(104) F. Planinsek and R. D. Rosenstein, *Amer. Crystallogr. Assoc. Meeting*, Minneapolis, Minn., August, 1967, Abstr. 70.

(105) I. N. Rabinowitz and J. Kraut, *Acta Crystallogr.*, **17**, 159–168 (1964).

(106) S.-H. Kim and R. D. Rosenstein, *Acta Crystallogr.*, **22**, 648–656 (1967).

(107) S. Takagi and R. D. Rosenstein, *Carbohyd. Res.*, **11**, 156–158 (1969).

(108) T. R. Lomer, A. Miller, and C. A. Beevers, *Acta Crystallogr.*, **16**, 264–268 (1963).

(109) S. S. C. Chu and G. A. Jeffrey, *Proc. Roy. Soc.* (London), **A285**, 470–479 (1965).

(110) R. Chandrasekaran and M. Mallikarjunan, *Z. Kristallogr.*, **129**, 29–49 (1969).

(111) G. N. Ramachandran, R. Chandrasekaran, and K. S. Chandrasekaran, *Biochim. Biophys. Acta*, **148**, 317–319 (1967).

(112) H. M. Berman, G. A. Jeffrey, and R. D. Rosenstein, *Acta Crystallogr.*, **B24**, 442–449 (1968).

(113) H. S. Kim, G. A. Jeffrey, and R. D. Rosenstein, *Acta Crystallogr.*, **B24**, 1449–1455 (1968).

(114) H. M. Berman and R. D. Rosenstein, *Acta Crystallogr.*, **B24**, 435–441 (1968).

$C_7H_{12}Cl_2O_4$	Methyl 4,6-dichloro-4,6-dideoxy-α-D-galacto-pyranoside[115]
	Methyl 4,6-dichloro-4,6-dideoxy-α-D-gluco-pyranoside[116]
$C_7H_{12}O_6 \cdot H_2O$	Sedoheptulosan monohydrate; 2,7-anhydro-β-D-*altro*-heptulopyranose monohydrate[117]
$C_7H_{13}BrO_5$	Methyl 6-bromo-6-deoxy-α-D-galactoside[118]
$C_7H_{13}ClO_5$	Methyl 2-chloro-2-deoxy-α,β-D-galactopyranoside[119]
$C_7H_{14}O_6$	Methyl α-D-glucopyranoside[120]
$C_7H_{14}O_7 \cdot H_2O$	D-*manno*-3-Heptulose monohydrate[121]
$C_8H_{15}NO_6$	2-Acetamido-2-deoxy-α-D-glucose[122]
$C_8H_{16}O_5S$	Ethyl 1-thio-α-D-glucofuranoside[123]
$C_9H_{17}O_{10}S^- \cdot Rb^+$	6-Deoxy-6-sulfo-α-D-gluco-pyranosyl-(1 → 1)-D-glycerol, rubidium salt[124]
$C_9H_{18}NO_6^+ \cdot I^-$	1,2-O-(Aminoisopropyli-dene)-α-D-glucopyranose hydroiodide[125]
$C_{10}H_{14}N_5O_4^+ \cdot Br^- \cdot H_2O$	Formycin hydrobromide, monohydrate[126]

(115) R. Hoge and J. Trotter, *J. Chem. Soc.*, A, 2165–2169 (1969).
(116) R. Hoge and J. Trotter, *J. Chem. Soc.*, A, 267–271 (1968).
(117) G. M. Brown and W. E. Thiessen, *Acta Crystallogr.*, A25, S195 (1969).
(118) J. H. Robertson and B. Sheldrick, *Acta Crystallogr.*, 19, 820–826 (1965).
(119) R. Hoge and J. Trotter, *J. Chem. Soc.*, A, 2170–2174 (1969).
(120) H. M. Berman and S.-H. Kim, *Acta Crystallogr.*, B24, 897–904 (1968).
(121) T. Taga and K. Osaki, *Tetrahedron Lett.*, 4433–4434 (1969).
(122) L. N. Johnson, *Acta Crystallogr.*, 21, 885–891 (1966).
(123) R. Parthasarathy and R. E. Davis, *Acta Crystallogr.*, 23, 1049–1057 (1967).
(124) Y. Okaya, *Acta Crystallogr.*, 17, 1276–1282 (1964).
(125) J. Trotter and J. K. Fawcett, *Acta Crystallogr.*, 21, 366–375 (1966).
(126) G. Koyama, K. Maeda, and H. Umezawa, *Tetrahedron Lett.*, 597–602 (1966).

$C_{11}H_{15}BrN_2O_4$ L-Arabinose (p-bromo-phenyl)hydrazone[127]
D-Ribose (p-bromophenyl)-hydrazone[128]

$C_{12}H_{12}BrNO_8S$ 1,4:3,6-Dianhydro-2-O-(p-bromophenylsulfonyl)-D-glucitol 5-nitrate[129]

$C_{12}H_{16}BrFO_7$ 3,4,6-Tri-O-acetyl-2-bromo-2-deoxy-α-D-mannopyranosyl fluoride[130]
3,4,6-Tri-O-acetyl-2-bromo-2-deoxy-β-D-mannopyranosyl fluoride[131]

$C_{12}H_{17}BrN_2O_5$ D-Glucose (p-bromophenyl)-hydrazone[132]
D-Mannose (p-bromo-phenyl)hydrazone[133]

$C_{12}H_{20}O_6$ 1,2:4,5-Di-O-isopropylidene-β-D-fructopyranose[134]
2,3:4,5-Di-O-isopropylidene-β-D-fructopyranose

$C_{12}H_{22}O_{11}$ β-Cellobiose[103,135-137]

$C_{12}H_{22}O_{11}\cdot Na^+\cdot Br^-\cdot 2H_2O$ Sucrose·sodium bromide, dihydrate[138]

(127) S. Furberg and C. S. Petersen, *Acta Chem. Scand.*, **16**, 1539–1548 (1962).
(128) K. Bjamer, S. Furberg, and C. S. Petersen, *Acta Chem. Scand.*, **18**, 587–595 (1964).
(129) A. Camerman, N. Camerman, and J. Trotter, *Acta Crystallogr.*, **19**, 449–456 (1965).
(130) J. C. Campbell, R. A. Dwek, P. W. Kent, and C. K. Prout, *Carbohyd. Res.*, **10**, 71–86 (1969).
(131) J. C. Campbell, R. A. Dwek, P. W. Kent, and C. K. Prout, *Chem. Commun.*, 34–35 (1968).
(132) T. Dukefos and A. Mostad, *Acta Chem. Scand.*, **19**, 685–696 (1965).
(133) S. Furberg and J. Solbakk, *Acta Chem. Scand.*, **23**, 3248–3256 (1969).
(134) S. Takagi and R. D. Rosenstein, *Acta Crystallogr.*, **A25**, S197 (1969); but see also Ref. 18.
(135) R. A. Jacobson, J. A. Wunderlich, and W. N. Lipscomb, *Acta Crystallogr.*, **14**, 598–607 (1961).
(136) J. W. Moncrief and S. P. Sims, *J. Chem. Soc.*, D, 914 (1969).
(137) C. J. Brown, *J. Chem. Soc.*, A, 927–932 (1966).
(138) C. A. Beevers and W. Cochran, *Proc. Roy. Soc.* (London), **A190**, 257–272 (1947).

$C_{13}H_{18}Cl_2O_5$ — 3-Deoxy-3,4-C-(dichloro-methylene)-1,2:5,6-di-O-isopropylidene-α-D-galactofuranose[139]

$C_{13}H_{20}O_8$ — Pentaerythritol tetraacetate[140]

$C_{13}H_{22}NO_4^+\cdot Br^-$ — 2,5'-Anhydro-5-deoxy-3,4-dihydro-2',3'-O-iso-propylidene-1-N-methyl-showdomycin hydro-bromide[141,142]

$C_{13}H_{24}O_{11}\cdot H_2O$ — Methyl β-maltoside, monohydrate[143]

$C_{14}H_{19}ClO_9$ — Tetra-O-acetyl-α-D-gluco-pyranosyl chloride[144]

$C_{14}H_{26}N_3O_9^+\cdot Br^-\cdot H_2O$ — Kasugamycin hydrobromide, monohydrate[145]

$C_{17}H_{17}BBrNO_4$ — N-(p-Bromophenyl)-α-D-ribo-pyranosylamine 2,4-di-benzeneboronate[146]

$C_{17}H_{21}N_4O_6^+\cdot Br^-\cdot H_2O$ — Riboflavine hydrobromide, monohydrate[147]

$C_{18}H_{34}ClN_2O_5S^+\cdot Cl^-\cdot H_2O$ — 7(S)-Chloro-7-deoxy-lincomycin hydrochloride, monohydrate[148]

$C_{18}H_{35}N_2O_6S^+\cdot Cl^-\cdot H_2O$ — Lincomycin hydrochloride, monohydrate[149]

(139) J. S. Brimacombe, P. A. Gent, and T. A. Hamor, *J. Chem. Soc., B*, 1566–1571 (1968).
(140) T. H. Goodwin and R. Hardy, *Proc. Roy. Soc.* (London), **A164**, 369 (1938).
(141) Y. Tsukuda, Y. Nakagawa, H. Kano, T. Sato, M. Shiro, and H. Koyama, *Chem. Commun.*, 975–976 (1967).
(142) Y. Tsukuda, T. Sato, M. Shiro, and H. Koyama, *J. Chem. Soc., B*, 843–848 (1969).
(143) S. S. C. Chu and G. A. Jeffrey, *Acta Crystallogr.*, **23**, 1038–1049 (1967).
(144) M. N. G. James and D. Hall, *Acta Crystallogr.*, **A25**, S196 (1969).
(145) T. Ikekawa, H. Umezawa, and Y. Iitaka, *J. Antibiotics* (Tokyo), **A19**, 49 (1966).
(146) H. Shimanouchi, N. Saito, and Y. Sasada, *Bull. Chem. Soc. Jap.*, **42**, 1239–1247 (1969).
(147) N. Tanaka, T. Ashida, Y. Sasada, and M. Kakudo, *Bull. Chem. Soc. Jap.*, **42**, 1546–1554 (1969).
(148) D. J. Duchamp, *Amer. Crystallogr. Assoc. Meeting*, Minneapolis, Minn., August, 1967, Abstr. 27.
(149) R. E. Davis and R. Parthasarathy, *Acta Crystallogr.*, **A21**, 109–110 (1966).

$C_{19}H_{23}BrO_{10}S$ — 4,5,7-Tri-*O*-acetyl-2,6-anhy-dro-1-*O*-(*p*-bromophenyl-sulfonyl)-3-deoxy-D-*gluco*-heptitol[150]

$C_{42}H_{59}IO_{14}S$ — Fusicoccin A *p*-iodobenzene-sulfonate[151]

$C_{72}H_{120}O_{60} \cdot 3.8C_2H_3O_2^- \cdot 3.8K^+ \cdot 19.4H_2O$ — Cyclohexaamylose–potassium acetate hydrate complex[152]

2. Nucleosides and Nucleotides

$C_9H_{10}N_2O_8P^- \cdot C_6H_{16}N^+$ — Triethylammonium uridine 3′,5′-cyclic phosphate[153]

$C_9H_{11}BrN_2O_5$ — 5-Bromo-2′-deoxyuridine[154]

$C_9H_{11}BrN_2O_6$ — D-Arabinofuranosyl-5-bromouracil[155]
5-Bromouridine[154]

$C_9H_{11}BrN_2O_6 \cdot C_2H_6OS$ — 5-Bromouridine–methyl sulfoxide complex[156]

$C_9H_{11}FN_2O_5$ — 2′-Deoxy-5-fluorouridine[157]

$C_9H_{11}IN_2O_5$ — 2′-Deoxy-5-iodouridine[158]

$C_9H_{11}N_2O_9P^{2-} \cdot Ba^{2+} \cdot 8.9H_2O$ — Uridine 5′-(barium phosphate), hydrate[159]

$C_9H_{12}N_2O_5S$ — 4-Thiouridine[160]

$C_9H_{13}N_3O_5$ — Cytidine[161]

$C_9H_{14}N_3O_8P$ — Cytidine 3′-phosphate[162,163]

(150) A. Camerman and J. Trotter, *Acta Crystallogr.*, **18**, 197–203 (1965).
(151) A. Ballio, M. Brufani, C. G. Casinovi, S. Cerrini, W. Fedeli, R. Pellicciari, B. Santurbano, and A. Vaciago, *Experientia*, **24**, 635–638 (1968).
(152) A. Hybl, R. E. Rundle, and D. E. Williams, *J. Amer. Chem. Soc.*, **87**, 2779–2788 (1965).
(153) C. L. Coulter, *Acta Crystallogr.*, **B25**, 2055–2065 (1969).
(154) J. Iball, C. H. Morgan, and H. R. Wilson, *Proc. Roy. Soc.* (London), **A295**, 320–333 (1966).
(155) P. Tougard, *Biochem. Biophys. Res. Commun.*, **37**, 961–967 (1969).
(156) J. Iball, C. H. Morgan, and H. R. Wilson, *Proc. Roy. Soc.* (London), **A302**, 225–236 (1968).
(157) D. R. Harris and W. M. Macintyre, *Biophys. J.*, **4**, 203–225 (1964).
(158) N. Camerman and J. Trotter, *Acta Crystallogr.*, **18**, 203–211 (1965).
(159) E. Shefter and K. N. Trueblood, *Acta Crystallogr.*, **18**, 1067–1077 (1965).
(160) W. Saenger and K. H. Scheit, *Angew. Chem.*, **81**, 121 (1969).
(161) S. Furberg, C. S. Petersen, and C. Romming, *Acta Crystallogr.*, **18**, 313–320 (1965).
(162) M. Sundaralingam and L. H. Jensen, *J. Mol. Biol.*, **13**, 914–929 (1965).
(163) C. E. Bugg and R. E. Marsh, *J. Mol. Biol.*, **25**, 67–82 (1967).

$C_{10}H_{11}N_4O_8P^-\cdot Na^+\cdot 8H_2O$	Inosine 5'-(sodium phosphate) octahydrate[164]
$C_{10}H_{11}N_4O_8P^-\cdot Ba^{2+}\cdot 6H_2O$	Inosine 5'-(barium phosphate) hexahydrate[165]
$C_{10}H_{11}N_4O_8P^-\cdot 2Na^+\cdot 7.5H_2O$	Inosine 5'-(disodium phosphate) hydrate[165]
$C_{10}H_{11}N_5O_6P$	Adenosine 3',5'-cyclic phosphate[166]
$C_{10}H_{12}BrN_5O_5\cdot 2H_2O$	8-Bromoguanosine dihydrate[167]
$C_{10}H_{12}N_4O_4S$	D-Ribofuranosyl-6-thiopurine[168]
$C_{10}H_{12}N_4O_8P^-\cdot Na^+\cdot 8H_2O$	Inosine 5'-(sodium phosphate) octahydrate[169]
$C_{10}H_{13}BrN_2O_4$	5-Bromo-5-deoxythymidine (form B)[170]
$C_{10}H_{13}N_2O_8P^{2-}\cdot Ca^{2+}\cdot 6H_2O$	Calcium thymidylate, hexahydrate[171]
$C_{10}H_{13}N_4O_7P\cdot 3H_2O$	Guanosine 5'-phosphate, trihydrate[172]
$C_{10}H_{13}N_5O_3\cdot H_2O$	2'-Deoxyadenosine monohydrate[173]
$C_{10}H_{13}N_5O_4\cdot C_9H_{11}BrN_2O_6\cdot H_2O$	Adenosine–5-bromouridine monohydrate[174]
$C_{10}H_{13}N_5O_4\cdot C_9H_{14}BrN_3O_4$	5-Bromo-2'-deoxycytidine-2'-deoxyguanosine complex[175]
$C_{10}H_{14}N_2O_5$	Thymidine[176]
$C_{10}H_{14}N_2O_5S$	1-(2-Deoxy-α-D-*erythro*-pento-furanosyl)uracil-5-yl methyl sulfide[177]

(164) S. T. Rao and M. Sundaralingam, *Chem. Commun.*, 995–996 (1968).

(165) N. Nagashima and Y. Iitaka, *Acta Crystallogr.*, **B24**, 1136–1138 (1968).

(166) K. Watenpaugh, J. Dow, L. H. Jensen, and S. Furberg, *Science*, **159**, 206–207 (1968).

(167) C. E. Bugg and U. Thewalt, *Biochem. Biophys. Res. Commun.*, **37**, 623 (1969).

(168) E. Shefter, *J. Pharm. Sci.*, **57**, 1157–1162 (1968).

(169) S. T. Rao and M. Sundaralingam, *J. Amer. Chem. Soc.*, **91**, 1210–1217 (1969).

(170) M. Huber, *Acta Crystallogr.*, **10**, 129–133 (1957).

(171) K. N. Trueblood, P. Horn, and V. Luzzati, *Acta Crystallogr.*, **14**, 965–982 (1961).

(172) W. Murayama, N. Nagashima, and Y. Shimizu, *Acta Crystallogr.*, **B25**, 2236–2245 (1969).

(173) D. G. Watson, D. J. Sutor, and P. Tollin, *Acta Crystallogr.*, **19**, 111–124 (1965).

(174) A. E. V. Haschemeyer and H. M. Sobell, *Acta Crystallogr.*, **18**, 525–532 (1965).

(175) A. E. V. Haschemeyer and H. M. Sobell, *Acta Crystallogr.*, **19**, 125–130 (1965).

(176) D. W. Young, P. Tollin, and H. R. Wilson, *Acta Crystallogr.*, **B25**, 1423–1432 (1969).

(177) G. W. Frank, *Amer. Crystallogr. Assoc. Meeting*, Buffalo, N. Y., August, 1968, Abstr. 99.

$C_{10}H_{14}N_2O_6 \cdot 0.5H_2O$	5-Methyluridine hemihydrate[178]
$C_{10}H_{14}N_5O_7P \cdot H_2O$	Adenosine 5'-phosphate monohydrate[179]
$C_{10}H_{14}N_5O_7P \cdot 2H_2O$	Adenosine 3'-phosphate dihydrate[180]
$C_{11}H_{16}N_5O_3{}^+ \cdot Br^-$	Aristeromycin hydrobromide[181]
$C_{13}H_{15}N_5O_3S \cdot 0.33H_2O$	5',8-Anhydro-2',3'-O-isopropylidene-8-endo-mercaptoadenosine hydrate[182]
$C_{13}H_{16}N_5O_3{}^+ \cdot I^-$	3,5'-Anhydro-2',3'-O-isopropylideneadenosine iodide[183]
$C_{15}H_{21}N_5O_5$	2'-O-(Tetrahydropyranyl)-adenosine[184]
$C_{18}H_{22}N_4O_{10}S_2$	1-(2-Deoxy-α-D-erythro-pentofuranosyl)uracil-5-yl disulfide[185] 4-Thiouridine disulfide[186]
$C_{19}H_{24}N_7O_{12}P \cdot 4H_2O$	Adenosine 5'-(2'-deoxyuridin-2'-yl phosphate), tetrahydrate[187]
$C_{63}H_{88}CoN_{14}O_{14}P \cdot 22H_2O$	Vitamin B_{12}[188]
$C_{72}H_{100}CoN_{18}O_{17}P \cdot 17H_2O$	5'-Deoxyadenosine-5'-ylcobalamin hydrate, vitamin B_{12} coenzyme[189]

(178) D. J. Hunt and E. Subramanian, *Acta Crystallogr.*, **B25**, 2144–2152 (1969).

(179) J. Kraut and L. H. Jensen, *Acta Crystallogr.*, **16**, 79–88 (1963).

(180) M. Sundaralingam, *Acta Crystallogr.*, **21**, 495–506 (1966).

(181) T. Kishi, M. Muroi, T. Kusada, M. Nishikawa, K. Kamiya, and K. Mizuna, *Chem. Commun.*, 852–853 (1967).

(182) K. Tomita, N. Nishida, T. Fujiware, and M. Ikehara, *Acta Crystallogr.*, **A25**, S178 (1969).

(183) J. Zussman, *Acta Crystallogr.*, **6**, 504–515 (1953).

(184) O. Kennard, W. D. S. Motherwell, D. G. Watson, J. C. Coppola, and A. C. Larson, *Acta Crystallogr.*, **A25**, S88 (1969).

(185) E. Shefter, M. P. Kotick, and T. J. Bardos, *J. Pharm. Sci.*, **56**, 1293–1299 (1967).

(186) E. Shefter and T. I. Kalman, *Biochem. Biophys. Res. Commun.*, **32**, 878 (1968).

(187) E. Shefter, M. Barlow, R. A. Sparks, and K. N. Trueblood, *Acta Crystallogr.*, **B25**, 895–909 (1969).

(188) C. Brink-Shoemaker, D. W. J. Cruickshank, D. C. Hodgkin, M. J. Kamper, and D. Pilling, *Proc. Roy. Soc.* (London), **A278**, 1–26 (1964).

(189) P. G. Lenhert, *Proc. Roy. Soc.* (London), **A303**, 45–84 (1968).

AUTHOR INDEX FOR VOLUME 31

Numbers in parentheses are reference numbers and indicate that an author's work is referred to, although his name is not cited in the text.

structure bibliography, 361
Dulcitol, *see* Galactitol

E

Ebenaceae, polysaccharides from, 253, 254
β-Elimination
  in degradation of capsular polysaccharides, 239
  of oligosaccharides and polysaccharides, 212–229
Enzymes
  in biodegradative hydrolysis of galactomannans, 256
  coffee-bean, in galactomannan structure study, 271, 274
  in UDP-apiose biosynthesis, 169–172
Epi-isosaccharic acid, preparation of, 18
Epimerization, and acetolysis, 200
Erythritol, crystal structure bibliography, 363
—, 1,1-bis(acetamido)-1-deoxy-D-, preparation of, 83
—, 1,1-bis(acetamido)-1-deoxy-L-, preparation of, 82, 84, 112
—, 1,1-bis(benzamido)-1-deoxy-L-, preparation of, 100
—, 2,3-di-O-acetyl-1-amino-4-bromo-1,4-dideoxy-, hydrobromide, deamination of, 63
—, 2,3,4-tri-O-acetyl-1-amino-1-deoxy-D-, p-toluenesulfonate (salt), deamination of, 63
Erythrofuranose, 1,2:3,3′-di-O-isopropylidene-[3-C-(hydroxymethyl)-α-D-, preparation of, 183
—, 3-C-(hydroxymethyl)-D-, *see* Apiose
—, 3-C-(hydroxymethyl)-1,2-O-isopropylidene-α-D-, preparation of, 183
—, 3-C-(hydroxymethyl)-2,3-O-isopropylidene-β-D-, preparation of, 183
—, 2,3,3′-tri-O-methyl-3-C-(hydroxymethyl)-D-, identity with tri-O-methyl-D-apiose, 142
Erythrofuranoside, methyl 2,3-O-isopropylidene-[N-acetyl-4-amino-4-deoxy-3-C-(hydroxymethyl)-β-D-, preparation of, 184

—, methyl 2,3-O-isopropylidene-[3-C-(hydroxymethyl-4-thio-β-D-, preparation of, 183, 184
Erythrose, 2,3,4-tri-O-acetyl-*aldehydo*-L-, ammonolysis of, 84
Esters, carbohydrate, reaction with ammonia, 81–134
Ethanol, as solvent in ammonolysis of benzoyl groups, 102
Explosive industry, galactomannans in, 309, 311

F

Fenugreek, *see* *Trigonella foenum-graecum*
Ferric chloride, in degradation of polysaccharides, 239
Fetuin, deamination of, 236
Fire-fighting industry, galactomannans in, 311
Flavone, 3′,4′,5,7-tetrahydroxy-, *see* Luteolin
—, 4′,5,7-trihydroxy-, *see* Apigenin
—, 4′,5,7-trihydroxy-3′-methoxy-, *see* Chrysoeriol
Flavonoid, from *Digitalis purpurea*, 145
Flavor enhancers, apiose-containing nucleosides as, 183
Fletcher, Jr., Hewitt Grenville, obituary, 1–7
Food industry, galactomannans in, 310, 312
Formycin
  hydrobromide monohydrate, crystal structure bibliography, 366
  monohydrate, crystal structure bibliography, 357
Fourier transformation, pulse, in characterization of polysaccharide degradation products, 239
Foxglove, *see* *Digitalis purpurea*,
Frangulin B, from *Rhamnus frangula*, 147
Fructofuranose, 6-O-α-D-glucopyranosyl-α-D-, *see* Isomaltulose
Fructofuranoside, α-D-glucopyranosyl β-D-, *see* Sucrose
Fructopyranose, β-D-, crystal structure bibliography, 365
—, 2,6-anhydro-β-D-, crystal structure

acetal, crystal structure bibliography, 353

—, 6-O-(α-D-galactopyranosyluronic acid)-D-, preparation of, 194

—, 1,2,3,4,6-penta-O-benzoyl-α-D-, ammonolysis of, 86

—, 3,4,5,6-tetra-O-benzoyl-2-S-ethyl-2-thio-*aldehydo*-D-, ammonolysis of, 114

—, 3,4,6-tri-O-acetyl-2,5-anhydro-D-, formation by deamination, 22, 68, 69

β-D-Mannosidase, for biodegradative hydrolysis of galactomannans, 256

Mannosylamine, N-benzoyl-D-, preparation of, 83, 86

Mass spectrometry, of degradation products of polysaccharides, 239

Melanoidins, formation in ammonolysis, 89, 90, 91

Melibiitol, 1,1-bis(acetamido)-1-deoxy-, preparation of, 94

Melibiononitrile, octa-O-acetyl-, ammonolysis of, 98

Melibiose, octa-O-acetyl-, ammonolysis of, effect of solvent on, 100

—, octa-O-methyl-, ammonolysis of, 94

Melibiosylamine, N-acetyl-β-, configuration of, 105

Mercaptolysis, in degradation of polysaccharides, 190

Metabolism, of apiose, 153, 154

Methanol, as solvent in ammonolysis of benzoyl groups, 102

Methanolysis, in degradation of polysaccharides, 190

Methylation, Hakomori method of, for polysaccharides, 215

Milk, hexasaccharide from human, acetolysis of, 199

Molecular weights, of galactomannans, 260, 261, 277–279

Monosaccharides

acyl esters, reaction with ammonia, 81–92

ammonolysis products from acylated, 126–131

Mucilage, galactomannan polysaccharides, history of, 244

Mucopolysaccharides, deamination and structure of, 74

Murine myeloma, immunoglobulins, 313–346

Muscarine, synthesis of, deamination in, 71

Mycobacterium phlei lipopolysaccharide, degradation by Lossen rearrangement, 238

N

Naphthalene, 9,10-*trans,cis*-2-aminodecahydro-, deamination of, 16, 17

—, 9,10-*trans,trans*-2-aminodecahydro-, deamination of, 15, 17

Neobiosamine B, deamination and structure of, 75

Neuraminic acid

deamination of methyl glycoside of, 35

in polysaccharides and glycoconjugates, 187

—, N-acetyl-, *see* Sialic acid

Nicotinic acid, carbohydrate esters, ammonolysis of, 88, 89, 124

Nigeran, Barry degradation of, 203

Nigerose, from dextran by acetolysis, 197

Nitriles

acylated sugar, ammonolysis and free sugar formation, 84

ammonolysis of acetylated, mechanism of, 110

of acylated aldobionic acid, 97

of acylated aldonic acids, 82–85

Nomenclature

of D-apiose, 137

of cordycepin and cordycepose, 149

Nonulopyranosonic acid, 5-acetamido-3,5-dideoxy-β-D-*glycero*-D-*galacto*-, *see* Sialic acid

Nonulosonic acid, 5-amino-3,5-dideoxy-D-*glycero*-D-*galacto*-, *see* Neuraminic acid

Nuclear magnetic resonance spectroscopy, of degradation products of polysaccharides, 239

Nucleosides

apiose-containing purine, synthesis, 183, 184

crystal structure bibliography, 355–363, 369–371

deamination in synthesis of, 70

synthesis of, 4

Nucleotides
  apiose-containing sugar, 163
  apiosyladenine, deamination of, 153
  crystal structure bibliography, 355–363, 369–371

O

Obituary, Hewitt Grenville Fletcher, Jr., 1–7
Octulosonic acid, 3-deoxy-D-*manno*-, in polysaccharides and glycoconjugates, 187
Oligosaccharides
  acetolysis and anomerization of, 199
  degradation by oxidation with chromium trioxide, 229
  β-eliminative degradation of, 212
Ovarian-cyst blood-group H substance, Smith degradation of, 209
Ovomucoid, deamination of, 236
1,2-Oxazoline, carbohydrate derivatives, 4
Oxidation
  periodate, degration of polysaccharides by, 200–211
    and structure determination of carbohydrate amines, 102–105
Oxidizing agents, for alcohol to carbonyl oxidation, 223
Ozone, oxidative degradation of polysaccharides by, 232

P

Paint industry, galactomannans in, 311
Palatinose, *see* Isomaltulose
Palmae, galactomannans from, 253, 254
Paper industry, galactomannans in, 311
Paromobiosamine, deamination of derivatives of, 26–28
Parsley, *see Petroselinum crispum*
Pentaerythritol
  crystal structure bibliography, 363
  tetraacetate, crystal structure bibliography, 368
  tetranitrate, crystal structure bibliography, 363
Pentasaccharides, hydrolytic degradation of, 188

Pent-3-enosid-2-ulose, methyl 4-deoxy-3-*O*-methyl-, formation and hydrolysis of, 221
Pentitols, 1-amino-1-deoxy-, deamination of, 60–62
—, 1,4-anhydro-, preparation of, 60–62
—, 1,1-bis(acylamido)-1-deoxy-, conformations of, 110
Pentodialdo-1,4-furanose, 1,2-*O*-isopropylidene-α-D-*xylo*-, in L-apiose synthesis, 181
Pentofuranoside, methyl 2-deoxy-2-(hydroxymethyl)-, formation by deamination, 22
Pentonic acid, 3-deoxy-2-*C*-(hydroxymethyl)-D-*erythro*-
  calcium salt, crystal structure bibliography, 364
  strontium salt, crystal structure bibliography, 364
Pentose, 2-deoxy-D-*erythro*-, crystal structure bibliography, 363
—, 3-deoxy-D-*erythro*-, *see* Cordycepose
—, 4-*C*-(hydroxymethyl)-D-*threo*-, D-apiose from, 176
Pentosidulose, methyl 4-*O*-ethyl-3-*O*-methyl-β-D-*threo*-, degradation of, 221
Pent-1-ynitol, 1,2-dideoxy-D-*erythro*-, preparation of, and triacetate, 70
—, 1,2-dideoxy-D-*threo*-, preparation of triacetate, 70
Periodate oxidation
  degradation of polysaccharides by, 200–211
  and structure determination of carbohydrate amines, 102–105
Petroselinin, in parsley plant, 140, 141
Petroselinum crispum
  apiose from, 137, 138
  glycosides from, 140–143
pH, effect on deamination, 65
Pharmaceuticals, galactomannans in, 310, 312
Phosphoglycolipids, deamination and structure of, 77
Phosphoric esters, of carbohydrates, synthesis of, 4
Phytoglycolipids, deamination and structure of, 77

R

Ranuncoside monohydrate, crystal structure bibliography, 350

Rearrangement, in deamination reactions, 11–13

Rhamnofuranosylamine, N-acetyl-α-L-, configuration of, 109

Rhamnononitrile, tetra-O-acetyl-L-, reaction with ammonia, 82

Rhamnopyranose, L-, peracetate, ammonolysis of, 87

—, tetra-O-acetyl-α,β-L-, ammonolysis of, 123

—, 1,2,3,4-tetra-O-benzoyl-L-, ammonolysis of, 86

Rhamnopyranosylamine, N-benzoyl-L-oxidation and configuration of, 103 preparation of, 86

Rhamnus frangula, frangulin B from, 147

Ribitol, crystal structure bibliography, 364

—, 1-amino-1-deoxy-D-, deamination of, 60–62

—, 2-amino-2-deoxy-D-, deamination of, 63

—, 4-O-(2-amino-2-deoxy-D-glucopyranosyl)-D-, deamination and structure of, 76

Riboflavine, hydrobromide, monohydrate, crystal structure bibliography, 368

Ribofuranose, α-D-, glycosides and N-glycosyl derivatives, preparation of, 4

—, 1,2,3,5-tetra-O-acetyl-β-D-, crystal structure bibliography, 353

—, 1,3,5-tri-O-benzoyl-α-D-, preparation of, 4

Ribofuranosyl bromide, tri-O-benzoyl-D-, preparation of, 4

Ribofuranosyl halide, 3,5-di-O-benzoyl-D-, preparation of, 4

Ribopyranoside, methyl 4-amino-4-deoxy-2,3-O-isopropylidene-β-L-, deamination of, 42

—, methyl 2-C-(aminomethyl)-, deamination of, 53

—, methyl 1-thio-α-D-, crystal structure bibliography, 348

—, methyl 5-thio-α-D-, crystal structure bibliography, 349

—, methyl 5-thio-β-D-, crystal structure bibliography, 349

Ribopyranosylamine, N-(p-bromophenyl)-α-D-, 2,4-dibenzeneboronate, crystal structure bibliography, 368

Ribopyranosyl bromide, tri-O-benzoyl-D-, preparation of, 3, 4

Ribose,

D-, (p-bromophenyl)hydrazone, crystal structure bibliography, 367

derivatives, preparation of, 3

—, 2-amino-2-deoxy-L-, deamination of, 26

—, 3-O-(2,6-diamino-2,6-dideoxy-β-L-idopyranosyl)-D-, deamination of derivatives of, 26–28, 75

—, 2-C-(hydroxymethyl)-D-, isolation of, 136

Rubidium β-D-glucuronate dihydrate, crystal structure bibliography, 364

Ruthenium tetraoxide

oxidative degradation of polysaccharides by, 232

for oxidation of an alcohol to a carbonyl compound, 223

S

Saccharinic acids, analysis of, 213

Salmonella kentucky lipopolysaccharide, degradation of acetylated, by oxidation with chromium trioxide, 230

Salmonella lipopolysaccharides degradation by Lossen reaction, 211 periodate oxidation of, 211

Salmonella newport lipopolysaccharide, degradation of acetylated, by oxidation with chromium trioxide, 231

Salmonella strasbourg lipopolysaccharide, degradation of acetylated by oxidation with chromium trioxide, 230

Salmonella typhi lipopolysaccharide, degradation of acetylated by oxidation with chromium trioxide, 230

Salmonella typhimurium LT2 lipopolysaccharide, degradation of methylated, 225

Salmonella typhimurium 395 MS lipopolysaccharide, hydrolysis and structure of, 188

Sedoheptulosan, monohydrate, crystal structure bibliography, 366

L-Serinatocopper(II), di-O-β-D-xylo-

ERRATA

VOLUME 30

Page 124, line 17. For "1,3-di-N-acetyl-2,4-di-O-acetyl-" read "1,3-di-N-acetyl-4,6-di-O-acetyl-."

Page 151, formula **127.** For "$R^2 = NH_2$" read "$R^1 = NH_2$."

Page 159, formula **155.** For "$R = C_6H_3$-2,3-$(NO_2)_2$" read "$R = C_6H_3$-2,4-$(NO_2)_2$."

Page 163, line 2. For superscript "190" read "191."

Page 172, formula **198.** For "$\overset{|}{C}H_2NH_3$" read "$\overset{|}{C}H_2NHR$."

Page 176, line 2. For "neamine (**209**)" read "3′,4′-dideoxyneamine (**209**)."

Page 294, Table III, column 1, under Oligosaccharides. Insert the following formulas:

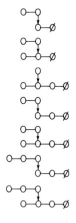

416

A 5
B 6
C 7
D 8
E 9
F 0
G 1
H 2
I 3
J 4